A PRICE GUIDE TO
VICTORIAN HOUSEWARE
HARDWARE & KITCHENWARE

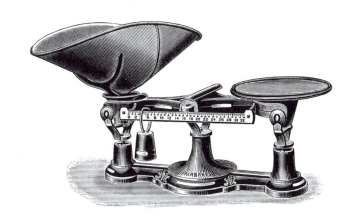

by Ronald S. Barlow

Linda Campbell Franklin
Contributing Editor

WINDMILL PUBLISHING COMPANY
El Cajon, California 92020

Copyright ©1992 by Ronald S. Barlow
All rights reserved under Pan American and
International Copyright Conventions.

Clip-Art Copyright Release:

Up to 10 illustrations may be used for graphic
design, advertising, or craft purposes in a single
publication without further written permission.
However, republication of any portion of this
book by a graphic archives, clip-art service,
or price guide publisher is strictly prohibited.

Additional copies of this book may be
ordered through your local museum shop,
independent bookstore, or directly from:

Windmill Publishing Company
2147 Windmill View Road
El Cajon, California 92020

Send a check or money order for $19.95 plus $2.00 postage.
(Please allow 3 to 4 weeks for delivery by Post Office.)

Canada, please remit money order only,
in U.S. funds, plus $1.00 additional postage.

Printed in the United States of America
International Standard Book Number:
0-933846-03-7

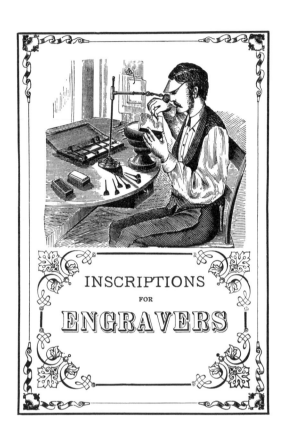

INSCRIPTIONS FOR ENGRAVERS

PREFACE

To quote the famous catalog historian and antiquarian bookseller, Lawrence B. Romaine, "One good 64-page illustrated catalog, showing a full line of products in any given field, will for such an undertaking be worth seven tons of manuscript ledgers, day books, copy books and correspondence." In *Victorian Houseware, Hardware & Kitchenware* we have cut, pasted, and assembled the best illustrations from seven important 19th century trade catalogs. Using Romaine's ratio of one to seven, we offer the equivalent of 49 tons of manuscript material and printed reports, covering the Hardware and Houseware business in the United States for half a century. For good measure, we are including a *price guide section* with current values listed for each item illustrated. To add further value, all of the illustrations in this book are copyright-free, and may be used in any way you choose, not to exceed ten per publication.

Although printing services were available in the United States as early as 1639, the first trade catalogs on record were folded advertising broadsides published in the 1790's. A typical catalog of those early days rarely exceeded a dozen or so pages, and precious few contained any illustrations. Most were issued by booksellers, and targeted at a small but growing market of people who could actually read.

Most artisans and tradesmen still relied on imported English pattern books, which were so well-illustrated that many items could be skillfully reproduced from the drawings presented. In 1853, Peck & Walter Manufacturing Company of New Britain, Connecticut, issued the first illustrated hardware catalog in America. Others soon followed, including Sargent & Company in 1856, and Hotchkiss & Sons in 1859. With the notable exception of Joseph Adams' outstanding 296-page catalog of housebuilder's hardware published in 1859, most of these were 30- to 50-page efforts with sparse illustrations.

Russell and Erwin's huge 486-page illustrated "American Hardware Catalog of 1865" set the standard for all that followed. Sargent and Company's hardware catalogs of 1874–1888 contained thousands of wonderful wood engravings — everything from door bells to barber's razor strops to buggy whips — plus the largest selection of builder's hardware ever assembled. In this price guide we have reproduced the best of Sargent's 1884 offering.

In 1885, the Wm. Frankfurth Hardware Company of Milwaukee, Wisconsin celebrated its 25th anniversary by incorporating the family-held company, and issuing a leatherbound 9 x 12 inch, 975-page trade catalog full of everything from coffee mills to fluting irons and oil stoves. In an effort to keep this expensive sales tool up to date and in circulation for as many years as possible, Mr. Frankfurth refused to publish a fixed discount schedule for subscribing merchants. Instead, he guaranteed to meet, or beat, any agent's prices, and offered prompt quotations by return mail.

Also reproduced in this book are dozens of pages from Russell & Erwin's 1875 catalog of builder's hardware and houseware. It is interesting to note by comparison how very few changes were made in products manufactured during these decades. An 1870's meat grinder looks very much like its 1900's counterpart, and cast iron cookware hardly changed at all.

Heavily represented on the following 372 pages are the products of Simmons Hardware Company of St. Louis. Organized in 1873 by E. C. Simmons, a former errand boy with the firm of Wilson, Levering & Waters, the firm rapidly grew to become one of the country's largest wholesale houses. Simmons did not limit his inventory to hardware alone, but branched into baskets, bicycles, wicker baby carriages, and golden oak furniture. He also owned and distributed the KeenKutter imprinted line of cutlery and carpenter's tools. By the early 1900's, the E. C. Simmons Hardware Catalog contained 4,200 pages and 21,000 product illustrations. The 25,000 KeenKutter dealer catalogs published in 1906 weighed in at the post office at half a million pounds, and had consumed 16 railroad cars of paper and 2,500 pounds of printer's ink.

Clocks and silverplated items in this guide are from the 1890's catalogs of Benjamin Allen & Co. and Marshall Field & Co., both venerable Chicago firms. Terms offered to 19th-century retailers were "six percent discount for cash, net payment due in thirty days. Old gold and silver taken in trade. Gold four cents per karat; silver ninety cents per ounce." The editor has made every effort to include silverplated items from these two firms that have not been reproduced in earlier reprints of this genre.

Woodburning cast iron stoves shown are from the 1877 and 1892 editions of "The Metal Worker," a weekly publication of the plumbing, tinware, and stove trades. Also included is a six-page selection of ranges and heaters from Bridge and Beach Manufacturing Company's 1905 catalog. The subtle evolution of stove designs during this 28-year period is interesting to behold. 4,275 trade names for various designs were recorded in the 1878 "Name Book of Stoves."

We don't wish to clutter up the preface with a complete bibliography of all the antique trade catalogs represented in this work, but these seven publications were our major resources, and reproductions of their inside-cover designs appear on the pages immediately following.

—1881—
House Furnishing Catalogue.

SIMMONS HARDWARE COMPANY,

ST. LOUIS, MO.

Sargent & Co.'s

ILLUSTRATED CATALOGUE

— OF —

HARDWARE.

—— 1884. ——

Sargent & Co.

NEW YORK, AND NEW HAVEN, CONN.

U. S. A.

Volume 3. 1875.

PRICE LIST

AND

DESCRIPTIVE CATALOGUE

OF

Russell & Erwin Mfg. Co.

MANUFACTURERS OF

GENERAL HARDWARE,

NEW BRITAIN, CONN.

WAREHOUSES:

NEW YORK, 45 & 47 Chambers Street.
PHILADELPHIA, 16 & 18 North Fifth Street.

Volume No. 1, contains illustrations, description and prices of Locks, Knobs, Padlocks, Builders and other Hardware.

Volume No. 2, now in press, will contain illustrations, description and prices of Real Compression Bronze, and other Hardware, for Public Buildings, Private Residences, etc.

ANNUAL ❖ ILLUSTRATED ❖ PRICE ❖ LIST

CLOCKS, SILVER and SILVER PLATED WARE,

141 · & · 143 · STATE · STREET · CHICAGO

BRIDGE & BEACH MFG. CO.

ST. LOUIS PRICE LIST.

JULY 1st, 1905.

"SUPERIOR"
Stoves and Ranges

503 South First Street,

Illustrated Catalogue of 1893.

Clocks, Lamps, and Silver-Plated Ware.

Marshall Field & Co.

Adams, Quincy, Franklin, Fifth Ave.

Chicago

How to Use This Catalog as a Price Guide

Find each item alphabetically in the index, or at random in the catalog. Note its stock number and the position of the object on the page, then turn to the Price Guide Section at the back of the book. Here you will find educated guesses by numerous antiquarians, including half a dozen guest experts, the editor, author, and publisher. If you are not entirely happy with our estimates, remember that prices do vary from shop to shop and state to state. And the real value of any collectible is what a buyer is willing to pay. Condition and rarity are also important factors in determining an antique's market value.

Your local auctioneer can be the most pragmatic appraiser in town because he has probably sold hundreds of similar household collectibles over the years. Getting cash offers from antique dealers is another inexpensive appraisal method. You can often double their quoted prices to determine an item's retail value. If you have time to travel to antique shops, shows, and major flea markets, take this catalog along and make detailed notes of prices and variations in quality and style. In a few years you may begin to understand the burden we price guide compilers bear.

Ronald S. Barlow

INDEX

INTRODUCTION	20
BIBLIOGRAPHY	362
PRICE GUIDE SECTION	363 to 375

A

Adzes —
   Carpenter's 229
Agricultural Implements —
   Axes to Ox Yokes .... 224 to 237
Animal Catchers —
   Plier on Pole 235
Andirons —
   Brass 24, 25
Apple Corers —
   Tin, Hollow Tube 106
Apple Parers —
   Clamp-on Style .... 77 to 80
Ash Sifters —
   Daisy 58
Augers —
   Fruit 86
Axes —
   Broad Axe 229

B

Bake Pans —
   Cast Iron 120, 121
   Granite Ironware 129
   Tinware 111
Balance Scales —
   Family .... 208 to 214
   Market .... 208 to 215
   Platform 213, 215, 216
Bath Tubs —
   Japanned Tin 206, 207
Beef Shavers —
   Enterprise 218
Bells —
   Call Bells 36, 37, 38
   Cow Bells 40
   Door Bells 262
   Hand 37
Benches —
   Wash Bench 95
Berry Dishes —
   Glass, Silverplated Frames 186, 187
Berry Spoons —
   Silverplated 167
Bird Cages —
   Cages and Accessories .... 194 to 197
Biscuit Beaters —
   Wooden, Crankstyle 91
Bon Bon —
   Trays, Silverplated 175
   Scoops 168
Bone Mills —
   Enterprise Poultry Mill 218, 232

Bootjacks —
   Cast Iron 203
Bowls —
   Cast Iron 118
   Granite Ironware .... 126 to 129
   Tin-plated 111
Bowl and Pitcher Sets —
   Granite Ironware 128
   Japanned Tin 205
Boxes —
   Billhead, Tin 188
   Cake, Tin 189
   Cash, Tin 189
   Flour, Tin 189
   Knife and Fork 94
   Nested, Wood 94
   Spice, Tin 188
   Salt, Wood 94
   Sugar, Wood 94
Brackets —
   Flower Pot 261
   Shelf, Cast Iron 238, 260
Brassware —
   Andirons 24
   Candlesticks 26
   Coal Hods 60
   Kettles 115
   Tea Kettles 29
Bread —
   Knives 160
   Plates, Wooden 90
   Rasps 101
Broilers —
   Meat 115
Buckets —
   Granite Ironware 127
   House Maid's 95
   Well Buckets 225
Bull —
   Rings, Snaps, Leaders 234
Butter Dishes —
   Granite Ironware 126
   Silverplated 171, 172
Butter Knives —
   Silverplated .... 156 to 162
Butter Prints/Molds —
   Woodenware 91
Butter Spades —
   Metal 86
Button Hole Cutters —
   Scissors .... 331 to 339

C

Cabbage —
   Borers 86
   Cutters 87
Cake Baskets —
   Silverplated 153, 185
Cake Knives —
   Silverplated 162

Cake Pans —
    Granite Ironware 127, 128
    Tin, Mold 108, 109, 111

Call Bells —
    Desk Top 36

Candlesticks —
    Brass Holders 26, 28
    Candle Moulds 113
    Candle Snuffers 26

Can Openers —
    American .... 96 to 98
    British 145

Carpet —
    Stretchers 193
    Sweepers 192

Carriage Jacks —
    Iron and Wood 198

Carving Knives and Forks —
    Rodgers & Sons, Silverplated 161

Caster Sets —
    Silverplated 149, 151, 153
    Pickle Casters 181

Cast Iron Holloware —
    Bake Pans, Bread and Rolls 120, 121
    Boilers, Bowls, Ovens 118
    Griddles, Kettles, Pots 118, 119
    Waffle and Wafer Irons 122, 123

Cattle Leaders —
    Spring-loaded, Nose 234

Celery Stands —
    Glass, Silverplated 173

Chafing Dish —
    Oyster, etc. 117

Chairs —
    Children's 311, 312
    Dining Room 316
    Arm Chairs 317
    Rockers .... 313 to 317
    Wicker .... 307 to 309
    Morris Chairs 315
    Office Chairs 318, 319

Chamber Pots —
    Granite Ironware 127

Champagne —
    Cork Screws .... 143 to 147
    Nippers 144
    Taps 142

Chandelier Hooks —
    For Hanging Lamps 278

Charcoal Irons —
    Family, Tailor's, Laundry 358

Cheese Scoop —
    Silverplated Server 166

Cheese Knives —
    Crankhandle Cutter 218

Cherry Stoners —
    Iron, Table Top 78, 218

Choppers —
    Meat 222-223
    Vegetable 222-223

Churns —
    Barrel, Dash, Swing 88
    Syllabub Churn 106

Cider Mills —
    Eagle, Junior and Senior 232

Cleaning Instructions —
    Kitchen and Utensils 75, 76

Clocks —
    Alarm 302
    Carved Wood 300, 301, 303
    Mantel, Bronze, etc. 296, 298
    Novelty 297, 302, 304
    Regulator 305

Coal —
    Hods, Brass 60
    Forks (shovel) 231
    Sifters 58

Coat Hooks —
    Coat & Hat Hooks .... 239 to 243

Coffee Makers —
    Granite Ironware 132
    Planished Tin 110

Coffee Mills —
    Lap .... 44 to 46
    Store 42, 43

Coffee Pots —
    Granite Ironware .... 125 to 133
    Silverplated 170, 178
    Tinware 114

Coffee Pot Stands —
    Trivets 353

Coffee Roasters —
    Home and Grocer's 42, 47

Coffee Urns —
    Large, Restaurant style 134

Colander —
    Stamped Tinware 109

Cookie Cutters —
    Tin Cake Cutters 106

Cookie Pans —
    8-Cup Mold, Tin 111

Copperware —
    Bailed Kettles 115
    Measures 348
    Tea Kettles 112, 115
    Wash Boilers 348

Coolers —
    Water, Galvanized Iron 134

Cork Presses —
    Cast Iron 142

Cork Pullers —
    Clamp-On 102

Corkscrews —
    American and British .... 143 to 147

Corn —
    Hooks, Huskers, Gloves 232
    Planters 233
    Shellers 232

Corn Cake Pans —
    Cast Iron 120, 121
    Stamped Tinware 116

Cream Pitchers —
    Granite Ironware 126
    Silverplated 170, 178, 183

Cream Ladles —
    Silverplated 155, 167, 168

Crumb Trays —
    Japanned Tin 191
    Silverplated 157

Cupboard Catches —
   Rural and Fancy 258, 259
Cups, Drinking —
   Granite Ironware 127
   Silverplated 176, 182
   Tinware 111
Curling Irons —
   Hair Frizzers, Pinchers, Curlers 359
Cuspadores —
   Brass and Japanned 190
   Granite Ironware 190
Cutlery —
   Carving Sets 161
   Knives, Forks .... 156 to 166
   Scissors .... 331 to 339

D

Dippers —
   Granite Ironware 128
   Tin 111
Dessert Sets —
   Silverplated 183
Dinner Plates —
   Granite Ironware 127
Dish Pans —
   Granite Ironware 128
Door Hardware —
   Handles, Store 251, 252
   Hinges, Butt 254, 255
   Knobs 256, 257
   Letter Box Plates 264, 265
   Push Plates 253
Door Bells —
   Nickel-plated 261, 262
Door Bell Pulls —
   Bronze-plated 263
Double Boiler —
   Granite Ironware 128
Drawer Handles —
   Pulls .... 272 to 277
Dust Pans —
   Japanned Tinware 191
Dutch Oven —
   Cast Iron 118

E

Egg —
   Beaters .... 82 to 84
   Fryers 128
   Poachers 113
   Timers 82
Embroidery Scissors —
   Sterling Silver Handles 338
Enterprise Mfg. Co. —
   Advertisement 41
Eraser —
   Eraser Knife, Ink 175
Escutcheons —
   Keyhole Surrounds .... 243 to 250

F

Farm Tools —
   Pumps to Posthole Diggers .... 224 to 237
Faucets —
   Molasses 41
   Wood 86

Filters —
   Water 135
Fire Dogs —
   Bronzed 34
Fire Irons —
   Sets, Stands, Tools 30, 35
Fish Knives/Forks —
   Silverplated Servers 157, 164
Flasks —
   Coffee, Tin 113
   Brandy, Silverplated 180
Flowerpot —
   Brackets 261
   Floor Stands 261
Fluters —
   Hand and Machine 360, 361
Fly Fans —
   Wind-Up 201
Fly Traps —
   Wire Screen 201
Fly Swatters —
   Longhandled 201
Food Grinders —
   Gem, with attachments 222
Foot Baths —
   Japanned Tin 205, 207
Foot Scrapers —
   Cast Iron 202
Forks, Table and Serving —
   Silverplated .... 158 to 168
Fork and Spoon Boxes —
   Japannedware 188
Freezers —
   Ice Cream 140, 141
Fruit Presses —
   Enterprise Mfg. Co. 81
Fruit Augers —
   Cast Steel 86
Fry Pans —
   Cast Iron Spider 119
   Granite Ironware 128
Fryers, Deep —
   Hotel Style 123
Funnels —
   Coffee, Tin 113
   Granite Ironware 128
   Jelly, Tin 113
Furniture —
   Children's Chairs 311, 312
   Dining Room Chairs 316
   Morris Chairs 315
   Office Chairs 318, 319
   Rocking Chairs .... 313 to 317
   Wicker, Assorted .... 306 to 311

G

Glass Cutters —
   Combination and Plain 84
Granite Ironware —
   Advertisement 124
   Pots, Pans, Holloware .... 125 to 133
Grain Cradles —
   With Scythes 231
Graters —
   Hand and Tabletop 89, 189

Gravy Boats —
   Silverplated 182
Gravy Strainers —
   Tin, Wood Handle 111
Griddles —
   Cast Iron 119
   Granite Ironware 128
   Soapstone 122
Grindstones —
   Mounted on Oak Frame 228

H

Ham Boilers —
   Cast Iron 118
Hardware —
   Household and Builders 238
Hatchets —
   Shingling, Lathing 229
Hay Knives —
   Spearpoint and Serrated 230
Hay Rakes —
   Wooden Teeth 231
Hods —
   Coal, Brass 60
Hog —
   Ringers and Tongs 235
Hooks —
   Bird Cage 56
   Coat and Hat .... 239 to 243
Hoosier —
   Cupboard by McDougall 67
Holloware —
   Silverplated .... 149 to 187
   Granite Ironware .... 125 to 133
   Stove, Cast Iron .... 118 to 121

I

Ice —
   Balance Scales 209
   Chippers, Picks, Saws 139
Ice Cream —
   Freezers and Spoons 140, 141
   Ice Cream Molds 138
Ink Blotters —
   Silverplated 175
Ink Erasers —
   Silverplated 175
Ink Wells —
   Silverplated 175, 182
Irons —
   Charcoal 358
   Curling 359
   Fluting 360, 361
   Sad Irons .... 350 to 357
Ironing Boards —
   Bosom and Skirt 94
   Ironing Tables (Boards) 95, 349

J

Jaggers —
   Pastry 101
Jelly Molds —
   Fancy Shapes 108
Jelly Presses —
   Enterprise Mfg. Co. 81

Jugs —
   Stoneware, undecorated 41

K

Kanakins —
   Sugar Box, Round 94
Kettles —
   Brass, Copper 115
   Cast Iron 118, 118
   Granite Ironware 129
Keys —
   Brass, Iron, Steel .... 244 to 247
   Padlock .... 266 to 270
Kitchen Cabinets —
   McDougall, Hoosier 67
Kitchens —
   How to Clean 75, 76
Knife Boxes —
   Wooden 94
   Tinware 188
Knife Rests —
   Silverplated 153, 177
Knives —
   Carving 161
   Bread 160
   Serving, Silverplated .... 156 to 164
   Table Knives and Forks .... 158 to 166
Knobs —
   Door 256, 257
   Cupboard 258, 259
Kraut Cutters —
   With Sliding Box 87

L

Ladders —
   Step, Pine 94
Ladles —
   Granite Ironware 128, 131
   Stamped Tin 116
Lamps, Oil-Burning —
   Hand Lamps 280, 285
   Library, Hanging 291
   Parlor Lamps 288, 289
   Piano 292
   Pittsburg 293, 295
   Student 286, 287
   Table 286, 290
Lamp Fillers —
   Oil Cans and Fillers 285
Lamp Shades —
   Silk, on frames 294
Lamp Trimmers —
   Wick Trimmers .... 26 & 330
Lanterns —
   Buggy, Hand, Railroad .... 280 to 284
Lard Presses —
   Heavy Duty 220
   Fruit and Lard 81
Laundry Equipment —
   Ironing Boards 94, 95
   Mangles 344
   Sad Irons .... 349 to 357
   Wash Boards 342, 343
   Wash Boilers 348
   Washing Machines .... 340 to 346
   Wringers .... 341 to 347

Lawn Mowers —
   Push-style 224
   Horsedrawn 224

Lemon Squeezers —
   Handheld and Countertop .... 99 to 103

Letter Box —
   Slot Plates 264, 265

Lobster Crack —
   Also used as a nutcracker 101

Locks —
   Door, Mortise 244, 251
   Padlocks .... 266 to 270
   Trunk 271

M

Mallets —
   Ice, Steak 139

Mangles —
   Clothes 344

Match Safes —
   Wallmounted .... 51 to 56

Measures —
   Copper 348

Meat Grinders, Cutters —
   Household Size 221, 222
   Starrett's Patent 223

Milk Cans —
   Iron Clad 88

Milk Skimmers —
   Tin 106

Milk Strainers —
   Retinned 109

Mills —
   Feed 218, 232
   Coffee .... 42 to 46
   Bone, Shell, Corn 218, 232
   Spice 42

Mincing Knives —
   Metal, Wood Handle 104, 105

Molasses Gates —
   Enterprise 41

Molds —
   Berlin 108
   Butter 91
   Cake 109
   Candle 113
   Ice Cream 138
   Jelly, Melon 108
   Pudding, Rice 108, 113

Mountain Cake Pans —
   Granite Ironware 128

Mouse Traps —
   Wood and Wire .... 198 to 200

Mustard Spoons —
   Silverplated 156

Mutton Holder —
   Screw-Clamp style 101

N

Napkin Rings —
   Silverplated 180, 184

Nutcrackers —
   Handheld 101, 169
   Tabletop 142

Nut Picks —
   Silverplated 156, 169

Oil Lamps —
   Assorted Styles .... 280 to 295

Oil Stoves —
   Monitor 58

One-Arm-Man Knives —
   Silverplated, Bone Handle 159

Oyster Ladles —
   Silverplated 155, 167

Ox Bows —
   Bows, Shoes, Yokes 234

P

Padlocks —
   Assorted Styles .... 266 to 270

Pails —
   Granite Ironware 127
   Wooden 95

Parers —
   Apple Parer & Corer .... 77 to 80
   Fruit & Vegetable .... 77 to 80

Paste Jagger —
   Brass 101

Patty Pans —
   Granite Ironware 127
   Pieced Tinware 106

Pepper Boxes —
   Tin Shaker 108

Pickle Casters —
   Glass, Silverplated 181
   Pickle Spear 168

Pie Knives —
   Silverplated Server .... 156 to 166

Pincers —
   Coal Tongs 35
   Hair Irons 359

Pitcher & Bowl Sets —
   Granite Ironware 128
   Japanned Tin 205
   Stands for above 204

Pitchforks —
   Barley Forks, Wood 237
   Steel Pitch Fork 237

Plate Lifters —
   Sherwood's Triumph 82

Plug Cutters —
   Tobacco 57

Pokers —
   Fire .... 30 to 35

Post Hole Diggers —
   Eureka, Samson, Giant 227

Potato —
   Parers and Slicers .... 77 to 80
   Ricer/Masher 81
   Mashers (all styles) 85

Potato Scoops —
   Steel Wire Shovel 231

Pots —
   Stove Pot, Cast Iron 119
   Stove Pot, Granite Iron 129

Pot Cleaners —
   Interlocked Rings 82

Presses —
   Fruit 81
   Jelly 81
   Meat 220

Pudding Molds —
  Plain Tin 113
Pumps, Well —
  Chain 226
  Garden, Brass 225
  Kitchen, Pitcher 225
  Town, Curb 226
Push Plates —
  Door 253

R

Rakes —
  Hay Rakes, Wooden 231
Refrigerators —
  Wooden, Zinc Lined 136, 137
Rice Boiler —
  Granite Ironware 128
Rolling Pins —
  Glass and Wood 85, 92

S

Sad Irons —
  Adult .... 350 to 357
  Children's 351, 355
  Heaters 118, 353
  Stands (Trivets) .... 349 to 357
Salad Fork —
  Silverplated 167
Salt Boxes —
  Maple, Walnut, etc. 93
Salt Cellars —
  Silverplated 175, 177
Salt and Pepper Sets —
  Glass and Silverplated 153, 177
Salt Spoons —
  Silverplated 156
Sardine —
  Knives 97
  Scissors 96
Sauce Pans —
  Granite Ironware 128
Sausage Stuffers —
  Butcher's 219, 220
  Family Size 219, 221
Scales —
  Family, Farm & Market 208–217
Scissors —
  Assorted Styles and Sizes .... 331 to 339
Scoops —
  Cooks, Wood Handle 117
  Solid Wood 91
  Grocers', Tinned 117
Screens —
  Window Cloth 201
Scythes —
  With Grain Cradles 231
Seed Strippers —
  Wooden Gatherer 232
Sewing Machines —
  Introduction to Collecting 320
  Price Guide to Rare Models 321
  History of Invention 322
  Illustrations (1850–1915) 21, 323, 329
Sewing Novelties —
  Sterling Silver Sets, etc. 339

Shelf Brackets —
  Fancy Iron 238
  Corner Brackets 260
Shovels —
  Coal Forks 231
  Potato Scoops 231
  Snow Shovels, Wood 94
Silverplated Ware —
  Baskets .... 185 to 187
  Desk Accessories 175, 180, 182
  Flatware .... 155 to 169
  Holloware .... 149 to 187
  Napkin Rings 177, 180, 184
  Tea Sets 170
Skewers —
  Polished Steel, Wire 82
Skillets —
  Cast Iron 119
  Granite Ironware 129
Skimmers —
  Metal, Flat Handle 116
  Metal, Wood Handle 116
Slates —
  School, Noiseless 93
Slaw Cutters —
  One and Two Knives 87
Slop Bowls —
  Granite Ironware 126
  Silverplated 178
Slop Jars —
  Granite Ironware 127
  Japanned 204, 205
Smoothing Irons —
  Mrs. Pott's and others .... 349 to 357
Soup Ladles —
  Granite Ironware 131
  Stamped Tinware 116
Spiders —
  Cast Iron 119
  Granite Ironware 129
Soap Holders —
  Wall-mount, Graniteware 128
  Wall-mount, Wire Basket 204
Soapstone —
  Griddles & Footwarmers 122
Spice Cabinets —
  Wooden, 8-Drawer 93
Spice Jars —
  Porcelain Canisters 67
Spice Tins —
  In Japanned Boxes 188
Spittoons —
  Brass and Japanned 190
  Granite Ironware 190
Spinning Wheels —
  German, Factory-made 93
Spoon Holders —
  Granite Ironware 126
  Silverplated 173, 174
Sponge Baskets —
  Wire, Hanging 204
Spoons —
  Basting, Graniteware 128
  Mixing, Wooden 91
  Silverplated 155–169

Spring Balances —
    Assorted Scales .... 208 to 217
Sprinklers —
    Garden, Green 189
Steamers —
    3 Pc. Set, with Teakettle 93
Steak Pounders —
    Metal, Comb. Ice/Steak 139
    Wooden Maul 85
Stoves —
    Coal and Wood .... 61 to 74
    Oil Stoves 58
    Pocket Stoves 59, 117
    Portable Cook 59
String Holders —
    (Also called Twine Boxes) .... 48 to 50
Sugar Bowls —
    Granite Ironware 126
    Silverplated 170, 178, 183
Sugar Shells —
    Silverplated .... 155 to 169
Sugar Tongs —
    Silverplated .... 155 to 169
Syllabub Churn —
    Tin Cylinder 106
Syrups —
    Granite Ironware 126
    Silverplated 171, 182

T

Tables —
    Kitchen and Ironing 95
    Settee, Table/Bench 95
Tailor's Shears —
    Heinisch's 330
Tap Borers —
    Common, Wood Handle 86
Tea Canisters —
    Japanned Tinware 189
Tea Kettles —
    Brass 26
    Cast Iron 118
    Copper 112, 115
    Granite Ironware 127
    Nickle-plated 112
    Tin 111, 113
    Range 112, 113
Tea Pots —
    Granite Ironware .... 125 to 133
    Silverplated 170, 178
    Tinware .... 110 to 114
Tea Sets —
    Silverplated 170, 178
Thermometers —
    Home, Office, and Dairy 279
Thimbles —
    Gold and Silver 336, 337
Tinware —
    Molds 108, 109
    Tea & Coffee Pots .... 110 to 115
    Dippers 116
    Scoops 117
Tobacco Cutters —
    Countertop 57
Toddy Warmer —
    Tinware Etna 108

Toothbrush Holders —
    Wire Rack 204
Toothpick Holders —
    Silverplated 180
Towel Stands —
    Japanned 204
Traps —
    Fly 201
    Mouse .... 198 to 200
    Rat 200
    Roach 201
Trays —
    Tea, Silverplated 170, 179
Trivets —
    Coffee Pot 353
    Sad Iron .... 349 to 357
Trunk —
    Locks and Keys 271
Tumblers —
    Granite Ironware 127
Twine Boxes —
    (Also called String Holders) .... 48 to 50

U

Umbrella Stands —
    Bronzeplated 31
    Japanned 204

V

Vegetable —
    Cutters 87
    Presses 81

W

Waffle Irons —
    Cast Iron 122, 123
Wash Benches —
    Wooden 95
Wash Boards —
    Brass, Glass, Enamelled 342, 343
Wash Boilers —
    Copper and Tin 348
Wash Bowls & Pitchers —
    Granite Ironware 128
    Tinware, Japanned 205
    Stands 204
Washing Machines —
    Advertisements 345, 346, 349
    History of 341, 342
Water Carriers —
    Granite Ironware 127, 129
    Japanned Tin 204, 205
Water Coolers/Filters —
    Galvanized Iron 134, 135
    Stands, Drawers 134
Water Pitcher Sets —
    Tilting, Silverplated 152, 154
Wheelbarrows —
    Garden 236
Wicker Furniture —
    Chairs, Settees, Stands .... 306 to 311
Wine Coolers —
    Granite Ironware 127
Wringers —
    Clothes .... 341 to 347

Graniteware showroom circa 1900. — Smithsonian Institution

VICTORIAN FACTORY-MADE ANTIQUES

by Linda Campbell Franklin, Contributing Editor

The official "Victorian" period lasted 64 years. Alexandrina Victoria (1819-1901) succeeded to the throne of Great Britain and Ireland in 1837, and was crowned queen the next year. The jubilee of her reign came in 1887, and her diamond jubilee in 1897, a few years before her death.

Queen Victoria's many artistic talents and her ideas about moral behavior affected social and family life, literature, and the decorative arts in Great Britain, as well as in the United States. The useful arts — in function and appearance — were also the beneficiaries of this period of expansion.

The Industrial Revolution was rushing full steam ahead by the middle of the 19th century. The first steam railways had been operating in England for only 15 years when Victoria ascended the throne. In the 1830's, these railways expanded all over Britain, helping to create and interconnect the vast, noisy, smoking manufacturing districts specializing in iron founding, pottery making, cloth weaving, shoe making, brass casting, and other trades. The monstrous iron locomotives were more than symbolic proof of machinery's successful race in to the future.

In North America, steam railways developed at the same time, with short runs in place in the Eastern United States in the 1830's, and completion of the first transcontinental line in 1869. A total of 50,000 miles of wrought iron track had been laid. In Canada, the first railroads were operating by 1836. Both developments meant a rapid expansion of iron foundries, brass mills, and other factories making tools and the giant lathes and presses that made even more tools. By the time Victoria was queen, everything was in place in England and North America to start turning the figments of inventors' imaginations into real, mass-produced objects by the hundreds of thousands, if necessary.

As for demand, there seemed to be room for thousands of inventions in the United States alone, where a numbered patent system didn't start until 1836 (just one year before Victoria succeeded to the throne). There were 98,459 patents granted by the end of 1869, when the western and eastern train tracks met at Promontory Point, Utah. Isaac M. Singer had retired three years earlier, a multimillionaire, and his factory produced 127,833 sewing machines in 1870. In 1871, the Metropolitan Washing Machine factory was shipping 600 tubs a day. The population, as counted by the 1870 census, was 39,818,449 people. By the end of the Centennial Year in 1876, which begins the period for most of the catalog pictures in this book, there had been 185,812 patents granted. By 1901, when Victoria died, the United States Patent Office had granted 690,384 patents. However, the million mark wasn't reached in the United States until 1911.

The death of Victoria was nominally the end of the "Victorian Period," but older household tools and decorations such as the ones depicted here didn't suddenly disappear, to be replaced by new forms. Many utilitarian items were made the same way, the same shape, even with the same decoration to some extent, from the 1870's into the 1930's. If it ain't broke, don't fix it; and if it sells, don't redesign it.

PLATE 2.

SEWING MACHINES.

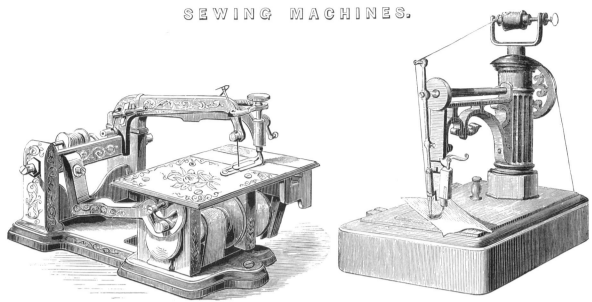

Fig. 1.—THE WANZER MACHINE.   Fig. 2.—WHIGHT & MANN'S MACHINE.

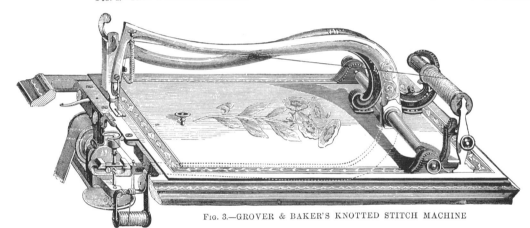

Fig. 3.—GROVER & BAKER'S KNOTTED STITCH MACHINE

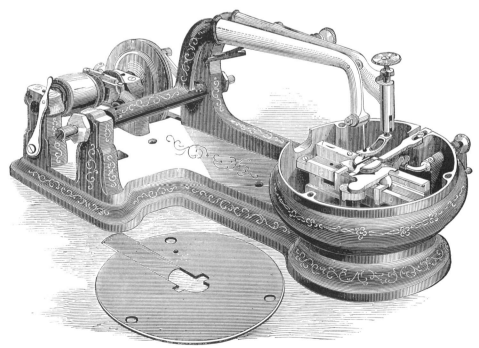

VOL. XI.   Fig. 5.—THE 'FLORENCE' MACHINE.

Gossip at the pump, circa 1899. — *Author's Collection*

An 1890's General Store. — *Author's Collection*

Many of the objects from the so-called "Victorian Period" so avidly collected today were simple necessities of life. Such things as tin bathtubs, enameled kitchenware, locks, keys, tools, scales, irons, and washboards changed very little over a fifty-year period. Fashionable items, such as lamps and furniture, were made pretty much the same way — it was just the decorations or the shapes that changed, or perhaps the maker utilized a newly invented or improved part. The same is true of silverplated flatware and serving pieces. The plating process itself (which democratized silver by making an inexpensive substitute), was a triumph of new scientific methods, and so were the intricate decorations. Popular repousse' designs (raised in relief, usually by stamping from the other side) were possible because of powerful steam-driven stamping machines which made the same patterns over and over on hundreds of pieces an hour.

To our jaded eyes, the mass-produced goods of the second half of the 19th century seemed positively handcrafted — real works of art! But this isn't true. Almost every single thing depicted in this book was basically machine-made. Manufacturers were geared to produce vast quantities of attractive, inexpensive goods which would be bought by hundreds of thousands of people. The pool of customers appeared to be bottomless. These consumer goods were not made with the idea of their breaking down, wearing out, and being replaced. Manufacturers counted on a rapidly growing population, built by large families and a steady stream of immigrants, plus an increasing demand created by illustrated periodicals of the day.

Today's collectors are reaping rewards from this booming manufacturing period. It was a time when large banks frequently encouraged speculation and greedy corporations printed bogus stock, both of which greatly unbalanced the economy. Panics and crashes were frequent, and consumer prices remained relatively low for longer periods than they do today.

Most of the utilitarian objects in the book have few claims to being decorative or ornamental in the traditional sense of the word "antique." These are the mechanical laborsaving devices which occupied most of the inventive genuis of a hundred and twenty years ago. Some of these objects had decorative cast iron parts incorporating hearts, flowers, and bunches of grapes. However, their main selling point was an efficient mechanical design, heightened (or hidden) by these decorations, which were part of the Victorian mind set. It seems naive, even charming, now, to find this evidence of an almost sentimental approach to the design of a household tool. And in many ways these decorative elements were integrated into products, not just tacked on as are the decals applied to some of today's housewares. To some art world experts this naivete of design qualifies it as a form of folk art. Handcranked farm machinery is beginning to appear in commercial galleries, and some 19th century sewing machines are bringing more than $1,000 each.

All the 19th-century laborsaving devices in the world couldn't actually free the housewife from her chores. They simply meant that she was expected to do more work in less time. Advertisements in the 1880's did the same job they do today. They sold lifestyles, morals, appearances and values, and said in so many words that to have an ideal home life you had to cook, clean, iron, sew, serve, entertain, bathe, wash, dust, preserve, and decorate to a very high standard.

"Mary, I just came down stairs to see where this delightful aroma of Coffee comes from. It perfumes the whole house."

We who love to collect (that is, to form select assemblages of objects that look or function in similar ways) can comfort any nagging doubts we have about the value of what we do by reassuring each other that we are recycling in a very educational way. We are giving landfills an extended life by not cluttering them with old wagon wheels, worn-out stoves, and wicker rockers. We might also, along the way, reflect on the overwhelming waste of throwaway societies, and try to rearrange some small part of our lives to be more in keeping with the simple pleasures of our Victorian heritage.

NOTE: Contributing editor Linda Campbell Franklin (who lives in Virginia and New York) has been writing about antique and collectible household tools and utensils for almost twenty years. She is responsible for most of the kitchen and houseware pricing in this book.

The third edition of her "300 Years of Kitchen Collectibles," which sold more than 150,000 copies in its second edition, has just been published, and I recommend it highly. It has over 7,000 detailed descriptions including value ranges, and pictures of several thousand objects. It also includes many old recipes, and hundreds of quotations from 19th century sources about all kinds of objects used in the kitchen. If you are looking for a complete history of kitchen tools you will want to add this fascinating volume to your collection.

Book may be ordered from the publisher, Books Americana, through your bookstore, or directly from the author, Linda C. Franklin, 2716 Northfield Road, Charlottesville, VA 22901. When requesting information, please include a self-addressed stamped envelope.

BRASS ANDIRONS.

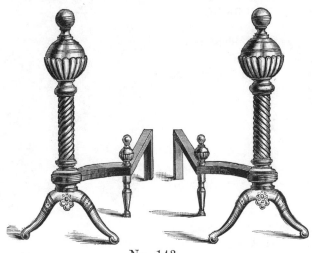

No. 143.

No. 143—Circular, 20 inches high, ... per pair, $17.50

No. 132.   No. 140.

No. 132—18 inches high, ... per pair, $9.00
No. 140—18 inches high, ... " 8.50

No. 139.   No. 115. Queen Anne Pattern.

No. 139—16½ inches high, ... per pair, $ 6.50
No. 115—20½ inches high, ... " 15.00

BRASS ANDIRONS.

No. 135.

No. 135—Circular, 26 inches high, ... per pair, $24.00

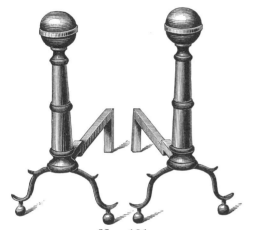

No. 101.

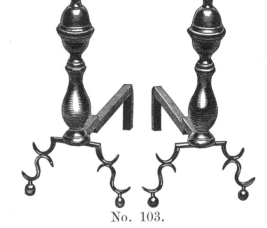

No. 103.

No. 101—16 inches high, ... per pair, $5.50
No. 102—20 inches high, ... " 6.50
No. 103—18 inches high, ... " 7.00

No. 107.

No. 104.

No. 107—18 inches high, ... per pair, $12.00
No. 104—22 inches high, ... " 8.75

BRASS CANDLESTICKS.

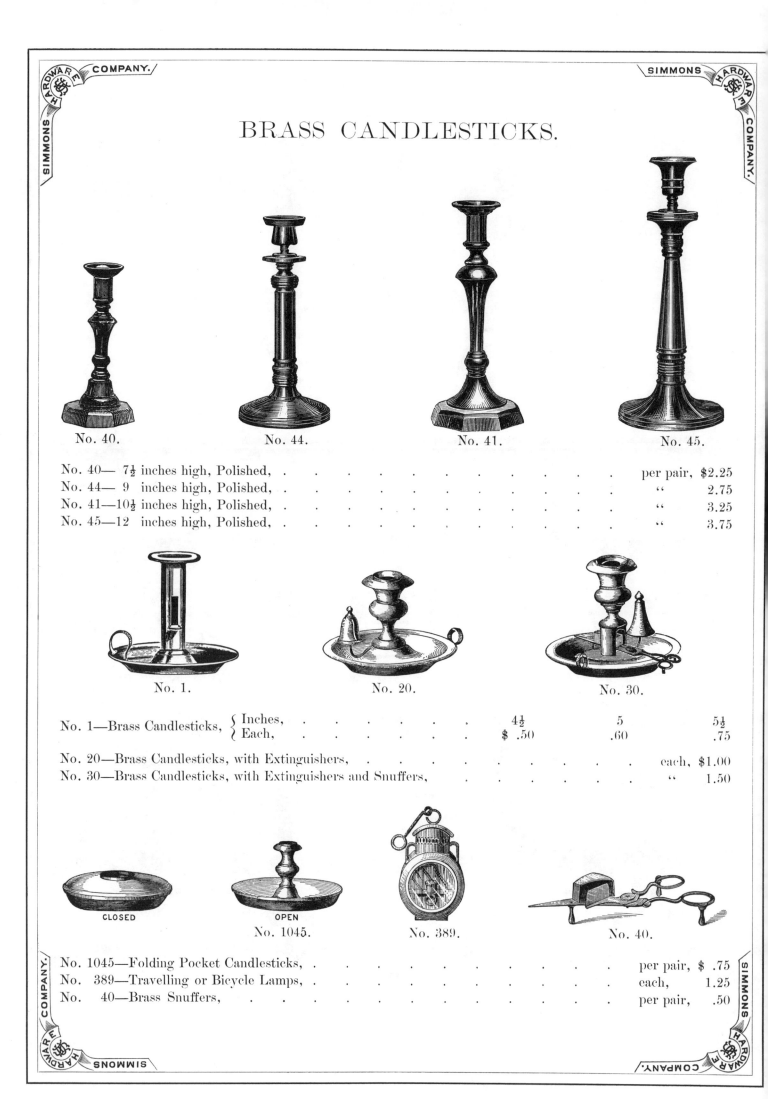

No. 40.   No. 44.   No. 41.   No. 45.

No. 40— 7½ inches high, Polished, ... per pair, $2.25
No. 44— 9 inches high, Polished, ... " 2.75
No. 41—10½ inches high, Polished, ... " 3.25
No. 45—12 inches high, Polished, ... " 3.75

No. 1.   No. 20.   No. 30.

| No. 1—Brass Candlesticks, | Inches, | 4½ | 5 | 5½ |
| | Each, | $ .50 | .60 | .75 |

No. 20—Brass Candlesticks, with Extinguishers, ... each, $1.00
No. 30—Brass Candlesticks, with Extinguishers and Snuffers, ... " 1.50

CLOSED   OPEN
No. 1045.   No. 389.   No. 40.

No. 1045—Folding Pocket Candlesticks, ... per pair, $ .75
No. 389—Travelling or Bicycle Lamps, ... each, 1.25
No. 40—Brass Snuffers, ... per pair, .50

BRASS CANDLESTICKS.

No. 449.   Cut two-thirds size.

No. 429.   Cut two-thirds size.

No. 392.   Cut two-thirds size.

No. 449—Polished Brass Candlesticks, ... per pair,	$9.00
No. 429—Polished Brass Candlesticks, ... "	8.00
No. 392—Polished Brass Candlesticks, ... "	5.00

BRASS CANDLESTICKS.

No. 414.   Cut two-thirds size.

No. 387.   Cut two-thirds size.

No. 447.   Cut two-thirds size.

No. 414—Polished Brass Candlesticks,	per pair, $ 4.00
No. 387—Polished Brass Candlesticks,	" 12.00
No. 447—Polished Brass Candlesticks,	" 4.50

BRASS TEA KETTLES.

	Brass.	Bronzed.
—Capacity, 1 pint, each,	$2.00	$2.75
—Capacity, 1 quart, "	3.00	4.00
—Capacity, 2 quarts, "	3.50	4.75
—Capacity, 3 quarts, "	4.00	5.25

complete with Alcohol Lamp,

FIRE IRONS.

Polished Steel.

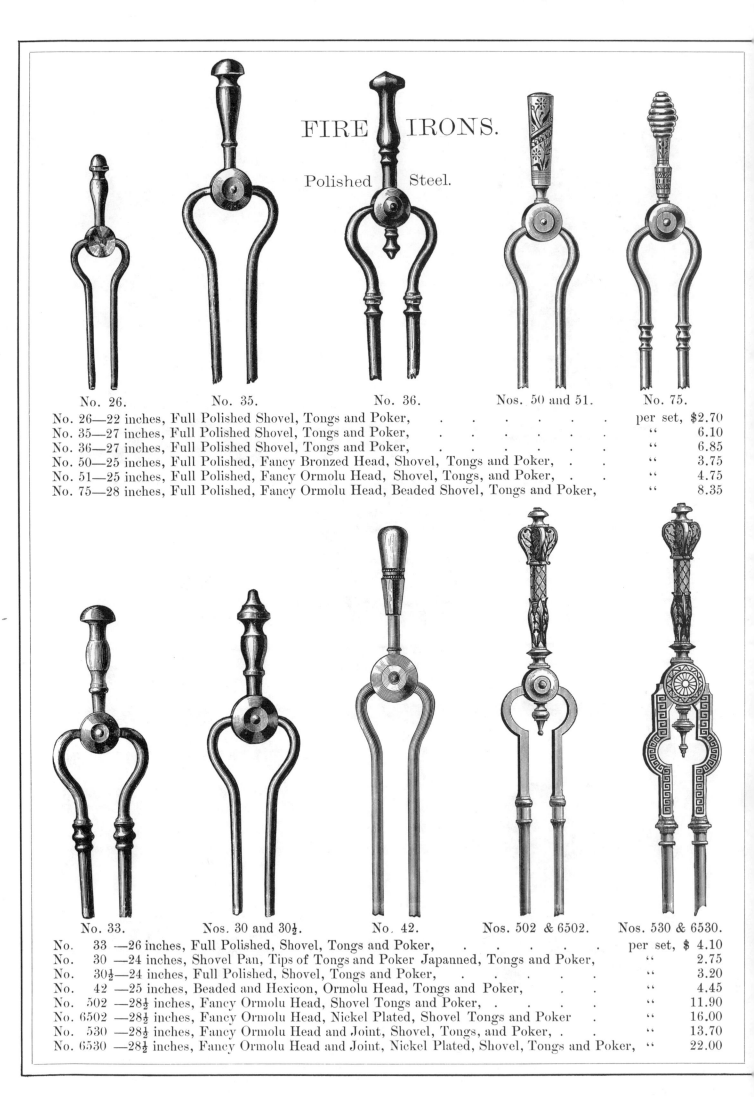

No. 26. No. 35. No. 36. Nos. 50 and 51. No. 75.

No. 26—22 inches, Full Polished Shovel, Tongs and Poker,	per set, $2.70
No. 35—27 inches, Full Polished Shovel, Tongs and Poker,	" 6.10
No. 36—27 inches, Full Polished Shovel, Tongs and Poker,	" 6.85
No. 50—25 inches, Full Polished, Fancy Bronzed Head, Shovel, Tongs and Poker,	" 3.75
No. 51—25 inches, Full Polished, Fancy Ormolu Head, Shovel, Tongs, and Poker,	" 4.75
No. 75—28 inches, Full Polished, Fancy Ormolu Head, Beaded Shovel, Tongs and Poker,	" 8.35

No. 33. Nos. 30 and 30½. No. 42. Nos. 502 & 6502. Nos. 530 & 6530.

No. 33 —26 inches, Full Polished, Shovel, Tongs and Poker,	per set, $ 4.10
No. 30 —24 inches, Shovel Pan, Tips of Tongs and Poker Japanned, Tongs and Poker,	" 2.75
No. 30½—24 inches, Full Polished, Shovel, Tongs and Poker,	" 3.20
No. 42 —25 inches, Beaded and Hexicon, Ormolu Head, Tongs and Poker,	" 4.45
No. 502 —28½ inches, Fancy Ormolu Head, Shovel Tongs and Poker,	" 11.90
No. 6502 —28½ inches, Fancy Ormolu Head, Nickel Plated, Shovel Tongs and Poker	" 16.00
No. 530 —28½ inches, Fancy Ormolu Head and Joint, Shovel, Tongs, and Poker,	" 13.70
No. 6530 —28½ inches, Fancy Ormolu Head and Joint, Nickel Plated, Shovel, Tongs and Poker,	" 22.00

FIRE IRON STANDS.

No. 8.   No. 556—Umbrella Stands.   No. 134.

No.   8—Fire Iron Stands, Gold Bronzed on Assorted Colors, 22½ inches high, . . . each, $ .75
No. 134—Fire Iron Stands, Berlin Bronzed, 27 inches high, ... " 1.50

No. 91.   Nos. 1278 and 1280.   No. 9.

No. 91—Fire Iron Stands, Green and Gold Bronzed, 22 inches high, ... each, $1.25
No.   9—Fire Iron Stands, Gold Bronzed on Assorted Colors, ... " 1.25
   No. 1280—Fire Iron Stands, Nickel and Blue, 22 inches high, . " 6.00
   No. 556—Umbrella Stands, Green and Gold Bronze, 32 inches high, " 2.75

Bronzed Fire Iron Stands.

Quarter Size Cut of No. 32. Patented.

Nos. 35, 36 and 38. Quarter Size Cut of Nos. 35 and 36.

Berlin Bronzed.

No. 32, Height 28½ Inches, . per dozen, $26 00

Tuscan Bronzed.

No. 35, Height 24 Inches, . per dozen, $21 00

Berlin Bronzed.

No. 36, Height 24 Inches, . per dozen, $23 00
No. 38, " 26 " . . " 24 00

Bronzed Fire Iron Sets.

Brass

No. 106.

per set, 8 50

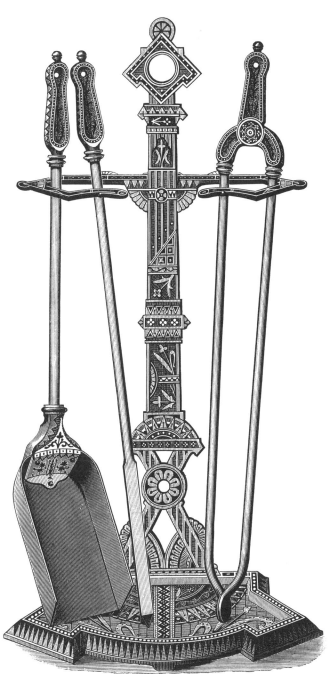

Nos. 33 and 34. Quarter Size Cut.

No. 94.

Tuscan Bronzed.
No. 33, Stand, Poker, Shovel and Tongs, complete as per cut, . . . per dozen, $40 00

Berlin Bronzed.
No. 34, Stand, Poker, Shovel and Tongs, complete as per cut, . . . per dozen, $43 00

Bronzed Fire Dogs.

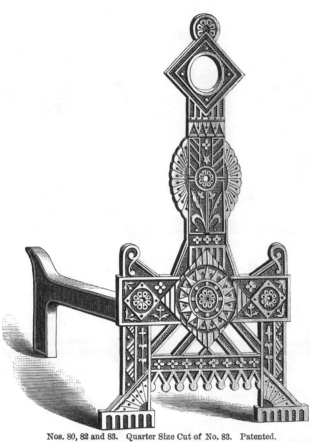

Nos. 80, 82 and 83. Quarter Size Cut of No. 83. Patented.

Berlin Bronzed.

				Per pair.
No. 80, Height 15¼ In.,	Length of Leg 12 In.,	$2 90
No. 82, " 15¼ "	" " " 15 "	3 00
No. 83, " 17¼ "	" " " 18½ "	3 50

Nos. 62 and 63. Quarter Size Cut of No. 62. Patented.

Nos. 72 and 73. Quarter Size Cut of No. 73. Patented.

Berlin Bronzed.

No. 62, Height 15½ Inches, . per pair, $2 80	No. 72, Height 16 Inches, . per pair, $3 25
No. 63, " 18 " . . " 3 50	No. 73, " 18 " . . " 3 75

Cottage Fire Iron Sets.

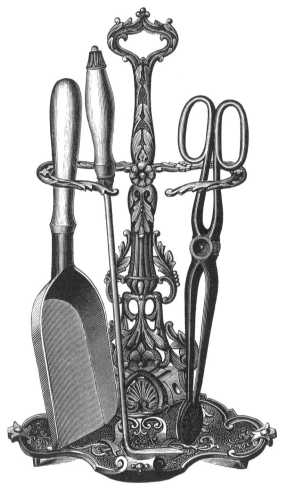

Stands only No. 21.
Complete, as per cut, No. 23.

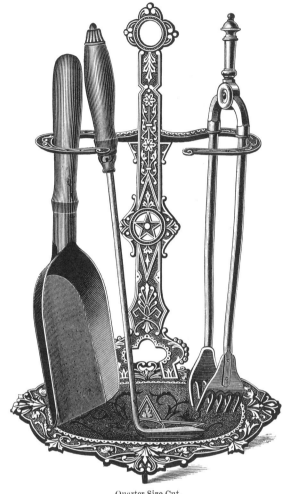

Quarter Size Cut.
Complete, as per cut, Nos. 222 and 226.

TUSCAN BRONZED.

No. 21, Tuscan Bronzed Stands (Stands only), ... per dozen, $14 85
No. 23,   "   "   " with Poker, Shovel and Tongs, . . . " 25 50
No. 126, Stand, Poker, Shovel and Tongs, as per cut, . . . per dozen, $19 60
No. 226,   "   "   "   " Spring Claw Tongs, as per cut, . . . " 20 85

ALL BERLIN BRONZED.

No. 122, Stand, Poker, Shovel and Claw Tongs, as per cut, . . . per dozen, $28 00
No. 222,   "   "   "   " Spring Claw Tongs, as per cut, . . " 29 25

CALL BELLS.

No. 6450.

Nos. 3000 and 3000½.

No. 8050.

No. 6450 —Call Bells, Nickel and Verde, Fancy Base, ... each, $1.65
No. 3000 —Call Bells, Bronze, Fancy Base, ... " 1.00
No. 3000½—Call Bells, Gilt, Fancy Base, ... " 1.25
No. 8050 —Call Bells, Nickel and Verde, Fancy Base, ... " 1.15

No. 3400.

No. 3700.

No. 5900.

No. 3400—Call Bells, Bronze, Fancy Base, ... each, $1.15
No. 3700—Call Bells, Gilt, Fancy Base, ... " 2.25
No. 5900—Call Bells, Gilt, Fancy Base, ... " 1.65

No. 2800.

No. 75.

No. 2900.

No. 2800—Call Bells, Bronze, Fancy Base, ... each, $1.00
No.   75—Call Bells, Bronze, Fancy Base, ... " .85
No. 2900—Call Bells, Bronze, Fancy Base, ... " 1.15

CALL BELLS.

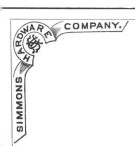
No. 6600.

No. 8800.

No. 2400.

No. 6600—Call Bells, Electric Alarm, Bronze, Fancy Base, ... each, $3.50
No 8800—Call Bells, Revolving Gong, Black Base, ... " 4.50
No. 2400—Call Bells, Hotel, Marble Base, ... " 5.00

Brass Hand Bells.

Silver Chime Hand Bells.

Tea Bells.

Brass Hand Bells,	Nos.	0	1	2	3	4	5	6	7	8	9	10	12	14
	Diam. at Mouth,	$2\frac{1}{8}$	$2\frac{1}{2}$	$2\frac{5}{8}$	3	$3\frac{1}{2}$	$3\frac{3}{4}$	4	$4\frac{3}{4}$	$5\frac{1}{4}$	$5\frac{1}{2}$	6	$6\frac{1}{2}$	$6\frac{3}{4}$
	Each,	$ .15	.20	.25	.35	.45	.50	.65	.75	1.00	1.35	1.75	2.00	2.65

Silver Chime Hand Bells,	Nos.	202	204	206	208	209	212
	Diameter at Mouth,	$3\frac{1}{4}$ in.	$3\frac{3}{4}$ in.	$4\frac{1}{2}$ in.	5 in.	$5\frac{3}{4}$ in.	7 in.
	Each,	$ .50	.75	1.00	1.65	2.00	3.25

Tea Bells, White Metal Bell, Plain Handle,	Nos.	4	6	7
	Each,	$ .10	.15	.20

Call Bells.

No. 8050. Bell, $1.00. Bronze Base.

No. 9416. Call Bell. $1.75 each.

No. 2200. Call Bell.

No. 9426. Electric Call Bell. $2.25 each.

NEW DEPARTURE.

No. 9418. Call Bell. $1.00 each.

NO. 213.

No. 385, Nickel Plated.

NO. 214.

No. 713, Silver and Gold.
No. 718, All Gold Plated.

No. 743, Silver and Gold.
No. 748, All Gold Plated.

No. 733, Silver and Gold.
No. 738, All Gold Plated.

No. 753, Silver and Gold.
No. 758, All Gold Plated.

TWO STORES CONTRASTED.

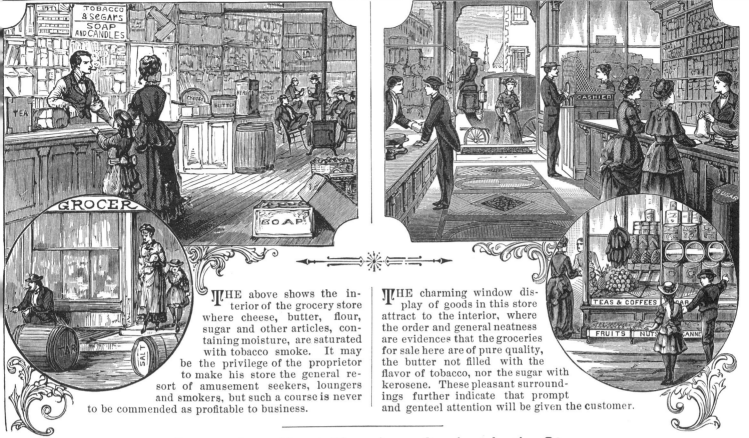

THE above shows the interior of the grocery store where cheese, butter, flour, sugar and other articles, containing moisture, are saturated with tobacco smoke. It may be the privilege of the proprietor to make his store the general resort of amusement seekers, loungers and smokers, but such a course is never to be commended as profitable to business.

THE charming window display of goods in this store attract to the interior, where the order and general neatness are evidences that the groceries for sale here are of pure quality, the butter not filled with the flavor of tobacco, nor the sugar with kerosene. These pleasant surroundings further indicate that prompt and genteel attention will be given the customer.

Suggestions About Shopping. Conduct in the Store.

PURCHASERS should, as far as possible, patronize the merchants of their own town. It is poor policy to send money abroad for articles which can be bought as cheaply at home.

Do not take hold of a piece of goods which another is examining. Wait until it is replaced upon the counter before you take it up.

Injuring goods when handling, pushing aside other persons, lounging upon the counter, whispering, loud talk and laughter, when in a store, are all evidences of ill-breeding.

Never attempt to "beat down" prices when shopping. If the price does not suit, go elsewhere. The just and upright merchant will have but one price for his goods, and he will strictly adhere to it.

It is an insult to a clerk or merchant to suggest to a customer about to purchase that he may buy cheaper or better elsewhere. It is also rude to give your opinion, unasked, about the goods that another is purchasing.

Never expect a clerk to leave another customer to wait on you; and, when attending upon you, do not cause him to wait while you visit with another. When the purchases are made let them be sent to your home, and thus avoid loading yourself with bundles.

Treat clerks, when shopping, respectfully, and give them no more trouble than is necessary. Ask for what is wanted, explicitly, and if you wish to make examination with a view to future purchase, say so. Be perfectly frank. There is no necessity for practicing deceit.

The rule should be to pay for goods when you buy them. If, however, you are trusted by the merchant, you should be very particular to pay your indebtedness when you agree to. By doing as you promise, you acquire habits of promptitude, and at the same time establish credit and make reputation among those with whom you deal.

It is rude in the extreme to find fault and to make sneering remarks about goods. To draw unfavorable comparisons between the goods and those found at other stores does no good, and shows want of deference and respect to those who are waiting on you. Politely state that the goods are not what you want, and, while you may buy, you prefer to look further.

If a mistake has been made whereby you have been given more goods than you paid for, or have received more change than was your due, go immediately and have the error rectified. You cannot afford to sink your moral character by taking advantage of such mistakes. If you had made an error to your disadvantage, as a merchant, you would wish the customer to return and make it right. You should do as you would be done by. Permanent success depends upon your being strictly honest.

Say "No" Politely.

A COMMON saying is, "A man's manners make his fortune." This is a well-known fact, and we see it illustrated every day. The parents who considerately train a child amid kindness and love, rear a support for their declining years. The teacher that rules well and is yet kind, is beloved by his pupils. The hotel proprietor, by affability and an accommodating spirit, may fill his hotel with guests. The railway conductor who has a pleasant word for the lonely traveler, is always remembered with favor. The postoffice clerk who very carefully looks through a pile of letters and says, "not any" very gently, pleasantly adding a word of hope by saying, "it may come on the afternoon train," we always gratefully recollect. When the time comes that we can return the kindness, we take great pleasure in doing so.

The man who shows himself to be a gentleman, even though he may not buy what we have to sell when we solicit him, we always know will get his reward. His affability, when he declined, demonstrated that he could say "no" with a pleasant word. The very fact of his impressing us so favorably, even when he did not purchase, clearly indicated that he was thoroughly schooled in the ways of politeness, and that he lived up to the golden rule of doing to others as he desired others to do to him.

Kentucky Cow Bells.

Half Size Cut of No. 1½.

Trade Mark.

Nos.	0	1	1½	2	3	4	5	6	7
Height (without loop),	7	6½	6	5¼	4½	4	3½	2½	2¼ Inch.
Per dozen,	$13 00	11 00	9 00	8 00	6 00	5 00	3 50	2 50	2 25

Western Cow Bells.

Half Size Cut of No. 3.

Trade Mark.

Nos.	1	2	3	4	5	6	7	8	9
Height (without loop),	6½	6	5¾	5½	5¼	4¾	4	3½	2½ Inch.
Per dozen,	$11 50	10 00	9 00	8 00	6 50	5 50	4 50	3 50	2 50

...ranted to measure correctly, and draw a gallon of the heaviest molasses in winter in one minute.

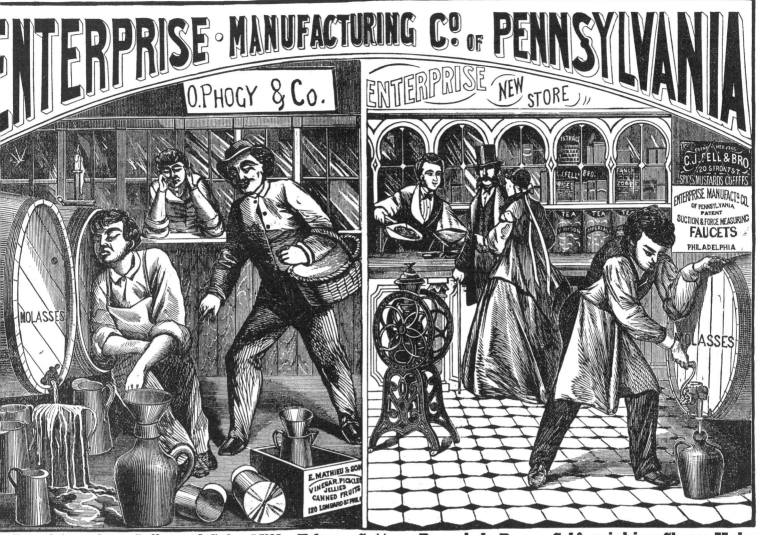

...uf'rs of American Coffee and Spice Mills, Tobacco Cutters, Bung-hole Borers, Self-weighing Cheese Knives,

ENTERPRISE PATENT COLD HANDLE SAD-IRONS, &c., &c.

...were awarded the **ONLY CENTENNIAL MEDAL** for **COFFEE MILLS.** } **THIRD AND DAUPHIN STREETS PHILADELPHIA.**

...atented Hardware Manufacturers and Iron Founders.

...TERPRISE COLD HANDLE, DOUBLE POINTED SMOOTHING IRONS,
(Mrs. Potts' Patent.)

...ee, Drug and Spice Mills, Measuring Faucets, Sausage Stuffers, Fruits, Lard and Jelly Presses, Cheese Knives, Tobacco Cutters, Bung Hole Borers.

ENTERPRISE MEASURING FAUCETS.

"Enterprise" Coffee, Drug and Spice Mills.

Fancy Iron Hopper.			Nickeled Dome Hopper.			Iron Hopper, on Stand.		
Nos.	Each.	Capacity per minute.	Nos.	Each.	Capacity per minute.	Nos.	Each.	Capacity per minute.
1	$2 00	6 oz.	2½	$4 00	6 oz.	15	$32 00	2 lbs.
2	3 00	6 "	4	8 00	½ lb.	16	37 00	2 "
3	5 00	½ lb.	6	12 00	¾ "			
5	8 00	¾ "	8	16 00	1 "	Nickeled Dome Hopper, on Stand.		
7	10 00	1 "	10	23 00	2 "	17	$40 00	2 lbs.
9	16 00	2 "	13	26 00	2 "	18	45 00	2 "
11	18 00	2 "	14	30 00	2 "	19	75 00	3 "
12	22 00	2 "				20	100 00	3 "

The "Swift" Mills—Coffee, Drug and Spice.

Highly Ornamented and Finished.

Swift Mill No. 12½.

Swift Mill No. 13½.

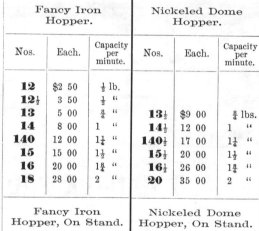

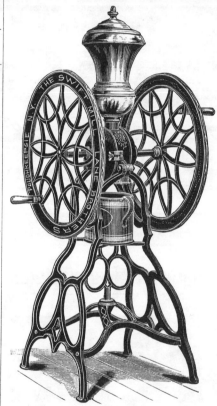

The Swift Mill No. 26.

Fancy Iron Hopper.			Nickeled Dome Hopper.		
Nos.	Each.	Capacity per minute.	Nos.	Each.	Capacity per minute.
12	$2 50	½ lb.			
12½	3 50	½ "			
13	5 00	¾ "	13½	$9 00	¾ lbs.
14	8 00	1 "	14½	12 00	1 "
140	12 00	1¼ "	140½	17 00	1¼ "
15	15 00	1½ "	15½	20 00	1½ "
16	20 00	1¾ "	16½	26 00	1¾ "
18	28 00	2 "	20	35 00	2 "

Fancy Iron Hopper, On Stand.			Nickeled Dome Hopper, On Stand.		
24	$37 00	2 lbs.	26	$45 00	2 lbs.
28	65 00	3 "	30	75 00	3 "

"Swift" Stand Mills.

For Corn, Coffee and Spices.

No. 3, Stand Mills, 20 Inch Fly Wheel, each, $6 75
No. 4,   "   "   24   "   "   "   "   8 00
No. 4½, "   "   30   "   "   "   "   9 00
No. 5,   "   " Two 24 In. Fly Wheels," 10 00

Can furnish also without Stand.

Lane Brothers' Patent Coffee Roasters.

For Grocers, etc.

No. 6, Size is about 3 feet 4 inches high, and weighs about 400 pounds, boxed.
It will roast thirty pounds at once, or sixty pounds an hour. . . Price, each, $50 00
Fitted for power, ... " 60 00

ENTERPRISE COFFEE AND SPICE MILLS.

No. 0.

No. 14.

SHOWING No 1 MILL CLOSED.

No. 3.

No. 5.

No. 2.

No. 214.

No. 9 Mill open.

No. 218.

Wm. Frankfurth Hardware Co.

COFFEE MILLS.

PARKER'S SIDE.

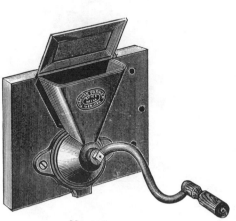

No. 135.

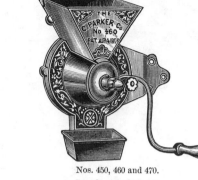

Nos. 450, 460 and 470.
Per dozen, $10 50

No. 50.

No. 135, Wilson's Improved, Wood Back,	- - - - -	per dozen, $6 00
50, Eagle, Small Side, "	- - - - -	" 7 50
60, " Medium " "	- - - - -	" 8 00
131, Old No. 0, Large, "	- - - - -	" 15 00
130, California Extra Large, "	- - - - -	" 25 00

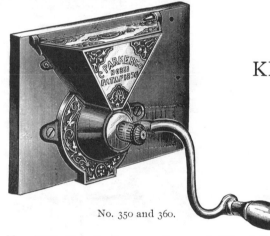

No. 350 and 360.

No. 350, Small, Copper Bronzed, Wood Back,
360, Medium, " " "

KENTUCKY HINGED

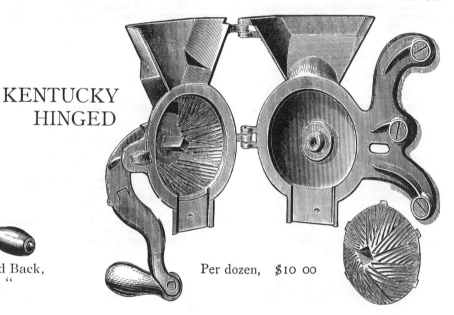

Per dozen, $10 00

NORTON'S CANNISTER.

Diamond Mill.
Copper Bronzed, . per dozen, $4 00

IRON BRACKET.

Anchor Mill and Cannister Combined, - per dozen, $9 00
No. 5, Iron Bracket Mill, - - - " 7 50

Milwaukee, Wisconsin.

COFFEE MILLS.

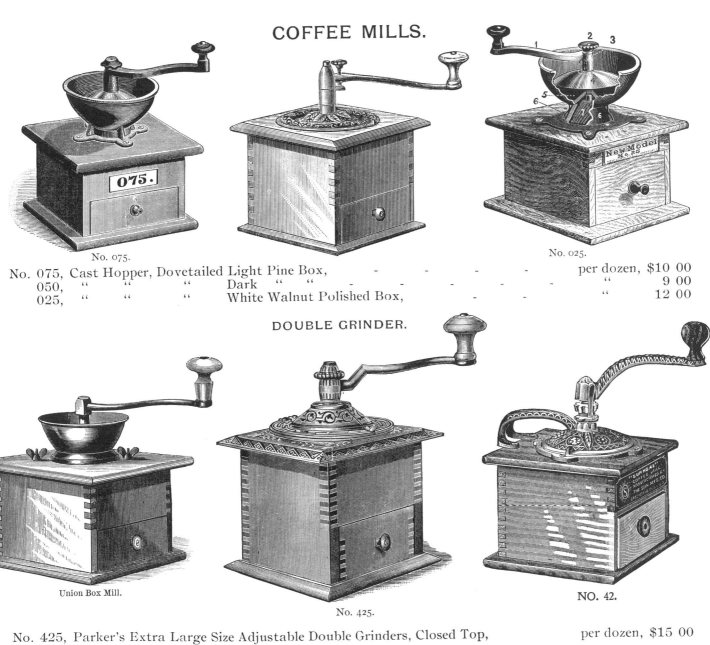

No. 075.   No. 025.

No. 075, Cast Hopper, Dovetailed Light Pine Box, - - - - per dozen, $10 00
   050,   "   "   "   Dark   "   " - - - - - "   9 00
   025,   "   "   "   White Walnut Polished Box, - - "   12 00

DOUBLE GRINDER.

Union Box Mill.   No. 425.   NO. 42.

No. 425, Parker's Extra Large Size Adjustable Double Grinders, Closed Top,   per dozen, $15 00

No. 55.   Parker's Iron Hopper,   No. 65.

No. 55, Parisian Pattern, with Patent Imported Steel Grinders and Patent Regulators, Wood Box, - - - - - - - - - - - - - - per dozen, $12 00
   65, Parisian Pattern, with Patent Imported Steel Grinders and Patent Regulators, Japanned Sheet Iron Box with Screw Cannister in the Bottom, - - - per dozen, 13 50

LOS ANGELES, CALIFORNIA.

COFFEE MILLS.

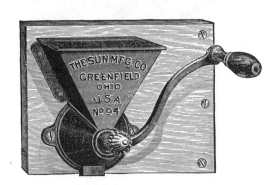

NO. 94.

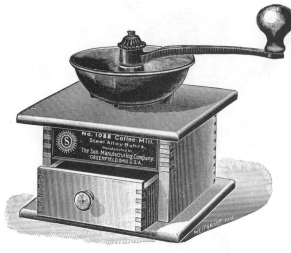

NO. 1088.

NO. 050.

NO. 1070.

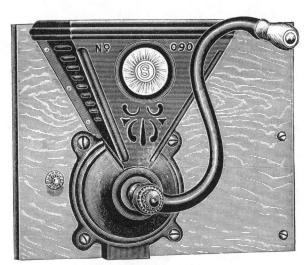

NO. 090.

NO. 1080.

Coffee Roasters.

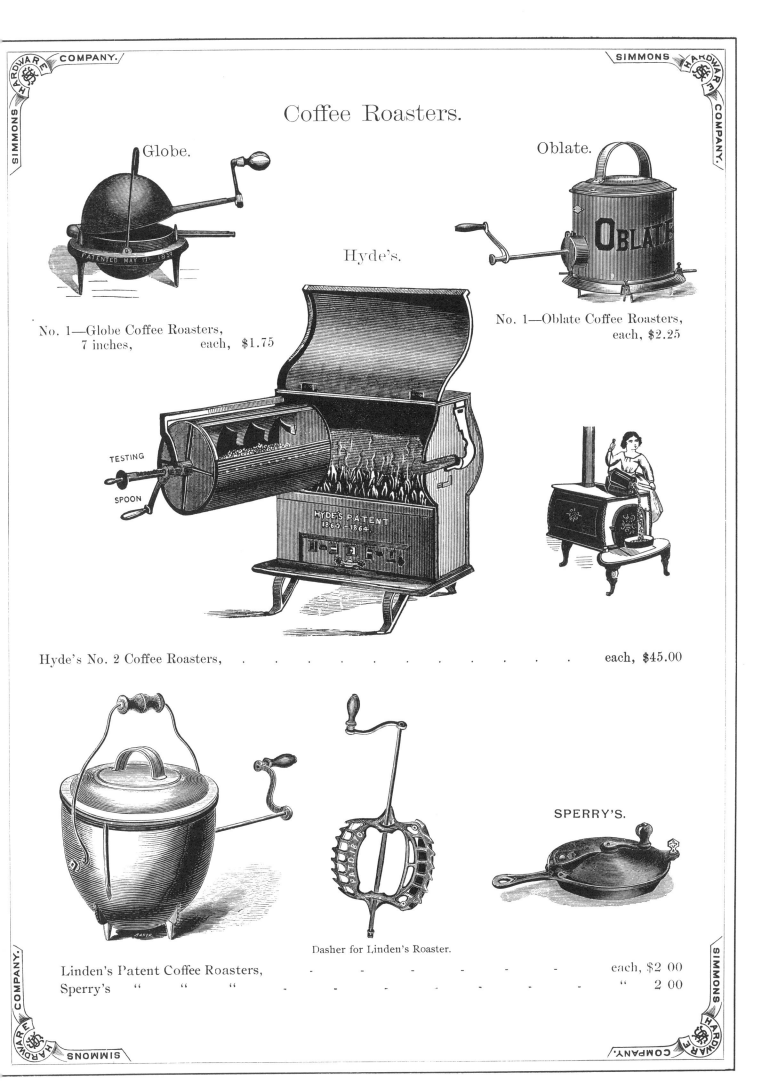

Globe.

No. 1—Globe Coffee Roasters,
7 inches, each, $1.75

Hyde's.

Oblate.

No. 1—Oblate Coffee Roasters,
each, $2.25

Hyde's No. 2 Coffee Roasters, ... each, $45.00

Dasher for Linden's Roaster.

SPERRY'S.

Linden's Patent Coffee Roasters, - - - - - - - - each, $2 00
Sperry's " " " - - - - - - - - " 2 00

SEYMOUR'S PATENT TWINE HOLDERS.

Nos. 1, 2, 3 and 4.

No. 1, Dark Bronzed, to hang, . per doz., $4 25
No. 2, " " large size, " 5 00
No. 3, Light Bronzed, " . " 4 25
No. 4, " " large size, " 5 00

Above numbers have Patent Brakes, as shown in Cut.

No. 01, Dark Bronzed, to hang, . per doz., $3 50
No. 02, " " large size, " 4 00
No. 0, Light Bronzed, " . " 3 50
No. 00, " " large size, " 4 00

These numbers are same style as Nos. 1, 2, 3, and 4, but without the Patent Brakes.

Nos. 0, and 3, are the most saleable numbers, and are large enough for ordinary size twine.

SMITH'S PATENT TWINE HOLDERS.

Nos. $12\frac{1}{2}$, $13\frac{1}{2}$, $14\frac{1}{2}$, and $15\frac{1}{2}$.

No. 12, Dark Bronzed, per doz., $5 00
No. 13, Light " " 5 00
No. 14, " " large size, " 6 00
No. 15, Dark " " " 6 00
No. $12\frac{1}{2}$, " " " 5 50
No. $13\frac{1}{2}$, Light " " 5 50
No. $14\frac{1}{2}$, " " large size, " 6 50
No. $15\frac{1}{2}$, Dark " " " 6 50

Nos. $10\frac{1}{2}$, and $11\frac{1}{2}$.

No. $10\frac{1}{2}$, Dark Bronzed, per doz., $5 00
No. $11\frac{1}{2}$, Light " " 5 00

These numbers have a patent twine cutter on the side, as shown in cut.

No. 10, Dark Bronzed, per doz., $4 50
No. 11, Light " " 4 50

RUSSELL & ERWIN MANUFACTURING CO.

WEBB'S PATENT
CAM-FASTENING HEMISPHERICAL TWINE BOXES.
IMPERIAL BRONZE.

No. 45, Regular Size,	per dozen,	$6 50	No. 65, Regular Size,	per dozen,	$8 50
No. 55, Large "	"	7 00	No. 75, Large "	"	9 00

Nos. 20 and 30.

No. 20, Bronzed, to Hang, per doz., $3 60
No. 30, Japanned,   "   "   3 25
No. 25, Bronzed, to Stand,   "   4 75
No. 35, Japanned,   "   "   4 00

Nos. 25 and 35.

Wm. Frankfurth Hardware Co.

TWINE BOXES.

Half Size Cut of No. 23.

Half Size Cut of No. 63.

No. 23, Coppered Bronzed, Drop Catch Hinge, - - - - - per dozen, $3 75
63, " " - - - - - - - " 7 00

No. 23, One-third Dozen in a Box; No. 63, One-sixth Dozen.

Bee-Hive Twine Boxes.

No. 31, Size of Twine Chamber, 4x4½ inches, Japanned, per dozen, $7 00

No. 76, Berlin Bronzed, per dozen, $8 50

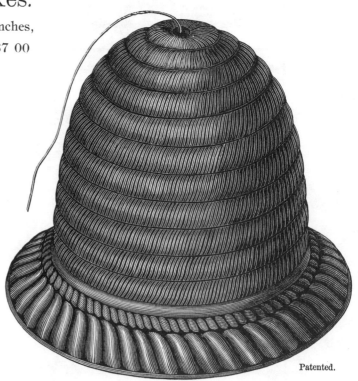

Half Size Cut of No. 31.

MATCH SAFES.

Verde Antique.

No. 80½. Cut full Size.

No. 10.   Nos. 2 and 3.

No. 80½—Match Safes, Berlin Bronzed,	each, $	.25
No. 10 —Match Safes, Bronzed, for Ordinary Matches,	"	.20
No. 2 —Match Safes, Bronzed, for Ordinary Matches,	"	.20
No. 3 —Match Safes, Bronzed, for Long Matches,	"	.25

MATCH SAFES.

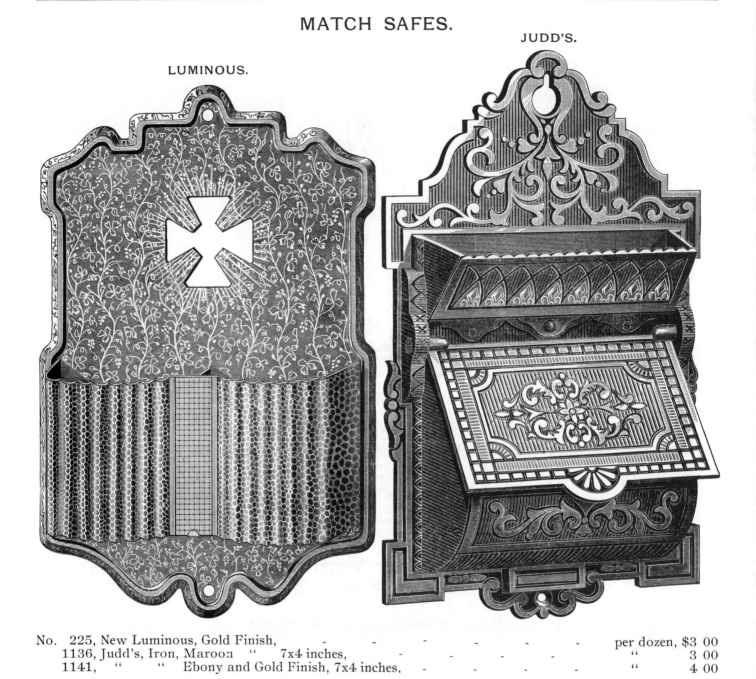

No. 225, New Luminous, Gold Finish, - - - - - - per dozen, $3 00
    1136, Judd's, Iron, Maroon " 7x4 inches, - - - - " 3 00
    1141, " " Ebony and Gold Finish, 7x4 inches, - - - - " 4 00

FRICTION MATCH PLATES.

No. 2, Adamantine Bronze, - - - - - - - - per dozen, $1 60

Ornamental Match Safes.

Berlin Bronzed. | **Imperial Bronze.**

No. 40, Berlin Bronzed, . per dozen, $1 70 | No. 48, Imperial Bronze, . per dozen, $9 00

Berlin Bronzed. | **Imperial Bronze.**

No. 50, Berlin Bronzed, . per dozen, $1 85 | No. 58, Imperial Bronze, . per dozen, $10 00

MATCH SAFES.

SQUARE.
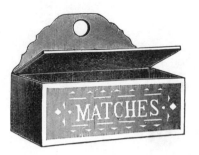

ROUND.

Square, Large, Japanned Tin, 3¾x2½x1½ inches, - - - - - per dozen, $0 75
Round   "   "   2½x4½   "   - - - - -   "   1 25

DAISY.

TWIN.

Daisy, Japanned Tin, Self-closing, - - - - - - - per dozen, $1 25
Twin,   "   "   - - - - - - - -   "   1 00

Full Size Cut of No. 1184.

One-half Size Cut of No. 30.

No. 1184, Judd's, Iron, Bronzed Maroon Finish, - - - - per dozen, $1 50
   30, Parker's Berlin Bronzed, Self-closing, - - - - -   "   1 75

Ornamental Match Safes.

 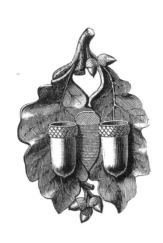 

All Fancy Bronzed.

No. 34.	No. 35.	No. 37.
Single Oak Leaf, per doz., $3 00	Double Oak Leaf, per doz., $5 50	Game Pattern, per doz., $4 00

Bronzed in Imitation of Black Walnut.

No. 30.	No. 31.	No. 33.
Rustic Pattern, per dozen, $3 60	Basket Pattern, per dozen, $3 10	Double, . per dozen, $6 00

Hanging Basket or Bird Cage Hooks.

One-third Size Cut of No. 20.　　　　　Nos. 71 and 72.　One-third Size Cut of No. 71.　Patented.

Extra Heavy.

Fancy Bronzed.		Berlin Bronzed.	
	Per dozen.		Per dozen.
No. 20, Assorted Green, Red and Blue,	$5 75	No. 71, Berlin Bronzed,	$4 50
		No. 72, Base not quite as wide as No. 71,	4 50

Ornamental Match Safes.

Full Size Cut of No. 60.

No. 60, Bronzed Match Safes, ... per dozen, $6 00

"CLIPPER" Tobacco Cutters.

One-third Size Cut. Whole length is 16 inches.

No. 12, "Clipper" Tobacco Cutters, Size of opening $\frac{7}{8} \times 3\frac{1}{4}$ inches,   per dozen, $24 00

Tobacco Cutters.

Bronzed and Ornamented.

Morse's Patent, .   per dozen, $12 00

One-third Size Cut of No. 2.

No. 2, Tuscan Bronzed, Polished Blade, Pattern as per Cut, .   .   .   per dozen, $10 00
No. 1, Old Pattern, Japanned, Wood Bottom (not same pattern as No. 2), .   "   14 00

A full case contains one dozen.

Wm. Frankfurth Hardware Co.

MONITOR OIL STOVES.

IMPROVED MONITOR.

No. 1.

IMPROVED MONITOR.

No. 2.

No. 1, Improved Monitor, Two 3¾-inch Burner, with Tray, - - each, $6 00

The No. 1 Stove is abundantly large for families of four or five persons. This Stove is handy for ironing, for bath-rooms, for druggists, book binders, restaurants, picnic parties, or any place where a quick reliable heat is required.

No. 2, Improved Monitor, Four 3¾-inch Burners, with Tray, - each, $10 75

This Stove will do all the cooking required for families of one dozen persons or more, making it a complete, neat and handy cook Stove, free from dust, ashes, smoke or smell.

The Monitor is the only absolutely safe Oil Stove in the world. The Stove can be filled with oil, if required, during the preparation of a meal, without extinguishing the fire. The wicks will last from four to eight months.

DAISY ASH SIFTER.

IMPROVED MONITOR.

No. 12.

No 12, Improved Monitor, with Four 5-inch Patent Geared Ratchet Burners, each, $14 00

This Stove having two back reservoirs may be separated if desired into two independent Stoves.

The Monitor has an unobstructed circulation of oxygen to its burning wicks; giving perfect combustion, The Daisy prevents the dust from arising, and can be used in all places. It has a strong galvanized wire screen and dovetailed frame.

Daisy, - - - - - - - - - - - - per dozen, $6 00

PORTABLE COOK STOVES.

Portable Cook Stoves, ... each, $6.00

Houchin's Pocket Cook Stoves.

No. 666. Closed.   No. 666. Open.

No. 666—Pocket Cook Stoves, with Square Grid Irons and Boilers, with Folding Handles, Holding nearly one Pint, Packed in a Box, 4 inches square, 1½ inches high, . . . each, $1.00
No. 667—Same as No. 666, Nickel Plated, ... " 1.75

No. 662.   No. 675.

No. 662—New Pocket Cook Stoves, no Cups, ... each, $ .50
No. 663—New Pocket Cook Stoves, with Cups, see No. 666, open, ... " .75
No. 675—New Pocket Cook Stoves, 3 Pint Boilers, packed in a box 5×5 inches, 6 inches high, " 1.50
No. 676—Same as No. 675, Nickel Plated, ... " 2.25

BRASS COAL HODS.

No. 50.

No. 60.

Funnel.

No. 59.

No. 52.

No. 54.

Nos.	50	60	59	52	54
Each,	$4.50	$12.00	$10.50	$4.75	$7.00

COOK STOVES.

Cook Stove.

Laundry Stove.

Parlor Cook.

Sam S. Utter's Ranges have no equal in Style, Durability or Price

Our Best Girl.
With Skirts.
With or without

Reservoir Boiler,
Hot Closet,
High or Low.

She is just lovely and takes on sight. In a variety of sizes.

SADIE.
With or without Skirts.
A beauty.

Nos. 6, 7 and 8.

SEE CUT.

SAM.
Nos. 60, 70, 80, 90, 6, 7 and 8.

EXTRA SAM.
7 and 8.

With or without

Hot Closets,
Reservoir Boilers
and Water Backs.

August 4, 1877.

Mr. Sam. S. Utter, 113 Beekman street, New York, has just "uttered" the "Sam" range, shown in the accompanying cut. Our only excuse for indulging in a pleasantry in a practical stove description is the fact that the eccentricity displayed in naming this range seems to invite it. We do not know, however, that "Sam" is not as good a name for a stove as "Dewdrop," or "Snowflake," or "Twinkling Star," or any other of the pretty names now so much in request. "Sam" is an honest, straightforward name, and it belongs in this case to an honest, practical stove. However, as the name is on a nickel plate, the dealer can substitute "Tom" or "Jack," or any other name he happens to like better. The "Sam" is new, and has been only a few weeks on the market. The aim of the manufacturer has been to make a large, substantial range at a low price. The trimmings are nickel plated.

THE NEW ARGAND OF 1877

Graceful in Form.

Chaste and Beautiful in Design.

The Stove that Revolutionized the Market in 1873.

Nearly every Anti-Clinker Base Burner now manufactured is copied from it.

All persons not licensed by Perry & Co. who manufacture, sell or use stoves of this character, are liable to them for the profits and damages.

Seventy-five thousand sold.

Adapted for Hard Coal, and also as Specially Constructed for Soft Coal.

MANUFACTURED UNDER THE ORIGINAL ANTI-CLINKER PATENTS

"The Housekeeper" Range.

The cut upon this page represents the latest design in elevated oven ranges from the firm of Messrs. Charles Noble & Co., of Philadelphia. It is intended by the manufacturers to be, in every respect, first-class. It has nickel-plated knobs and ornaments, front illumination, nickel-plated shifting guard-rail (in order that a tin kitchen can be used), shaking and dumping clinkerless grate, ash flue, hot-air arrangements to heat any number of rooms, and many other desirable characteristics. One feature to which attention is especially called by the manufacturers, is the fact that the water-back can be removed at any time, without disturbing any of the brickwork. In setting the range there are no flues to be found in the brickwork. An award was made to the firm by the Centennial Commission at the Exhibition last year.

The Fall Trade.

The Philadelphia *North American* says: There is really a fair prospect of a good fall trade, and the pioneer indications have already been felt in the dry goods line, which always leads the van. With the reopening of railway travel there has come an influx of mercantile visitors from the interior, as well as unusually large receipts of produce. The deadness of the summer is fast giving way to a brisk movement. But there are frequent and desperate efforts to prolong the stagnation by reports of fresh railroad strikes, and by the propagation of absurd rumors of all kinds of trouble, apparently set in circulation by some central authority engaged in working the wires far and near in some speculative scheme at the expense of the general prosperity. Undeniably the success of these aims is aided by the sensitiveness of the public mind and the absence of any commanding progressive movement. So long as the attention of the people remains so wide awake to every appearance of possible or probable disaster, active business enterprise will trade. As the renewed agitation for a free-trade tariff must again unsettle foreign imports and induce importers to hold off and await developments before ordering fresh cargoes, as well as to sell off as fast as possible the stocks in hand to avoid possible loss consequent upon a low tariff, the field for domestic goods is better than it was. Indeed, the same feeling of uncertainty just alluded immigrants have arrived at New York, and there is a manifest increase also at Philadelphia. The through ticket arrangement from Europe to the Western frontier, however, doesn't appear to be as actively at work as could be desired, and the consequence of the influx is an overflow of the crowded labor markets of the seaboard. In any Western region where there is a large demand for cal agitation to mining and comm ing ground that products of indust profitably worked kets, and that co be justifiable to se of the trade. The of Southern railw

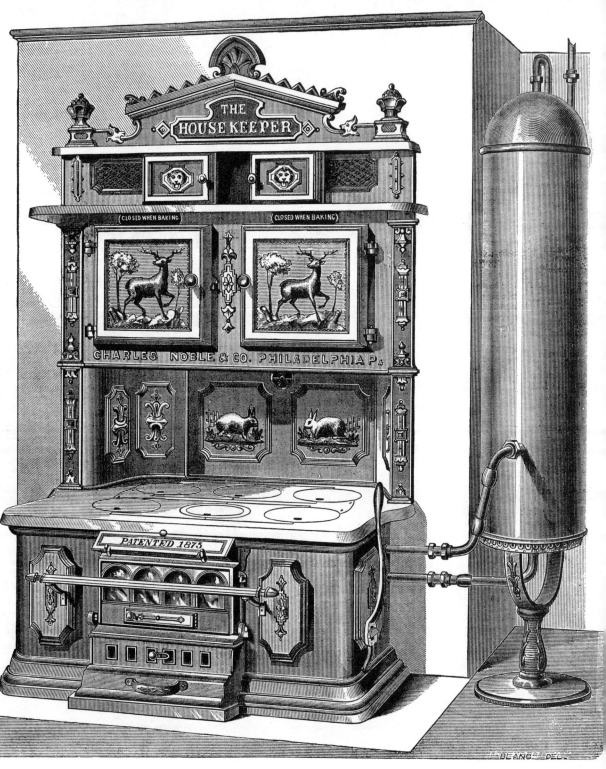

"THE HOUSEKEEPER" RANGE.

The "Uncle Sam" Range of 1877.

We present herewith a cut of the "Uncle Sam" range for 1877, manufactured by Abendroth Brothers, 109 and 111 Beekman street, and 282 Pearl street, this city. This range, in addition to substantial excellence, good form and tasteful decoration, has many points of especial merit, prominent among which is the reservoir. This is the Abendroth patent, for which many advantages are claimed. It is placed on the end of the range, and is so constructed that the water can be boiled almost as quickly as if placed directly over the fire. A cast iron water back of large capacity can also be used when desired, giving an abundant supply of hot water for bath, laundry and cooking purposes. It is furnished, when ordered, with all the couplings complete. We also notice the illuminated fire box; clinker clearing, shaking and dumping grate; sliding grate between the fire grate and ash pit, intended to free the cinders from ashes; large oven with flues so arranged as to secure equal heat on all sides; smoke collar, for either perpendicular or horizontal pipe; spacious hot closet, for warming dishes and articles of food; cast iron shelf, for plates; doors lined.

FRANKLIN,

THE "UNCLE SAM" RANGE OF 1877.

A Model Kitchen of the 1870's.

Majestic

You can cook and heat water for the entire house, with either coal or gas, or both at the same time, with the MAJESTIC Combination Coal and Gas Range.

The Highest Economy

of fuel using of either kind. Economy of kitchen space, compared to two separate stoves. One plumbing connection.

Our book, "Cost Saving," tells what you save over buying a coal and gas range separately, what you save over using them separately, what you save over buying or using any other combination range, and gives full description and prices. Address St. Louis office for it.

We make a full line of Majestic Family Coal or Wood Ranges, also Majestic Family Gas Ranges, and are the largest makers of Hotel Ranges in the world. Descriptive circular, showing style and price of any of these Majestic Ranges together with name of nearest dealer handling them, sent on application.

MAJESTIC MFG. CO., 2016 Morgan St., St. Louis, Mo.

Popular from 1890–1920.

McDougall Kitchen Cabinets

Every mother wants her daughters to have economical housekeeping ideas. Your children will appreciate the object lesson in kitchen economy that is taught by the McDougall Kitchen Cabinet. The McDougall Idea is to lighten the work of the housewife and to make the kitchen more attractive. This idea is the foundation upon which McDougall Kitchen Cabinets are built.

30 Days' Trial in Your Own Home

You can only appreciate the immense saving in time, energy and food supplies that a McDougall Kitchen Cabinet will effect for you, by putting it to the actual test in your kitchen. Any dealer is authorized to place a McDougall Kitchen Cabinet in your home on this plan.

Ask Your Dealer To Show You The McDougall Kitchen Cabinets

or write for handsomely illustrated catalogue, showing styles ranging in price from $14.90 to $90.00.

Look for the name-plate "*McDougall, Indianapolis.*" It is the maker's guarantee of quality, your protection against imitation.

G. P. McDougall & Son, 527 Terminal Building, Indianapolis, Ind.

The McDougall cabinet, model of 1905.

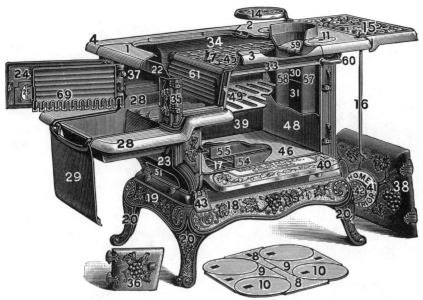

HOME SUPERIOR No. 8-20, SQUARE, 1891.

WOOD COOK.

THE SUNK HEARTH (No. 28 ON CUT) IS CONSIDERED "FRONT" OF COOK STOVE IN ORDERING REPAIRS.

1	Front Section Top.	22	Upper Front.	41	Tile Frame.
2	Back Section Top.	23	Lower Front.	43	Oven Door Kicker.
3	Right Section Top.	24	Front Door.	45	Top Oven Plate.
4	Left Section Top.	25	Front Door Slide.	46	Bottom Oven Plate.
7	Centre Rest.	28	Sunk Hearth.	48	Back Oven Plate.
8	Cut Long Centre.	29	Hearth Slide.	49	Oven Slide.
9	Short Centre.	30	Upper Back.	51	Flue Stopper.
10	Cover.	31	Lower Back.	54	Bottom Flue Strip.
11	Flue Cover.	33	Right Side.	57	Right Back Flue Strip.
14	Tea Shelf.	34	Left Side.	58	Left Back Flue Strip.
15	Back Shelf.	35	End Fire Door Frame.	59	Direct Damper.
16	Towel Rod.	36	End Fire Door.	60	Direct Damper Handle.
17	Bottom Plate.	37	Left Fire Box Side.	61	Fire Back.
18	Base Side.	38	Right Oven Door.	69	Wood Fender.
19	Base End.	39	Left Oven Door.		
20	Base Leg.	40	Outside Oven Shelf.		

VICTOR SUPERIOR No. 8-20, SQUARE, 1891.

COAL COOK.

THE SUNK HEARTH (No. 28 ON CUT) IS CONSIDERED "FRONT" OF COOK STOVE IN ORDERING REPAIRS.

NAMES OF REPAIRS.

1	Front Section Top.		
2	Back Section Top.		
3	Right Section Top.		
4	Left Section Top.		
7	Centre Rest.		
8	Cut Long Centre.		
9	Short Centre.		
10	Cover.		
11	Flue Cover.		
14	Tea Shelf.		
15	Back Shelf.		
16	Towel Rod.		
17	Bottom Plate.		
18	Base Side.		
19	Base End.		
20	Base Leg.		
22	Upper Front.		
24	Front Door.		
25	Front Door Slide.	48	Back Oven Plate
26	Front Feed Door.	49	Oven Slide.
27	Front Feed Door Slide.	51	Flue Stopper.
28	Sunk Hearth.	54	Bottom Flue Str
29	Hearth Slide.	57	Right Back Flue
30	Upper Back.	58	Left Back Flue S
31	Lower Back.	59	Direct Damper.
33	Right Side.	60	Direct Damper F
34	Left Side.	61	Fire Back.
36	End Fire Door.	62	Right End Lining
38	Right Oven Door.	63	Left End Lining.
39	Left Oven Door.	64	Front Grate.
40	Outside Oven Shelf.	65	Bottom Grate.
41	Tile Frame.	66	Grate Rest.
43	Oven Door Kicker.	67	Grate Clamp.
45	Top Oven Plate.	68	Shaker.
46	Bottom Oven Plate.	71	Ash Pan.
47	Front Oven Plate.		

THE OVEN DOOR FRAME (SEE No. 23 ON CUT) IS CONSIDERED "FRONT" OF SQUARE STEEL RANGE IN ORDERING REPAIRS.

NAMES OF REPAIRS.

1. Front Section Top.
2. Back Section Top.
3. Right Section Top.
4. Left Section Top.
5. Top Number Plate.
6. Top Plate Lugs.
7. Centre Key Plate.
8. Cover Key Plate.
9. Cover.
10. Reducing Ring Cover.
11. Short Centre.
12. End Shelf.
13. Long Base Side.
14. Base End.
15. Front Rail.
16. Door Frame.
17. Wood Feed Door.
18. "   "   " Number.
19. "   "   " Lining.
20. "   "   " Hinge.
21. Grate Door.
22. Ash Door.
23. Oven Door Frame.
24. Oven Door Panel.
25. Oven Door Handle.
26. Oven Door Latch.
27. "   " Latch Frame.
28. "   " Catch.
29. Right Oven Door Hinge.
30. Left Oven Door Hinge.
31. Clean Out Door and Hinge.
32. Clean Out Door Frame.
33. Name Plate.
34. Flue Box.
35. Direct Damper.
36. Direct Damper Frame.
37. Direct Damper Handle.
38. Direct Damper Knob.
39. Fire Box Extension.
40. Bottom Flue Strip.
41. Bottom Oven Brace.
42. Key Plate Support.
43. Centre Rest.
44. Oven Slide.
45. Oven Slide Bracket.
46. Reservoir Plate (See Page 123).
47. Left End Frame.
48. Pouch Feed Door.
49. Pouch Feed Door Slide.
50. Left End Draft Door.
51. "   "   " Slide.
52. Fire Back (Right Fire Lining).
53. Left Fire Lining.
54. Front Wood Lining.
55. Front Coal Lining.
56. Back Fire Lining.
57. Fire Lining Frame.
58. Right Duplex Grate.
59. Left Duplex Grate.
60. Grate Rest.
61. Grate Clamp.
62. Grate Gear.
63. Shaker.
64. Right Grate Rest Support.
65. Left Grate Rest Support (One Piece).
66. Front Section Left Grate Rest Support.
67. Back Section Left Grate Rest Support.
68. Front Ash Chute.
69. Back Ash Chute.
70. Left Ash Chute.
92. Right Closet Bracket.
93. Left Closet Bracket.
94. Right Closet Corner.
95. Left Closet Corner.
96. Right Closet Door Hinge.
97. Left Closet Door Hinge.
98. Pipe Register Frame.
99. Pipe Register Slide.
100. Closet Door Panel-Handle.
101. Closet Tea Shelf,

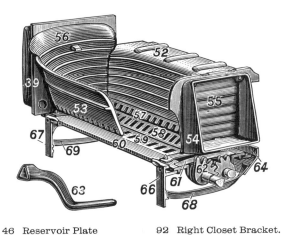

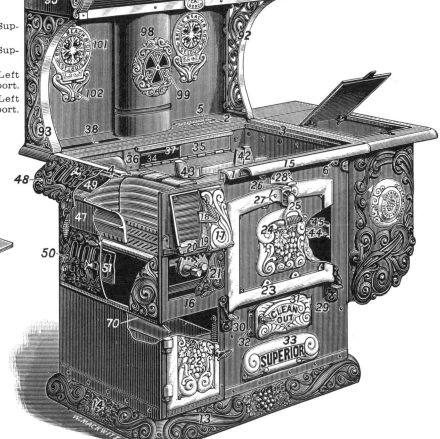

SUPERIOR STEEL RANGE No. 468-18 RESERVOIR AND HIGH CLOSET.

STEEL RANGE.

"FLORA."

"LAUNDRY."

"VICTOR."

"NOVEL SUPERIOR."

"CLIO."

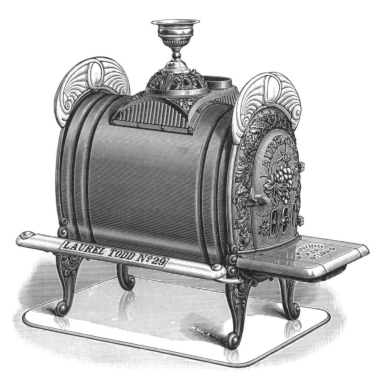

"LAUREL TODD."

"GEM OAK."

"PIONEER."

"SUPERIOR RADIATOR."

"MYRTLE."

"ALAMO."

"SLIGO."

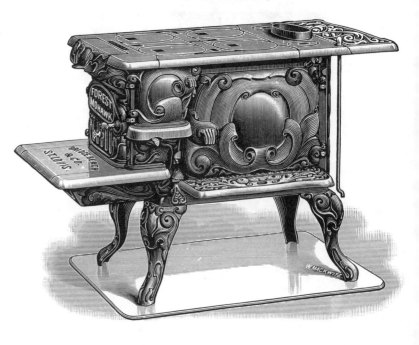

"SUPERIOR AIR TIGHT."

"GLEN MOHAWK."

"OSAGE."

"SUPERIOR RADIATOR."

Modern cooking methods, circa 1900. — *Library of Congress*

Hotel kitchen staff, circa 1890. — *Author's Collection*

DIRECTIONS FOR CLEANING THE KITCHEN AND UTENSILS,

FROM MISS JULIET CORSON'S *1880*

Cooking School Text-Book and Housekeeper's Guide.

BY PERMISSION.

HOW TO CLEAN THE KITCHEN.

FIRST.—Dust down the ceiling and side walls with a feather duster, or a clean cloth tied over a broom.

SECOND.—Sweep the floor, setting the broom evenly upon the floor, and moving it with long, regular strokes, being careful not to fling the refuse about the room, or to raise much dust.

THIRD.—Wash the paint with a piece of clean flannel dipped in hot water, in which borax has been dissolved in the proportion of one tablespoonful to a gallon of water; if the spots are not easily rubbed off, use a little soap, rinsing it off thoroughly, and wiping the paint with the flannel wrung out of clean water.

FOURTH.—Wash the window glass with a soft cloth which does not shed lint, dipped in clean water and wrung out; polish the glass with a clean, dry cloth, or with newspaper.

FIFTH.—Scrub the tables with hot water, in which a little washing-soda and soap have been dissolved, using a stiff brush; then rinse them with a cloth wrung out of clean, hot water, and wipe them as dry as possible.

SIXTH.—Scrub the floor in the same manner, and wipe it quite dry.

SEVENTH.—Wash all the scrubbing brushes and cloths in hot water containing a little soda and soap.

EIGHTH.—Wash all the dish cloths, and kitchen towels in hot water, with soap and soda, or borax, every time they are used, and keep a clean, dry stock of them on hand.

HOW TO CLEAN THE STOVE.

FIRST.—Let down the grate and take up the cinders and ashes carefully to avoid all unnecessary dirt; put them at once into an ash sifter fitted into the top of a keg or pail with handles, and closed with a tight fitting cover; take the pail out of doors, sift the cinders, put the ashes into the ash-can, and bring the cinders back to the kitchen.

SECOND.—Brush the soot and ashes out of all the flues and draught-holes of the stove, and then put the covers on, and brush all the dust off the outside. A careful cook will save all the wings of game and poultry to use for this purpose. If the stove is greasy, wash it off with a piece of flannel dipped in hot water containing a little soda.

THIRD.—Mix a little black-lead or stove-polish with enough water to form a thin paste; apply this to the stove with a soft rag or brush; let it dry a little and then polish it with a stiff brush.

FOURTH.—If there are any steel fittings about the stove, polish them with emery paper; if they have rusted from neglect, rub some oil on them at night, and polish them with emery paper in the morning. A "burnisher" composed of a net-work of fine steel rings, if used with strong hands, will make them look as if newly finished.

FIFTH.—If the fittings are brass they should be cleaned with emery, or finely powdered and sifted bath brick dust rubbed on with a piece of damp flannel, and then polished with dry dust and chamois skin.

SIXTH.—Brush up the hearthstone, wash it with a piece of flannel, dipped in hot water containing a little soda, rinse and wipe it dry with the flannel wrung out of clean hot water.

HOW TO CLEAN COPPERS.

FIRST.—If the utensils are very much tarnished, let them stand fifteen minutes in scalding hot water, with a tablespoonful of soda dissolved in each gallon of water; then scour the inside with fine sand and soap well rubbed on with the hand, or with a soft rag, until perfectly clean and bright; wash and dry with a soft, dry cloth; clean the outside in the same way, and wash and dry with a clean cloth.

DIRECTIONS FOR CLEANING KITCHEN, etc., Continued.

SECOND.—Never use vinegar or lemon juice, unless to remove spots which can not be scoured off as directed above; if acid is used mix it with salt, apply it quickly, and thoroughly wash it off at once; this care will serve to prevent the possible collection of verdigris upon the copper, but the utensils tarnish more quickly when they are cleansed with acid, than when any other method is employed. Salt is sometimes substituted for sand in the above named process.

THIRD.—One of the best chefs belonging to the New York Cooking School always had the coppers cleaned with the following mixture, rubbed on with the hand, and then washed off with clean, cold water, and employed a soft towel to dry them thoroughly; the mixture consisted of equal parts of salt, fine sand and flour, made into a thick paste with milk or buttermilk.

HOW TO CLEAN IRON-WARE.

FIRST.—Wash thoroughly inside and out with hot water, soap and soda; rinse and wipe thoroughly with a clean towel, and finish drying near the fire. If the inside is coated with the remains of food, put the pot over the fire, fill it with hot water, dissolve a tablespoonful of soda in it, let the water boil for fifteen minutes; this will soften the dirt so it can be scoured off with sand and soap.

SECOND.—When about to use a fish kettle, set it over the fire empty to heat; if it is not perfectly clean, an odor will be perceptible; in that case clean it as directed above before using it.

THIRD.—If you are obliged to use the same gridiron for broiling steak that has been used for fish, wash it thoroughly with hot water, soap and soda, rinse it in clean, hot water, heat it over the fire, rub it thoroughly with clean brown paper, and then with an onion cut in two pieces.

HOW TO CLEAN ENAMELED WARE.

Put it over the fire, filled with hot water containing a tablespoonful of soda to every gallon of water, and let it boil fifteen minutes; then if it is not perfectly clean, scour it with a little soap and fine sand, wash it well and dry it with a clean cloth.

HOW TO CLEAN TINWARE.

FIRST.—Scour till bright with fine sand or powdered and sifted brick dust, moistened with water, and rubbed on with chamois skin; polish with dry whiting and chamois skin.

SECOND.—If the tin is very much tarnished, boil it in hot water with soda before scouring. If no whiting is at hand dry flour may be substituted for it.

HOW TO CLEAN JAPANNED WARE.

Wash it with a sponge dipped in clean, cold water, wipe it dry and polish it with dry flour well rubbed on with a soft cloth.

HOW TO CLEAN PLATED WARE.

FIRST.—Put it in hot soda water for five minutes, then wipe it on a clean, dry cloth, and polish it with chamois skin.

SECOND.—If it is much tarnished use a little very fine whiting or silver powder in polishing it, taking care to brush it all out of the crevices and chased work on the plate.

HOW TO CLEAN STEEL KNIVES.

Rub them on a leather-covered board with a little finely-powdered brick dust, and wipe them thoroughly on a clean, dry towel.

HOW TO CLEAN SILVER AND JELLY BAGS.

FIRST.—Put the sieves into hot water containing a little soda or borax, but no soap; scrub them well with a clean brush, rinse them thoroughly in plenty of boiling water, and shake them dry.

SECOND.—Never use soap for washing jelly bags; wash them in as hot, clean water as the hands can endure, rinse them in boiling water, and wring them as dry as possible; then dry them where no dust will reach them, and keep them in a clean place.

GENERAL KITCHEN CLEANLINESS.

FIRST.—Never cease to exercise the greatest care in keeping the kitchen clean; it is the best place in the house to recall to mind the proverb that "Cleanliness is next to Godliness."

SECOND.—After attention has been given to all the directions enumerated in this chapter, remember to watch the sinks and drains; flush them several times a day with boiling water.

THIRD.—Take care that no scraps of meat or parings of vegetables accumulate in them to attract vermin, or choke the traps.

FOURTH.—Never throw soapsuds into the sink without afterwards flushing it with clean, hot water.

FIFTH.—Run hot water, containing a little chloride of lime, into the drains at least once a day in summer, and once in every two or three days in winter, to counteract all unpleasant and unhealthy odors.

Remember that the best cook always has the cleanest kitchen.

APPLE AND POTATO PARERS.

Reading, 1878.

Reading Apple Parers, Covered Works, ... each, $1.00

White Mountain Potato Parers.

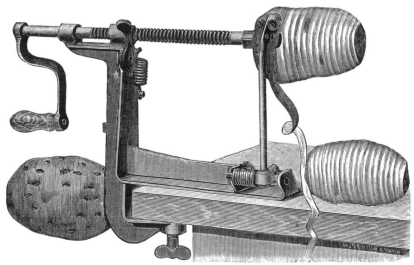

White Mountain Potato Parers, ... each, $1.25

POTATO PARERS AND SLICERS.

Saratoga.
Parer and Slicer.

Herring's Patent Ribbon Slicer.

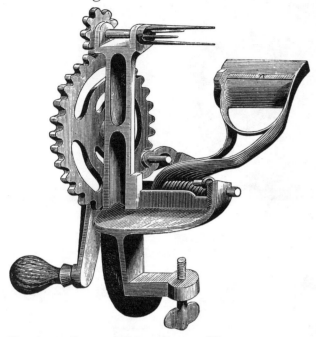

Herring's Patent Ribbon Potato Slicer, ... each, $ .75
Saratoga Potato Parer and Slicer, ... " .75

Cherry Stoners.

Family Cherry Stoners, ... each, $ .85

PARERS AND SLICERS.

Rotary Potato Parers.

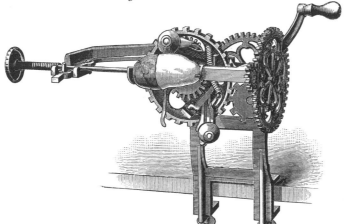

Rotary Potato Parers, ... each, $1.50

Rotary Peach Parers.

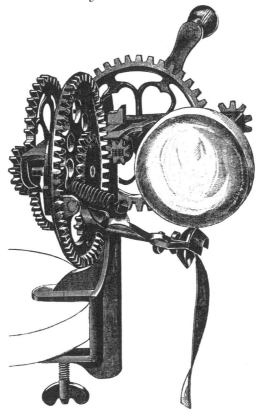

Rotary Peach Parers, ... each, $1.50

Family Parer, Corer and Slicer.

Family Parer, Corer and Slicer, ... each, $1.00

Wm. Frankfurth Hardware Co.

GOLD MEDAL APPLE PARER.

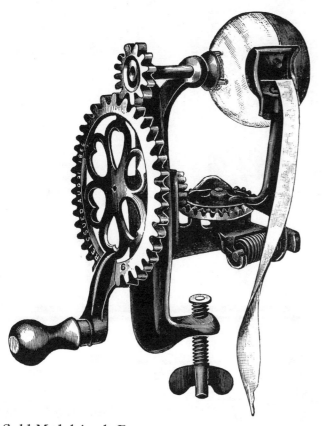

O. K. PARER AND SLICER.

Gold Medal Apple Parers, - - - - - - - - per dozen, $8 00
O. K. Parer and Slicers, - - - - - - - - - " 1 25

WHITE MOUNTAIN APPLE PARER, CORER AND SLICER.

No. 40.

White Mountain Apple Parer, Corer and Slicer, - - - - - per dozen, $9 00

FRUIT, LARD AND JELLY PRESS.

ENTERPRISE COMBINATION FRUIT PRESS.

(Rack Movement.)

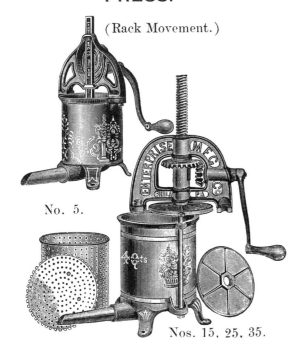

No. 15, 2-quart Japanned, Screw Movement,	each, $3 00
25, 4 " " " "	" 5 00
35, 8 " " " "	" 6 00
Combination Fruit Press, Galvanized,	" 3 00

FRUIT AND VEGETABLE PRESSES.

Enterprise   HENIS'S.

Save All Fruit, Lard and Jelly Presses,	per dozen, 10 00
Henis's " " " " "	" 6 00

Also used to Mash or Rice Potatoes.

Wm. Frankfurth Hardware Co.

SILVER'S EGG BEATER AND MEASURING GLASS.

SILVER'S PATENT EGG TIMER.

Dover.
Family Size.
Each, $.50

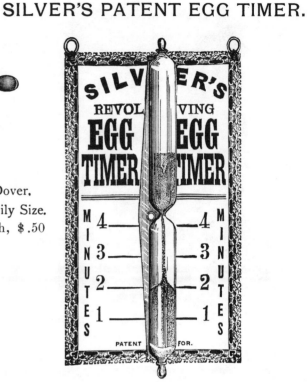

Silver's Patent Egg Beater and Measuring Glass, - - - - per dozen, $9 00
"   "   " Timer, Revolving, - - - - - - - "   3 50

POT CLEANERS.

SKEWERS.

LIGHTNING.

TRIUMPH PLATE LIFTERS.

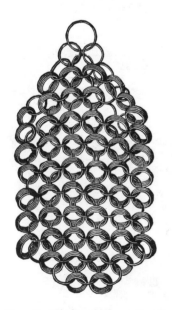
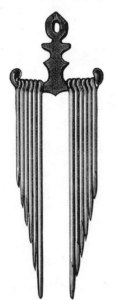
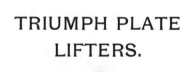

No. 2, Plain Wire, Large, Two Ring Pot Cleaner, - - - - - per dozen, $1 50
    2,   "   "   " Three   "   "   " - - - - - - "   1 75
    2, Retinned "   " Two   "   "   " - - - - - - "   2 00
    1, Polished Steel Wire, Meat Skewers, per set, (12), - - - - "   30
Triumph Plate Lifters, - - - - - - - - "   2 50

EGG BEATERS.

SPOON.

WHIP.

Wire Spoon Egg Beaters, Wood Handle, - - - - - - - per dozen, $1 00
"   "   "   "   Tin   "   - - - - - - -   "   75
No. 2, 15-inch Hotel Egg Whips, Wire Handle, - - - - -   "   9 00

SPIRAL.

for Tumblers, per dozen, $4 50

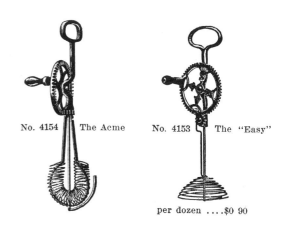

No. 4154   The Acme   No. 4153   The "Easy"

Spiral Egg Beaters, Wood Handle,

per dozen $0 90

Monroe's.   Peerless.   Dover

Mammoth. Each, 5 00

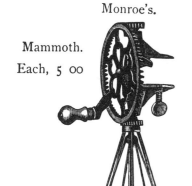

No. 3.

each, $3 00

P. D. & Co.'s Egg Beater, - - - - - - - - - per dozen, $3 00

EGG BEATERS.

Globe.

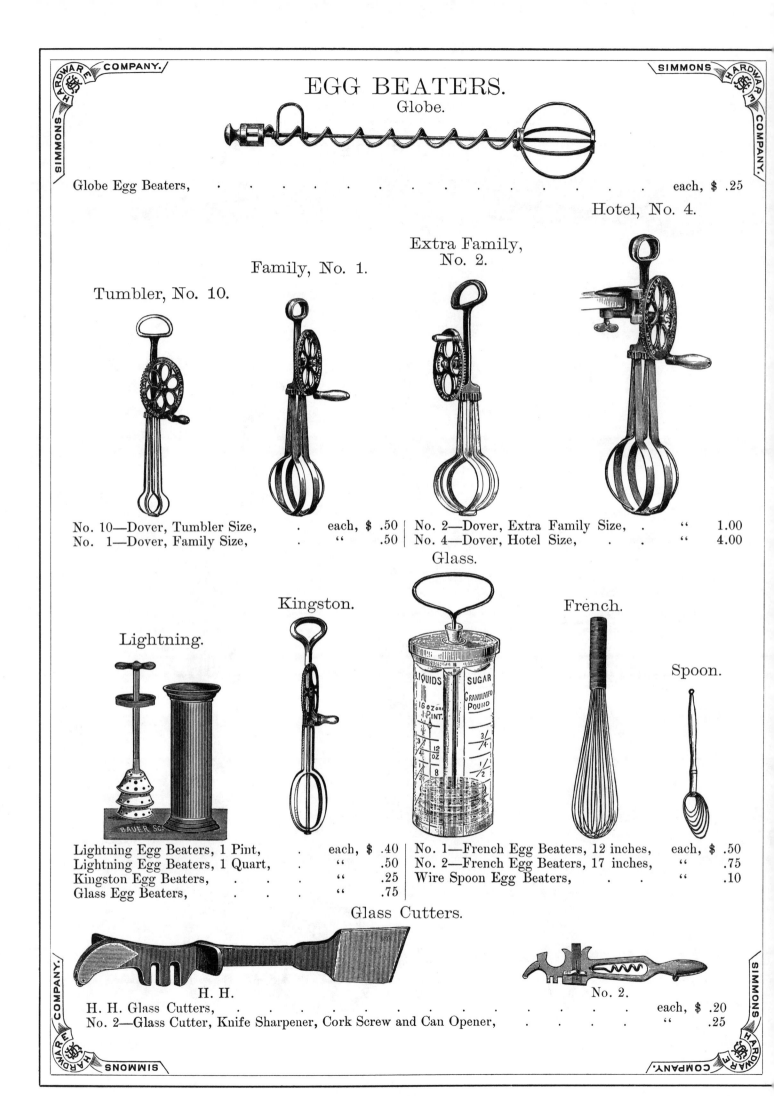

Globe Egg Beaters, ... each, $ .25

No. 10—Dover, Tumbler Size, . . each, $ .50	No. 2—Dover, Extra Family Size, . " 1.00
No. 1—Dover, Family Size, . . " .50	No. 4—Dover, Hotel Size, . . " 4.00

Glass.

Lightning Egg Beaters, 1 Pint, . each, $ .40	No. 1—French Egg Beaters, 12 inches, each, $ .50
Lightning Egg Beaters, 1 Quart, . " .50	No. 2—French Egg Beaters, 17 inches, " .75
Kingston Egg Beaters, . . . " .25	Wire Spoon Egg Beaters, . . " .10
Glass Egg Beaters, . . . " .75	

Glass Cutters.

H. H. Glass Cutters, ... each, $ .20
No. 2—Glass Cutter, Knife Sharpener, Cork Screw and Can Opener, ... " .25

WOODEN WARE.

Wood Steak Mauls.

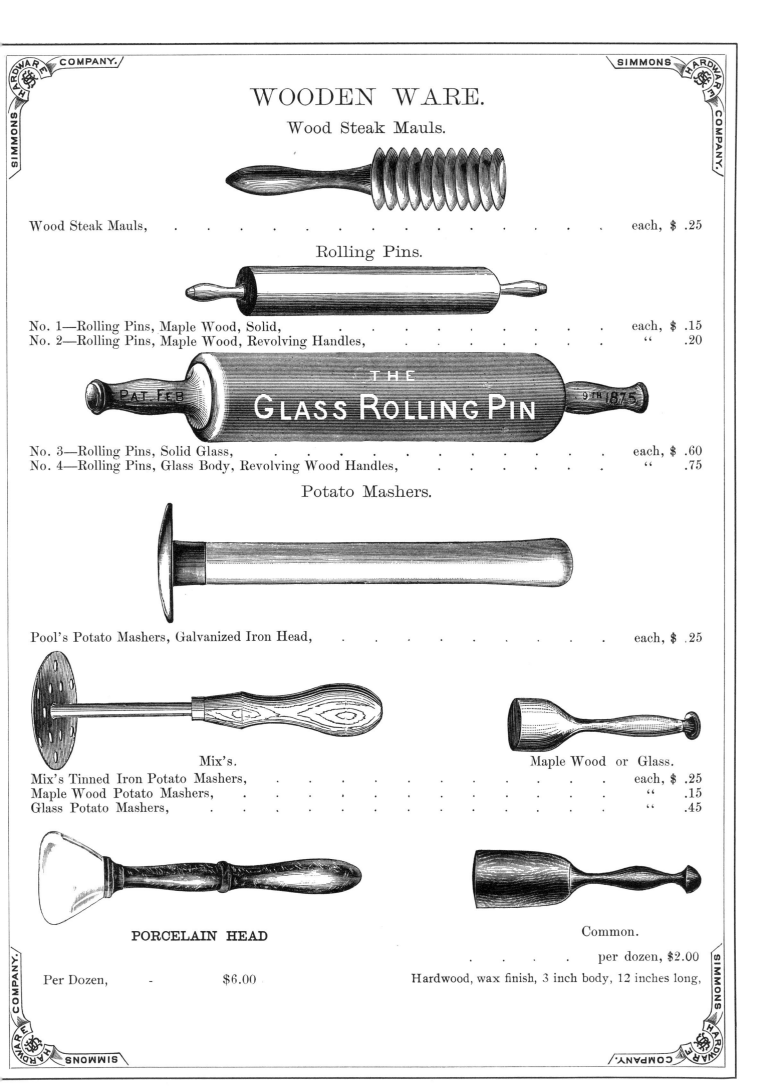

Wood Steak Mauls, ... each, $ .25

Rolling Pins.

No. 1—Rolling Pins, Maple Wood, Solid, ... each, $ .15
No. 2—Rolling Pins, Maple Wood, Revolving Handles, ... " .20

No. 3—Rolling Pins, Solid Glass, ... each, $ .60
No. 4—Rolling Pins, Glass Body, Revolving Wood Handles, ... " .75

Potato Mashers.

Pool's Potato Mashers, Galvanized Iron Head, ... each, $ .25

Mix's.   Maple Wood or Glass.
Mix's Tinned Iron Potato Mashers, ... each, $ .25
Maple Wood Potato Mashers, ... " .15
Glass Potato Mashers, ... " .45

PORCELAIN HEAD   Common.
                                          . . . . per dozen, $2.00
Per Dozen,   - $6.00   Hardwood, wax finish, 3 inch body, 12 inches long,

Wm. Frankfurth Hardware Co.

TAP BORERS.

ENTERPRISE.

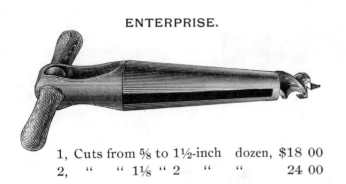

1, Cuts from ⅝ to 1½-inch   dozen, $18 00
2,   "   " 1⅛ " 2   "   "   24 00

COMMON.

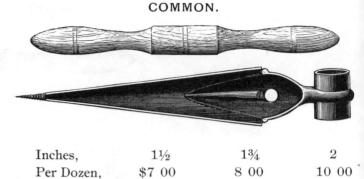

Inches,	1½	1¾	2
Per Dozen,	$7 00	8 00	10 00

WOOD FAUCETS.

CORK LINED.

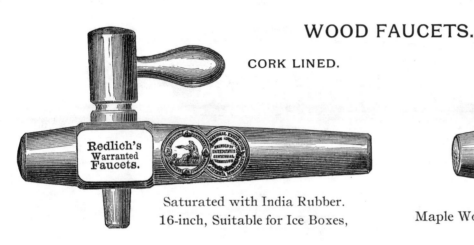

Saturated with India Rubber.
16-inch, Suitable for Ice Boxes,

METAL KEYED.

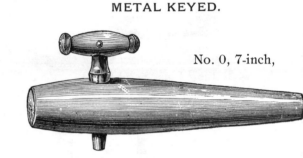

No. 0, 7-inch,

Maple Wood, with Block Tin Key, Leather Lining,

CABBAGE BORERS.

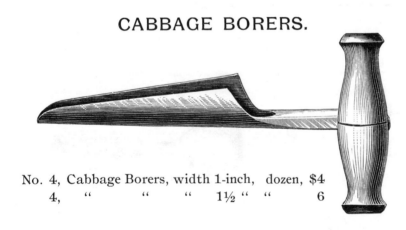

No. 4, Cabbage Borers, width 1-inch,   dozen, $4
   4,   "   "   "   1½ "   "   6

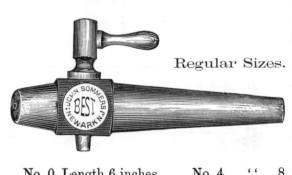

Regular Sizes.

No. 0, Length 6 inches,   No. 4, " 8
No. 2,   "   7 "   No. 6, " 9

FRUIT AUGERS.

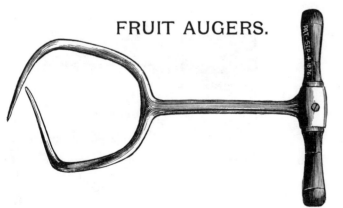

No. 1, Cast Steel Tinned Fruit Auger, Wood Handle,
7-inch Steel Tinned Butter Spade, Tinned Malleable Handle,

BUTTER SPADES.

- - -   per dozen, $18 00
        "   10 00

Milwaukee, Wisconsin.

VEGETABLE CUTTER.

Vegetable Cutter, Adjustable, - - - - - - - per dozen, $3 00

SLAW CUTTERS.

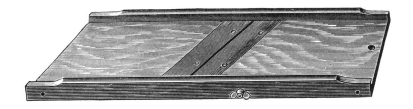

No. 1, One Knife, Adjustable, - - - - - - - per dozen, $4 00

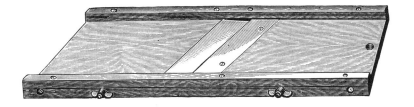

No. 2, Two Knives, Adjustable, - - - - - - - per dozen, $5 00

KRAUT CUTTERS.

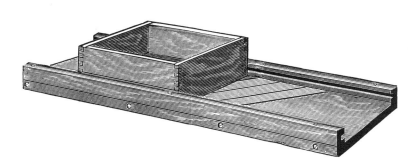

No. 3, Three Knives with Box, Size 8 x 26 inches, - - - - - -	each,	$1 50
5, " " " " " 9 x 30 " - - - - - - -	"	2 50
8, Four " " " " 12 x 36 " - - - - - - -	"	3 50

Wm. Frankfurth Hardware Co.

Improved Cedar Cylinder Churn.

CHURNS.
CHARM BARREL.

Swing Churn.

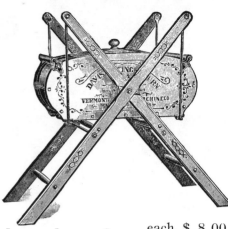

56869 This we consider by far the best small, cheap churn on the market. It is made from the best Virginia cedar; it has a double dasher, and the crank is locked to the churn with a clamp and thumb-screw, which prevents leakage. Lock cannot break. The top is large and dasher easily removed. The hoops are of galvanized iron and will not rust.

No	1	2	3	4
Will hold...	3	4	7	10 gallons
Will churn...	2	3	4	5 gallons

No. 0,	5 Gallons will Churn	1 to	2 Gallons Cream,	each,	$ 8 00
1,	9 " " "	1 "	4 " "	"	8 50
2,	15 " " "	2 "	7 " "	"	9 00
3,	20 " " "	3 "	9 " "	"	10 00
4,	25 " " "	4 "	12 " "	"	12 00
5,	35 " " "	5 "	16 " "	"	16 00

DASH.

Common.

LIGHTNING CHURN

MILK CAN

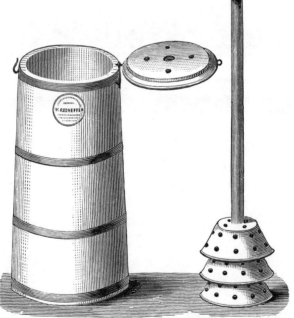

COMMON DASH CHURNS.

Nos.		1	2	3	4
Gallons,		6	5	4	3
Per Dozen,		$9 00	8 00	7 00	6 00

Gallons,		3	4	5	6
Tin, 3 Dash,		each, $2.35	2.75	3.00	3.40
Ash, 2 Dash,		" 2.00	2.25	2.50	2.75

Milwaukee, Wisconsin.

GRATERS.

BOX GRATERS

GEM NUTMEG

The rotary motion makes the grating continuous.
Per dozen. $1.20

Cut One-half Size.

Rajah Patent " " Wood Handle,

| TIN HANDLE | WOOD HANDLE | PLAIN. |

Pieced, panel back and lid, finished without solder, japanned.
No. 04—2 x 5 inches. . . . . Per dozen $0.80

Pat. Aug. 18, 1891.

Does not clog nor tear the fingers, nor drop the nutmeg, very simple, nicely finished.
No. 7—5 x 5 inches. . . . . Per dozen $2.70

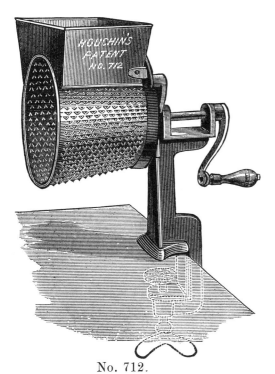

No. 712.

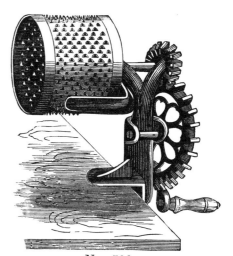

No. 709.

No. 712—Graters, Hotel Size, for Grating Cocoanut, Horse Radish, etc., ... each, $2.50
No. 709—Graters, Family Size, for Grating Cocoanut, Horse Radish, etc., ... " .75

WOOD CARVINGS.

Bread Plates.

No. 1—Carved Bread Plates, Diameter, Inches, . . . 10 11 12
Each, ... $1.25 1.50 1.75
Mottoes Assorted.

No. 2—Extra Carved Bread Plates, Assorted Mottoes and Designs, ... each, $2.50 to $3.50
Handsome Carvings for Wooden Weddings in Stock.

WOODEN WARE.

Wood Bowls.

Round.

Oblong.

Maple Wood Bowls, Round,	Inches,	12	13	15	17	19
	Each,	$ .15	.20	.25	.40	.50

Maple Wood Trays, Oblong,	Size,	Small,	Medium,	Large.
	Each,	$ .50	.75	1.00

Butter Pats.   Biscuit Beaters.   Wood Spoons.

Butter Pats,	each, $ .10
Biscuit Beaters,	" 3.50
Wood Spoons,	each, .10 to .25

Butter Prints.   Butter Moulds.

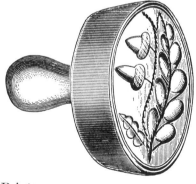 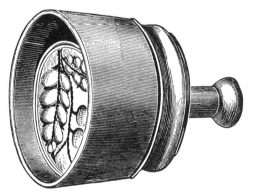

Butter Prints,	each, $ .20
Butter Moulds, Individual,	" .25
Butter Moulds, ½ pound,	" .40
Butter Moulds, 1 pound,	" .50

Butter Ladles.

Maple Butter Ladles,	each, $ .15	No. 2, For Rice, Crackers, etc.,	per dozen, $3 25
Wood Tooth Picks, Double per box of 2500,	.15	3, " Coffee, Sugar, etc.,	" 4 00
		4, " Salt, Flour, etc.,	" 5 30

UNION HARDWARE AND METAL COMPANY,

BUTTER MOULDS.
CALIFORNIA.

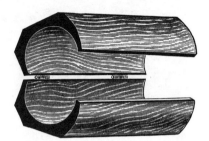

ROUND.

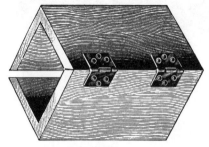

SQUARE.

Two pound round, brass hinges, weight per dozen, 22 pounds. Per dozen	$5.00
One pound square, brass hinges, weight per dozen, 15 pounds. Per dozen	4.50
Two pound square, brass hinges, weight per dozen, 21 pounds. Per dozen	5.00

ROUND.

Butter Moulds, ½ pound,

Made from select white maple.

Individual, one ounce, weight per dozen, 2 pounds. Per dozen	$1.50
Half pound, weight per dozen, 12 pounds. Per dozen	3.00
One pound, weight per dozen, 18 pounds. Per dozen	3.75

Butter Prints,

ROLLING PINS.

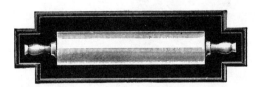
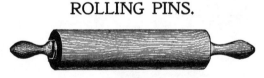

Plain handle.

Roller made of pure Opalite, snow white,

Plain handle rolling pin, 28 pounds, size 2⅝x12 inch. Per dozen	$1.80
Enameled revolving handle rolling pin, size 2⅝x12 inch. Per dozen	2.40

FANCY CAKE ROLLER AND PRINT.

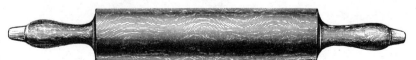

No. 1, Stained Handle, Maple Wood, Revolving Handles, per dozen,	$1 25
3, Black Enameled Handle, Maple Wood, Revolving Handles, "	1 75
5, Polished Cherry " " " " " "	2 00
Fancy Cake Roller and Print, Cherry " with Polished Handles, "	6 00

Size: Length, 17 inches; circumference, 12 inches; divided into 24 spaces, and rolls that number of cakes 2 inches square; on each a beautiful design carved by hand.

Milwaukee, Wisconsin.

SALT BOXES.

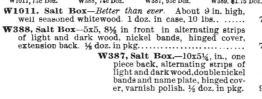

W1011. Salt Box—*Better than ever.* About 9 in. high, well seasoned whitewood. 1 doz. in case, 10 lbs.. 73
W388, Salt Box—5x5, 8½ in front in alternating strips of light and dark wood, nickel bands, hinged cover, extension back. ½ doz. in pkg...................... 74
W387, Salt Box.—10x5¼, in., one piece back, alternating strips of light and dark wood, double nickel bands and name plate, hinged cover, varnish polish. ½ doz. in pkg. 93

No. 1811, Varnished Maple and Walnut Wood, Striped, Nickel Plated Band and Brass Label,
    Holds 2½ Pounds of Salt, - - - - - per dozen, $3 50
  1992, Fine Polished Hardwood, Striped, Nickel Plated Band and Label, Holds 2½ pounds, " 5 00
  1993, " " " " " " " " " 3 " " 5 50
  7055, Extra " " " Tripple Stock, White Metal Label, " 2½ " " 7 00

SPINNING WHEELS.

German Spinning Wheels,
each, $4 00

Spice Cabinet, Ash Varnished, 8 Drawers, each, $2 00
Mustard Spoons, Boxwood,   per dozen, 50

SLATES.

STEAMER SET.

Steamer Set, - each, $2 00

The Steamer Set, which is a Teakettle and two Steamers combined, is the favorite method of cooking vegetables puddings, etc. Several kinds of vegetables may be cooked at the same time without flavoring each other in the least.

VICTOR NOISELESS.

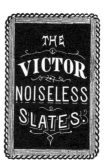

6x9   7x11   8x12

WOODEN WARE.

Knife Boxes.

No. 10.   No. 20.

No. 10—Striped Knife Boxes, Straight Sides, 3 Apartments,	each, $1.50
No. 20—Striped Knife Boxes, Straight Sides, 2 Apartments,	" 1.00
No. 30—Walnut Knife Boxes, Flaring Sides, 2 Apartments,	" .75
No. 40—Extra Whitewood Knife Boxes, Flaring Sides, 2 Apartments,	" .50
No. 50—Plain Whitewood Knife Boxes, Flaring Sides, 2 Apartments,	" .35

Knife Boards.   Skirt Boards.

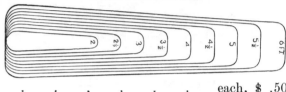

Knife Scouring Boards, ... each, $ .50

Skirt Ironing Boards,	Feet,	3	3½	4	4½	5	5½	6
	Each,	$ .50	60	.70	.80	.90	1.00	1.15

Pastry Boards.   Meat Blocks.   Bosom Boards.

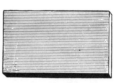
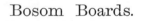

Pastry or Bread Boards, Plain,	Nos.	2	3	4	5
	Inches,	16×22	18×24	20×27	20×30
	Each,	$ .60	.65	.75	.85
Pastry or Bread Boards, Rimmed,	"		.75	.85	1.00

Meat Blocks, Hard Wood, . . each, $ .65 | Bosom Boards, . . . each, $ .35

Kanakins.   Nest Boxes.

Clear Norway Pine,

Sugar

Salt Boxes, Nest of 5, . . . each, $ .50

Nos.	1	2	3
Plain, Each,	$ .65	.50	.40
Varnished, Each,	.90	.75	.60

Snow Shovels.

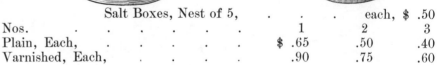

.60 Each

the Best and Strongest Ladder	Feet,	4	5	6	7	8	10	12
	Each,	$3 00	3 50	4 00	4 50	5 00	6 00	8 00

WOODEN WARE.

Settee Tables.

Settee Tables, Combined Wash Bench and Kitchen Table, 4 feet long, ... each, $5.00

Kitchen Tables.

Patent Ironing Tables.

Folding Kitchen Tables,	Length,	3½	4	4½	5
	Each,	$4.00	4.50	5.00	5.50

Patent Ironing Tables, ... each, $4.50

Patent Ironing Tables are supported wholly at one end, so that skirts or other garments can be readily slipped on for ironing. It can be placed at any desired height, from 21 to 30 inches.

Wash Benches.

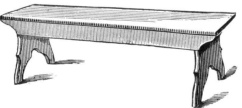

Wash Benches,	Length, feet,	4	5	6
	Each,	$1.50	1.75	2.00

Pine Pails. Cedar Pails. Horse Pails.

Extra Pine Pails, Oak Grained, 2 Hoops, ...	each,	$ .25
Extra Pine Pails, Oak Grained, 3 Hoops, ...	"	.30
Red Cedar Pails, Brass Bound, 3 Hoops, ...	"	1.00
Extra Pine Horse Pails, Oak Grained, ...	"	.45
Oak Horse Pails, ...	"	.85

CAN OPENERS.

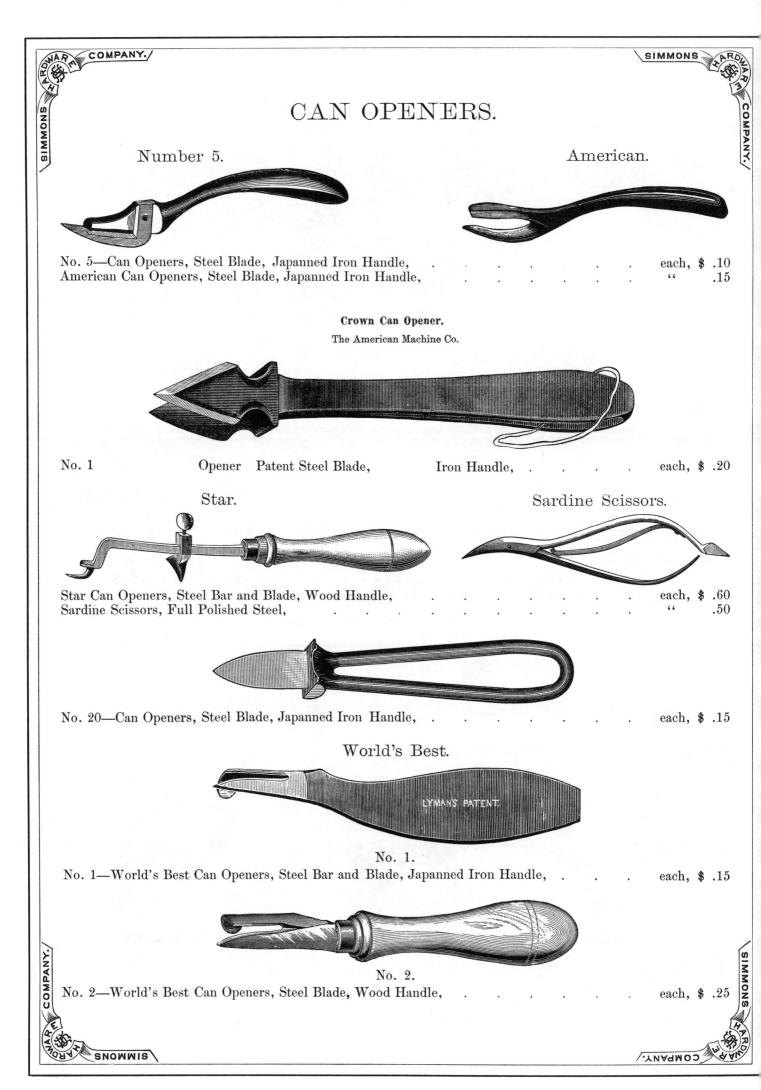

Number 5.   American.

No. 5—Can Openers, Steel Blade, Japanned Iron Handle, ... each, $ .10
American Can Openers, Steel Blade, Japanned Iron Handle, ... " .15

Crown Can Opener.
The American Machine Co.

No. 1   Opener Patent Steel Blade,   Iron Handle, . . . each, $ .20

Star.   Sardine Scissors.

Star Can Openers, Steel Bar and Blade, Wood Handle, ... each, $ .60
Sardine Scissors, Full Polished Steel, ... " .50

No. 20—Can Openers, Steel Blade, Japanned Iron Handle, ... each, $ .15

World's Best.

No. 1.

No. 1—World's Best Can Openers, Steel Bar and Blade, Japanned Iron Handle, . . . each, $ .15

No. 2.

No. 2—World's Best Can Openers, Steel Blade, Wood Handle, ... each, $ .25

THE ORIGINAL SPRAGUE PATENT.

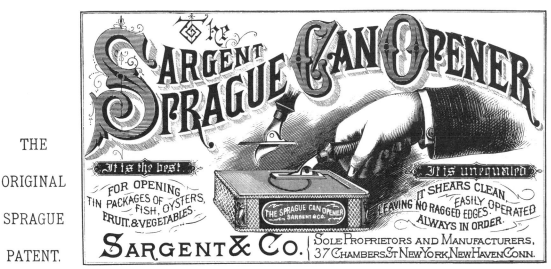

THE ORIGINAL SPRAGUE PATENT.

Sargent-Sprague Can Openers.

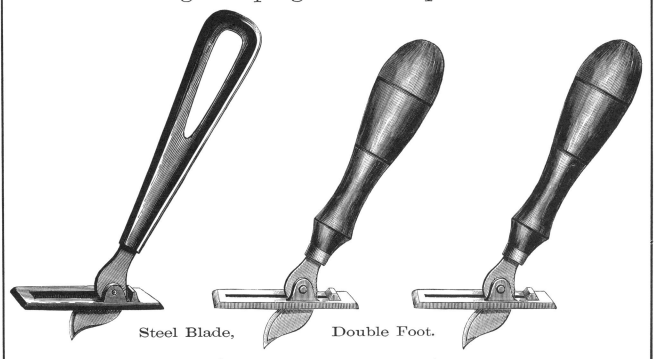

Steel Blade,　　　　　　Double Foot.

No. 1.	No. 2.	No. 3.
Jap'd Handle, per dozen, $2 00	Wood Handle, per dozen, $2 25	Imitation Ebony Wood Handle, per dozen, $2 50

No. 5, Can Openers.

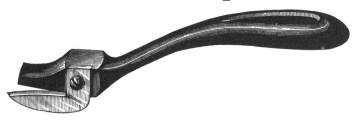

No. 5, Japanned Handle, Steel Blade, ... per dozen, $1 00

All on this page one dozen in a box.

Full cases contain twelve dozen. No charge for cases when ordered in original packages.

CAN OPENERS.

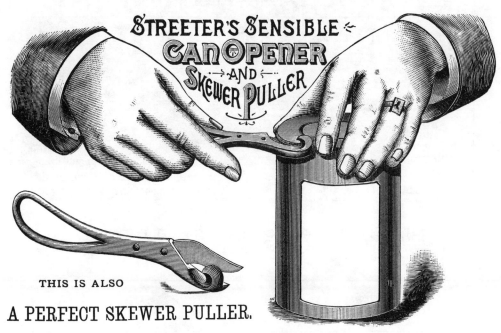

No. 1, Japanned Handle, - - - - - - - per dozen, $1 75

MESSENGER'S

Adjustable Cutter, ... per dozen, $3 00

LYMAN'S

No. 1. Half Size.

No. 1, Adjustable, Revolving Cutters, ... per dozen, $3 75

POOLE'S.

Steel Bar and Blade, Japanned Loop Handle, - - - - - per dozen, $3 50

All One Dozen in a Box.

LEMON SQUEEZERS.

DEAN'S.   DRUM.

No. 2.

No. 2, Dean's Fancy Japanned Frame with Glass Cup,	per dozen, $6 00
Extra Glass Cups for Same,	" 1 50
Drum, Porcelain Lined, Japanned Handles,	" 8 00

One Dozen in a Case.

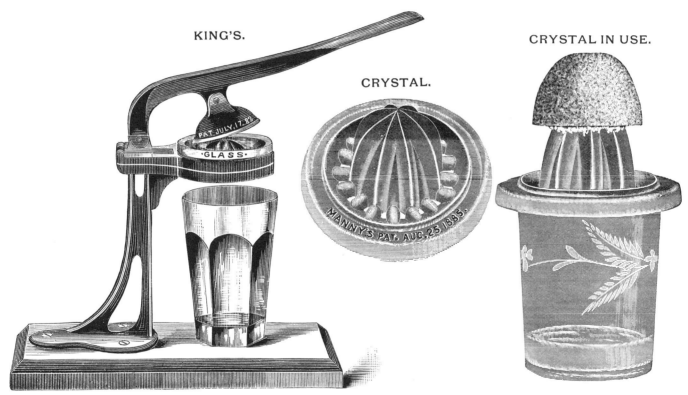

KING'S.   CRYSTAL.   CRYSTAL IN USE.

No. 3, King's,	per dozen, $15 25
Crystal, all Glass,	" 5 00

The King is especially adapted for Hotel, Restaurant and Bar use. It has a heavy perforated tinned metal bowl, and a hardwood plunge, so that in use the lemon juice cannot come in contact with the iron. Packed one dozen in a case. For convenience in packing, the frame and base are separate, and can readily be fastened by means of the bolt and nut, with which each article is provided.

Porcelain Lined,   per dozen, $7 00

MUDDLERS, OR "TODDY STICKS."

No. 1, All Wood, Fancy,   dozen, $1 25

LEMON SQUEEZERS.

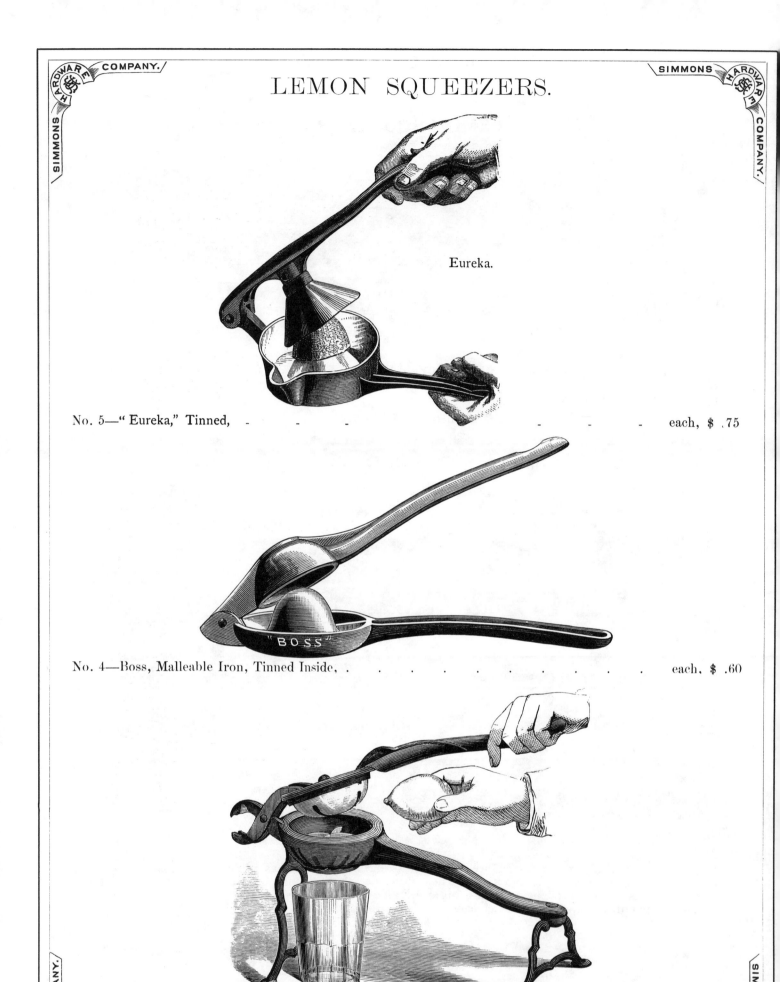

No. 5—"Eureka," Tinned, - - - - - - each, $ .75

No. 4—Boss, Malleable Iron, Tinned Inside, ... each, $ .60

No. 12—Patent, Cuts and Squeezes the Whole Lemon at once, ... each, $1.65

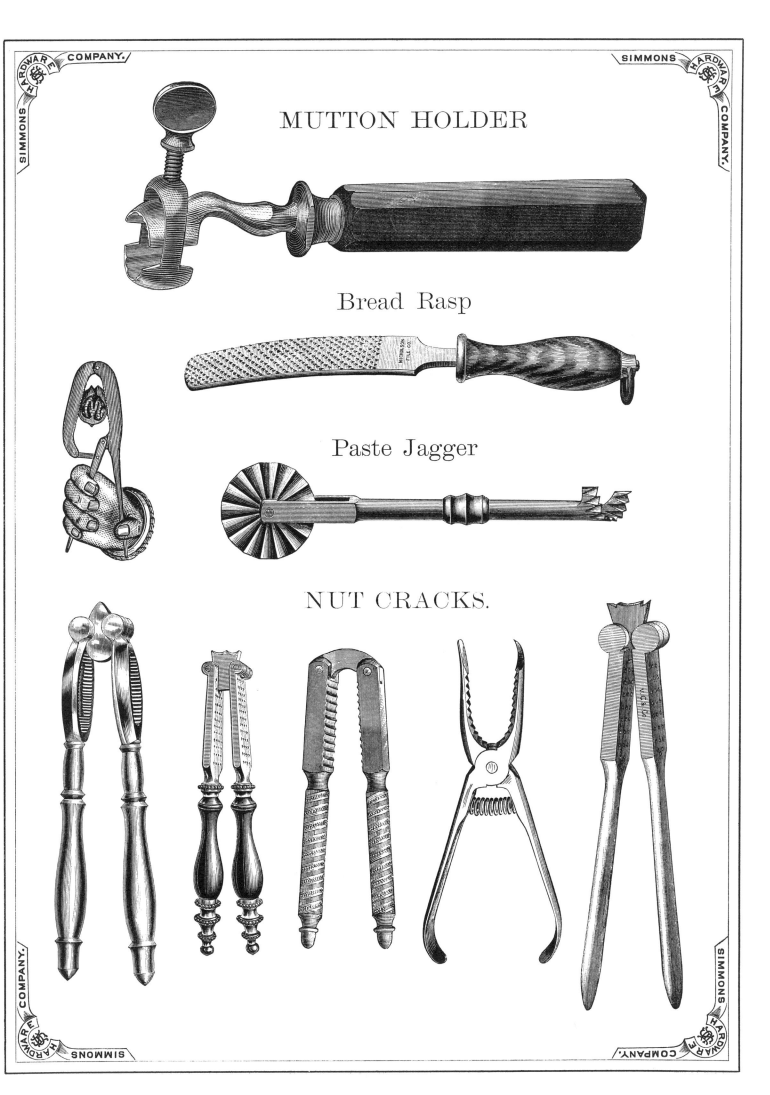

LIGHTNING BAR.

Nickel plated, lever handle, spiral drive. This lemon squeezer is particularly adapted, for rapid work, where many lemons are to be juiced. Weight, per dozen, 60 pounds. Each.................................. $ 2.50
One-twelfth dozen in a box.

CORK PULLERS.
MODERN.

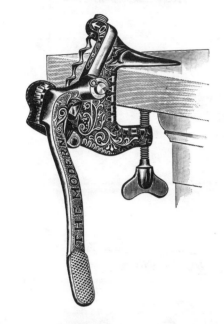

To clamp on table or bar, nickel plated, height above bar two inches. One lifting of the lever will extract the tightest cork. Weight, per dozen, 42 pounds. Each.. $ 3.00

CHAMPION.

No. 1, nickel plated, automatic pull, stands five inches above counter. Weight, each, 6 pounds. Each............................ $3.50

LEMON SQUEEZERS.

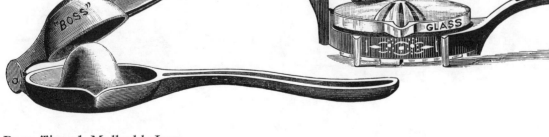

Boss, Tinned, Malleable Iron,	- - - - - - - - - per dozen,	$5 00
No. 2, King's, Japanned, with Glass Cup,	- - - - - - - - "	5 25

MANNY'S.

NO. 1.

NO. 2.

No. 1. Clear glass, to be used over a tumbler, weight per dozen, 4½ pounds. Per dozen.................... $1.00
No. 2. Clear glass, provided with a saucer which surrounds the cone, into which the clear juice is delivered, weight per dozen, 10 pounds. Per dozen.................... 2.00

One-third dozen in a paper box.

LEMON SQUEEZERS.

LITTLE GIANT.

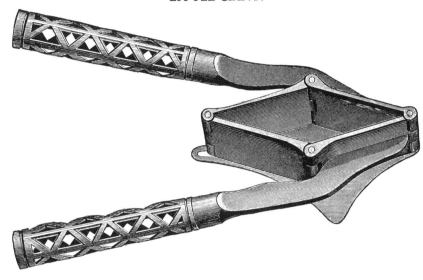

No. 22. Tinned, weight per dozen, 17 pounds, per dozen.................... $9.00

Half dozen in a box.

CROWN.

Crown, japanned, with hardwood bowl, weight per dozen, 13 pounds. Per dozen.................... $4.50

Half dozen in a box.

WOOD FRAME.

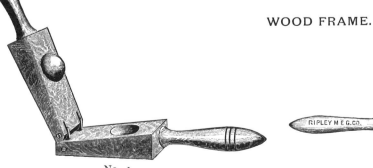
No. 1.

No. 2.

No. 1, Hard Wood Frame, Plain,   -   -   -   -   -   -   per dozen, $2 00
2,   "   "   "   Porcelain Bowl and Compress,   -   -   -   -   "   6 00

Mincing Knives.

No. 2.

Nos. 10 and 11.

No. 12.

No. 2, Single Shank,	per dozen, $1 50	No. 11, Braced Shank, Polished,	per dozen, $0 90
No. 10, Braced Shank, Unpolished,	" 80	No. 12, Double Shank, "	" 2 00

No. 33, Double Blade.

No. 51, Iron Handle. No. 55, Wood Handle.

No. 33, Double Blade, Polished,	per dozen, $2 15	No. 51, Iron Handle, Polished,	per dozen, $2 70
No. 21, Hammered Cast Steel,	" 3 50	No. 55, Wood Handle, "	" 3 40
No. 22, " " "	" 4 00		

Smith's Patent Wire Braced Mincing Knives.

Nos. 3 and 50. Style of Nos. 4 and 40.

Nos. 1 and 10, Single Blade.

Nos. 2 and 20, Double Blade.

Nos. 5 and 50, Double Blade.

No. 1, Plain Handle, Single Blade, ... per dozen, $2 00
No. 10, Black   "   "   "   . . . . " 2 50

No. 2, Plain Handle, Double Blade, ... per dozen, $3 00
No. 20, Black   "   "   "   . . . . " 3 50

No. 3, Black Handle, Large Single Blade, ... per dozen, $5 00
No. 4,   "   "   Extra Large Single Blade, Forged Blade, Hardened and Tempered, ... " 6 25
No. 30, Black Handle, Large Single Blade, Silvered Mountings, . " 10 50
No. 40, Black Handle, Extra Large Single Blade, Forged Blade, Hardened and Tempered, Silvered Mountings, . . " 12 00

No. 5, Black Handle, Large Double Blade, ... per dozen, $8 00
No. 50, "   "   "   "   "   Silvered Mountings, . " 16 00

MINCING KNIVES.

Lightning Mincing Knife.

We show in the accompanying cut a mincing knife of novel construction, manufactured by Logan & Strobridge, of Pittsburgh, Pa. The cutting edges are ground upon revolving discs, and the operation of the knife will be seen at a glance from the construction. The revolving discs cut each way, backward and forward, and the motion is very easy and regular. In construction this knife is light and strong, neat in appearance, and much more easily handled than the old style of chopping knife. This tool is an attractive novelty which promises the merit of utility in a larger degree than most articles of the same general class.

February 24, 1877.

No. 0.

Fine Polished Blade
Polished Maple Handle,

No. 6, Cast Steel Blade,   Varnished Handle,   - - per dozen, $3 00
15, Extra "   " Etched, "   "   - - - - " 5 00

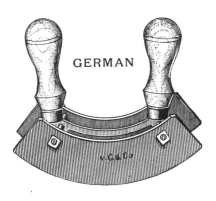

Double.

No. 675, 10-inch Cast Steel, Single Blade, Rocker Mincers,
675, 12 "   "   "   "   "   "
676, 10 "   "   Double "   "   "
676, 12 "   "   "   "   "   "

No. 5.

No. 5, Steel Blade, Japanned Iron Handle,

PIECED TIN WARE.

Syllabub Churns.

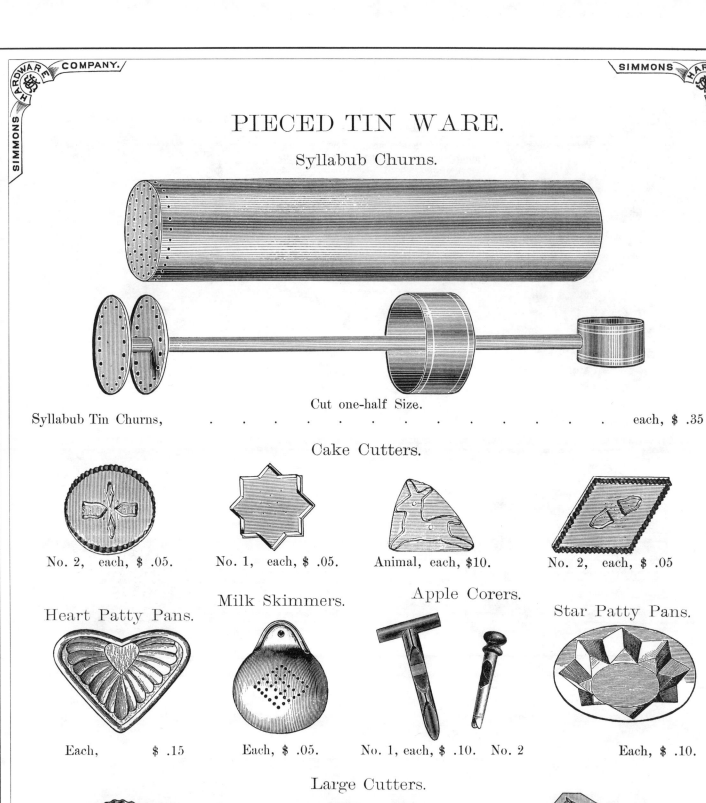

Cut one-half Size.

Syllabub Tin Churns, ... each, $ .35

Cake Cutters.

| No. 2, each, $ .05. | No. 1, each, $ .05. | Animal, each, $10. | No. 2, each, $ .05 |

Heart Patty Pans.   Milk Skimmers.   Apple Corers.   Star Patty Pans.

Each, $ .15   Each, $ .05.   No. 1, each, $ .10. No. 2   Each, $ .10.

Large Cutters.

Nos.   1   2   3   . . .
Each, .15   .20   .27   . . .

FANCY TIN CUTTERS.

Round.

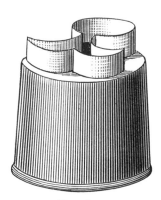

No. 3.

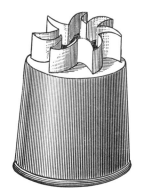

No. 1.

Tin Cutters, Round,	Nos.	0	1	2	3
	Each,	$ .15	.15	.20	.25

Oval.

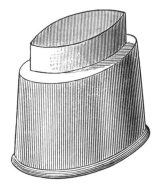

No. 1.

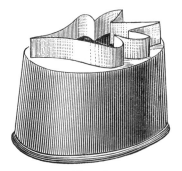

No. 3.

Tin Cutters, Oval,	Nos.	1	2	3
	Each,	$ .15	.20	.25

Tube Cutters.

No. 11.

No. 1.

No. 11—Scalloped Cutters,					per set, $2.75
Tube Cutters,	Nos.	1	2	3	4
	Per set,	$1.50	1.75	1.85	2.00

PLANISHED TIN WARE.

Shaving Etnas.

Pepper Boxes.

Flour Dredges.

Planished Tin Pepper Boxes, 3¾×2 inches,	each, $ .10
Planished Tin Shaving Etnas, 3¾×6 inches,	" .70
Planished Tin Flour Dredges, 4½×2½ inches,	" .15
Planished Tin Flour Dredges, 5 ×3 inches,	" .20

Oval Melon Moulds.

Berlin Moulds.

Oval Melon Moulds, 3 Pints, each, $ .70 ; 3½ Pints, each, $ .80 ; 4 Pints, each, $ .90 ; 5 Pints, each, $1.00
Berlin Moulds, 4 inches, . . per dozen, $1.50 | Berlin Moulds, 5 inches, . . per dozen, 2.00

Rice Moulds.

Jelly Moulds.

Confectioners' Moulds.

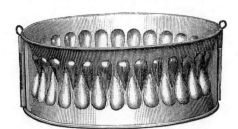

Oval Jelly Moulds.		Rice Moulds.		Confectioners' Moulds.	
½ Pint,	each, $ .35				
1 Pint,	" .40	3 Pints,	each, $ .95	1 Quart,	each, $1.50
1½ Pints,	" .45				
2 Pints,	" .60	5 Pints,	" 1.15	2 Quarts,	" 1.75
3 Pints,	" .75				
4 Pints,	" .85	7 Pints,	" 1.30	3 Quarts,	" 2.00

DEEP STAMPED WARE.

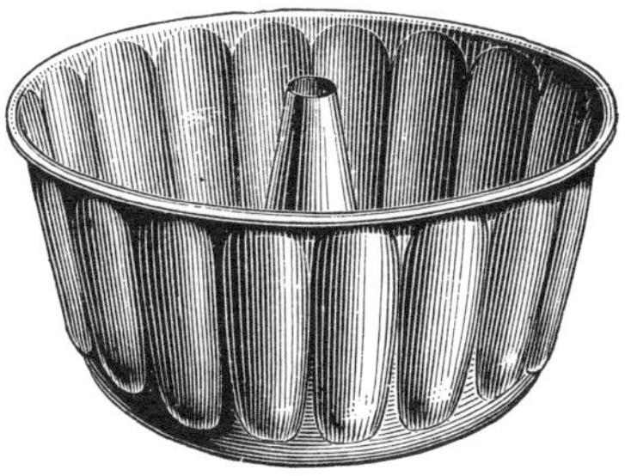

Turban Cake Moulds.   TURBAN.   Octagon Cake Moulds.

Retinned, with Tube.

Turban Cake Moulds.

Nos.	200	300	400	500	600
Quarts,	2	3	4	5	6
Inches,	8½×3	9×3	9½×3¼	10⅜×3½	11¾×11⅝
Each,	$ .30	.35	.40	.45	.50

Octagon Cake Moulds.

Nos.	720	730	740	750	760
Quarts,	2	3	4	5	6
Inches,	8⅛×3	9×3	9½×3¼	10⅜×3½	11¾×3⅝
Each,	$ .30	.35	.40	.45	.50

SCALLOPED CAKE PANS.   TURKS HEAD.

Plain, not Tubed.

Retinned, with Tube.

Plain, with Tubed.

Nos.									61	62	63	64
Quarts,									2	4	5	6
Inches,									7¼×3⅜	8½×3¾	9½×3⅞	11×4¼
Each,									$ .45	.55	.70	.80

MILK STRAINERS.   TURKS HEAD PANS.   Cullenders.

Retinned.

$6 85 Per Dozen

Improved Cullenders.

Nos.		42	43	44

104	205
9½×3	10½×3½
$ .25	.30

Wm. Frankfurth Hardware Co.

PLANISHED TEA AND COFFEE POTS.

Nos. 570 to 590.

Britannia Handle,
Cover and Spout,
Copper Bottom,

per dozen, $18 00

Nos. 405 to 420.

Etnas on Stand.

Tea Pots.

Planished.
Tin Bottoms.

Britannia Covers and Rings,

Good Morning Coffee Makers,

Textile Fabric Filter, used in all "Good Morning" Coffee Makers.

No. 12, 3 Pints, Planished Tin Coffee Biggins, Tin Bottom, - - - per dozen, $13 00

110

DEEP STAMPED WARE.
Tea Kettles.

Retinned Flat Bottom Tea Kettles.

GRAVY STRAINER

Wood Handle, Retinned.

Retinned

Nos.	16
Quarts,	1½
Diam. in Inches of Bottom	6¾
Per Dozen,	$18 75

WASH BOWLS

Nos.	31½
Quarts,	3
Diam. in Inches of Bottom,	5
Per Dozen,	$17 50

with Rings

$6 00 Per Dozen,

Retinned, Handled, Feet Fast.

Finished Water Dippers.

Straight Cup   Drinking Cups.

Finished Water Dippers.

		½	1	2	4	5	6
Pints,							
Plain,	each,	$ .08	.09	.12	.15	.17	.25
Tinned,	"	.09	.11	.15	.17	.20	.30

Cookey Pans.

Nos.	12
Retinned, per card,	.70

Patent Cookey Pans.

Sponge Cake Pans.

Sponge Cake Pans, Planished, 13½×3½ inches,
Soap Dishes, No. 01, Retinned, 4×1¼ inches,

Retinned.
TEA KETTLES.

Wm. Frankfurth Hardware Co.

COPPER WARE.

TEA KETTLES, SPUN COPPER.

| PLANISHED COPPER. | NICKEL PLATED. |

Pit Bottoms.

Nos.	7	8	9	Nos.	7	8	9
Per Dozen,	$27 00	30 00	33 00	Per Dozen,	$36 00	39 00	42 00

RANGE TEA KETTLES.

| Spun Copper, Flat Bottoms. | Nickel Plated, Spun Copper, Flat Bottoms. |

Inches,	6	7	8	Inches,	6	7	8
Per Dozen,	$15 00	16 50	20 00	Per Dozen,	$22 50	24 00	27 00

TEA KETTLES.

ALADDIN, NICKEL PLATED,

Chicago or Duck Spouts, Pit or Flat, Copper Bottoms.

Nos.	7	8	9
Per Dozen,	$56 00	60 00	66 00

PIECED TINWARE.

TEA KETTLES.

COPPER RIM.

Copper Rim Tea Kettles

JELLY FUNNELS

TEA STEEPERS

| Inches | - | - | 4½x3⅝ |
| Per Dozen, | - | - | $2 50 |

Duck Spout, Pit or Flat Copper Bottom.

Nos.	-	-	-	-	-	-	7	8	9
Per Dozen,	-	-	-	-	-	-	$21 50	24 00	26 50

RANGE

Range Tea Kettles

COFFEE FUNNELS

Duck Spout, Flat Copper Bottoms.

OIL STOVE

| Inches, | 8½ |
| Per Dozen, | $9 00 |

CANDLE MOULDS

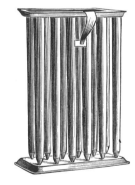

COFFEE FLASKS

1 Pint

PUDDING MOULDS

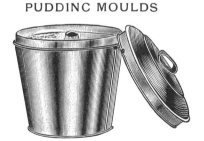

Plain.

Holes,	4	6	8	10	12	Quarts,	1½	2	3	4
Per Doz. Stands,	$4 80	7 20	9 60	12 00	14 00	Inches,	6¼x6	6¾x6½	7¾x7	8¼x7⅜
						Per Dozen,	$9 00	11 00	12 00	14 00

EGG POACHERS.

Three Hole.

Three Hole Poachers,   per dozen, $4 50

Six Hole.

Six Hole Poachers,   per dozen, $6 00

Wm. Frankfurth Hardware Co.

PIECED TINWARE.

COFFEE POTS

Bossed Handles.

Solid Lips.

Hinged Covers.

Quarts,	2	3	4	6
Per Dozen,	$7 75	9 25	10 50	12 00

Planished, Copper Bowl Bottoms

Tin, with Black Handles,

Copper Bowl Bottoms, Patent Enameled Handles and Hinged Covers.

Pints,	3	4	5	6		2	3	4	5
Per Dozen,	$8 50	10 00	11 50	12 50	Dozen,	$11 00	12 00	13 00	14 00

TEA POTS.

Planished, Copper Bowl Bottoms

Copper Bowl Bottoms, Patent Enameled Wood Handles and Hinged Covers.

Pints,	3	4	5	6	Quarts,	2	3	4	5
Per Dozen,	$9 00	10 50	12 00	13 00	Per Dozen,	$12 00	13 00	14 00	15 00

Brass Preserving Kettles.   American Broiler,   Spun Copper Kettles.

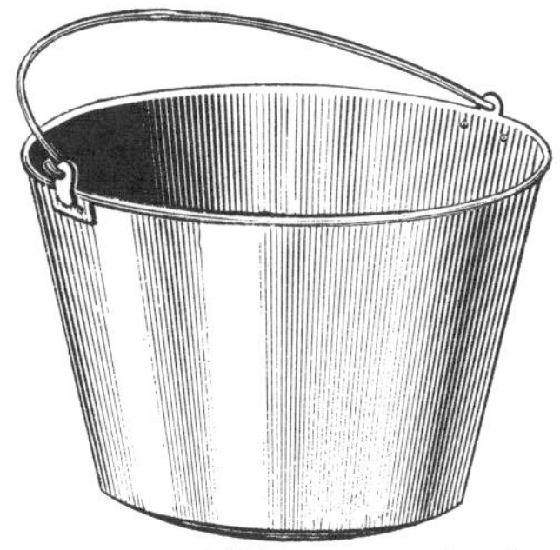

Brass Kettles,																	
Diameter, inches,	7	8	9	10	11	12	13	14	15	16	17	18	19	20	22	24
Weight, pounds,	1	1½	2½	3	3½	4	5	5¾	6½	7½	9	10½	12½	16½	20	27½
Capacity, gallons,	½	1	1½	2	2½	3	4	4½	5	6	8	10	12	14	18	25
Per pound, $																	

Spun Copper Kettles,					
Diameter, inches,	.	.	5	6	7	8
All Copper,	.	. each,	$1.75	2.00	2.25	2.50
All Copper, Nickel Plated,	"	2.50	2.75	3.00	3.25

Hammered Copper Kettles,									
Diameter, inches,	16	17¾	18	20	21¼	23¼	24½	25½	27½
Weight, pounds,	12	13½	17	21	24	29½	36½	41½	47
Capacity, gallons,	10	12	15	18	20	25	30	35	40
Per pound, $									

Broilers.

Epicure Broilers, ... each, $1.25

Self Basting Broilers.

No. 8—Round, fits No. 8 Stove, ... each, $ .75
No. 9—Round, fits No. 9 Stove, ... "   .90
No. 11—Oblong, 9×11 inches, ... " 1.00

DEEP STAMPED WARE.

Flat Handle Cup Dippers.

Wood Handle Cup Dippers.

Threaded Handle Soup Ladles.

Wood Handle Soup Ladles.

Flat Handle Ladles.

Wood Handle Ladles.

Flat Handle, Deep Skimmers.

Wood Handle, Deep Skimmers.

Flat Handle, Flat Skimmers.

Wood Handle, Flat Skimmers.

Flat Handle, Flat Skimmers.

Nos.	10	11	12	14	15
Inches,	4½	4⅝	5	5¾	6¼
Each,	$ .10	.13	.15	.18	.20

Patent Muffin Pans.

Retinned, No. 5, card, $ .50
Retinned, No. 5½, card, $ .75

Patent Corn Cake Pans.

DEEP STAMPED WARE.

Grocers' Scoops, Open.

Plain Tin.

Grocers' Scoops, Covered.

Grocers' Scoops, Open.

Nos.	1 Thumb.	2 Thumb.	3	4	5
Inches,	3×5	4¼×6¾	5¼×7¼	5¾×9	7×11
Each,	$ .10	.15	.30	.35	.50

Wood Handle, Retinned.

Grocers' Scoops, Covered.

Tin.

Solid Steel, Tinned, Wood Handle Scoops.

Cooks' Wood Handle Scoops.

Solid Steel, Wood Handle, Grocers' Scoops, 9½×6 inches, ... each, $1.50
Cooks' Wood Handle Scoops, ... " .20

Oyster Stands, Complete.

Pocket Oil Stoves for Alcohol . . . . each, $ .30
Oyster Stands, Complete, Retinned, ... " 1.25

CAST IRON HOLLOW WARE.

SCOTCH BOWL.

YANKEE BOWL.

Nos.	3	4	5	6	Nos.	3	4	5	6
Each,	$0 45	50	60	70	Each,	$0 55	65	75	90

TEA KETTLE.

HAM BOILER.

Nos.	7	8	9	10	Nos.	7	8	9
Each,	$0 80	90	1 00	1 25	Each,	$1 90	2 25	2 50

BAKE OVENS.

SHALLOW.

PATENT SAD IRON HEATERS.

Shallow.

Nos.	1	2	3
Inches,	12	11	10
Each,	$1 50	1 30	1 20

Deep.

Nos.	0	1	2	3
Inches,	14	12	11	10
Each,	$2 50	1 75	1 40	1 30

No. 7, Patent Sad Heater, - - each, $1 00
8, " " " - - " 1 25
9, " " " - - " 1 50

CAST STOVE HOLLOW WARE.

POT.

KETTLE.

SPIDER.

Nos.	7	8	9	10
Pots, Each,	$0 75	85	1 00	1 25
Kettles, Each,	$0 65	70	85	1 00
Spiders, Each,	$0 30	35	40	50
In Sets, Each,	$1 70	1 90	2 25	2 75

LONG PAN.

No. 7, Cast Iron Long Pans,	each, $0 50
8, " " " "	" 60
9, " " " "	" 75

HANDLED GRIDDLE.

BAILED GRIDDLE.

No. 7, Handled Griddles,	each, $0 25	12-inch, Bailed Griddles,	each, $0 50
8, " "	" 27	14 " " "	" 65
9, " "	" 30	16 " " "	" 80

CAST IRON BAKE PANS.

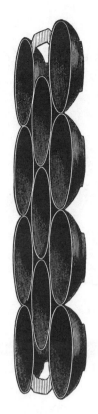
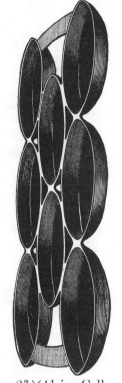
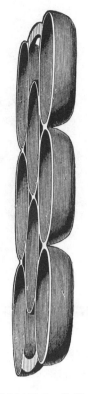

No. 1, 2¾ in ; No. 2, 3 in. Cells.   3 in Cells.   2⅜×4½ in. Cells.   2½×4½ in Cells.

No. 1—Round, small, $ .40 (Each.)
No. 2—Round, large,   .50

No. 3—Rd. Egg Fryer, $ .45 (Each.)

No. 4—French, Roll, $ .45 (Each.)

No. 5—Oval, $ .40 (Each.)

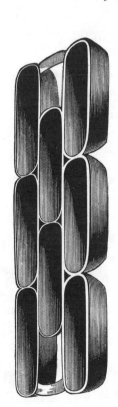

2×3 in Cells.   2½×3¾ in. Cells.   2¼×2¼ in. Cells.   2¼ in. Cells.

No. 6—Square, $ .40 (Each.)   No. 7—Square, $ .40 (Each.)   No. 8—Oval, $ .40 (Each.)   No. 9—R'd, Small, $ .40 (Each.)

CAST IRON BAKE PANS.

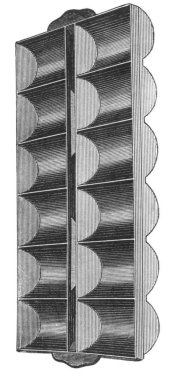

2×2¼ inch Cells.
No. 11—12 French Roll Pans,
Square, . . each, $ .60

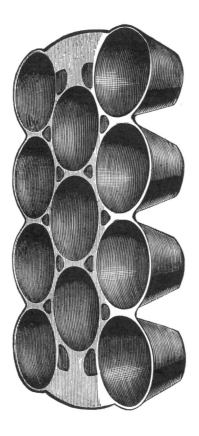

2¾ inch Deep Cells.
No. 10—12 Deep Corn Cake
Pans, . . each, $ .60

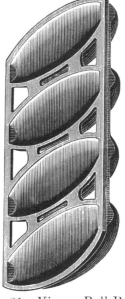

No. 21—Vienna Roll Pans,
each, $ .45

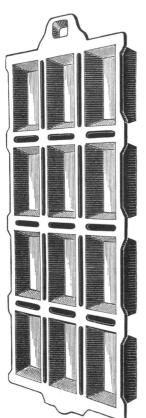

2×2⅝ inch Cells.
No. 12—12 Oblong Pans,
each, $ .45

No. 20—Vienna Bread Pans,
each, $ .45

WAFFLE IRONS.

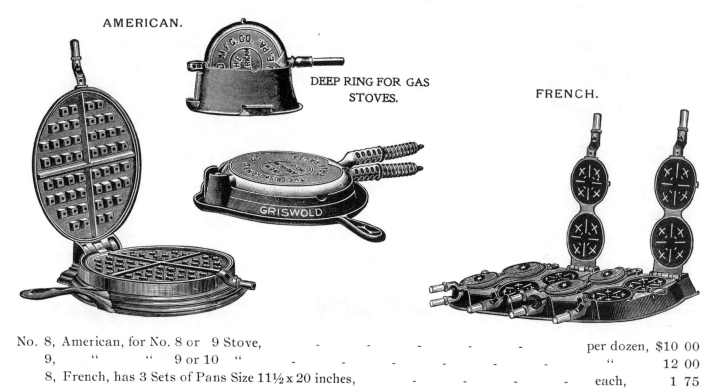

No. 8, American, for No. 8 or 9 Stove,		per dozen,	$10 00
9, " " 9 or 10 "		"	12 00
8, French, has 3 Sets of Pans Size 11½ x 20 inches,		each,	1 75
9, " " 4 " " " 11½ x 22½ "		"	2 00

SOAPSTONE GRIDDLES.

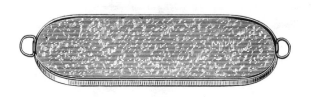

OVAL.

Inches,	9 x 18	10 x 20	11 x 22
Per dozen,	$15 00	18 75	22 50

ROUND.

Inches	10	12	14
Per Dozen,	$8 75	12 50	16 25

FOOT WARMERS.

Inches,	6 x 8	7 x 9	8 x 10
Per Dozen,	$3 75	4 37	5 00

PATENT FRYERS.

AMERICAN.
1865

Waffle Irons

Extra Large Fryer.

For Hotel Use.

Extra Large; Diameter, inches,	13	14	16
Each,	$4.50	6.00	8.00

Waffle Irons.

Excelsior Waffle Irons.

Square Waffle Irons.

No. 7—Excelsior Waffle Irons, 7 inches, each, $1.00	No. 0—Square Waffle Irons, 4 Cake, each, $1.75
No. 8—Excelsior Waffle Irons, 8 inches, " 1.15	No. 1—Square Waffle Irons, 3 Cake, " 1.50
No. 9—Excelsior Waffle Irons, 9 inches, " 1.25	

Long Handle Wafer Irons.

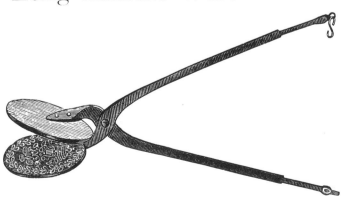

No. 3—Heavy English Wafer Irons, ... each, $1.75

GRANITE IRON WARE.

Pots.

					Metal.	Nickel Plated
No. 3410—4 Pints, Metal Trimmed,	each, $2.50		No. 04010—1½ Pints,	each,	$1.75	2.75
No. 3420—5 Pints, Metal Trimmed,	" 2.90		No. 4000—3 Pints,	"	2.25	3.25
No. 3430—6 Pints, Metal Trimmed,	" 3.35		No. 4010—4 Pints,	"	2.60	3.50
			No. 4020—5 Pints,	"	3.00	4.00

	Metal.	Nickel Plated.			Nickel Plated.			Nickel Plated.
No. 4810—4 Pints,	$2.60	3.75	No. 5200—3 Pints,	each,	$3.25	No. 3110—4 Pints,	each,	$3.75
No. 4820—5 Pints,	3.00	4.25	No. 5210—4 Pints,	"	3.75	No. 3120—5 Pints,	"	4.25
No. 4830—6 Pints,	3.50	4.75	No. 5220—5 Pints,	"	4.25	No. 3130—6 Pints,	"	4.75
			No. 5230—6 Pints,	"	5.00			

No. 3710—4 Pints, Metal Trimmed,	each, $2.20	No. 4910—4 Pints, Metal Trimmed,	each, $2.00
No. 3720—5 Pints, Metal Trimmed,	" 2.60	No. 4920—5 Pints, Metal Trimmed,	" 2.25
No. 3730—6 Pints, Metal Trimmed,	" 3.00	No. 4930—6 Pints, Metal Trimmed,	" 2.65

GRANITE IRON WARE.

Syrup Pitchers.

No. 3100.

Slop Bowls.

No. 3100.

Cream Pitchers.

No. 3100.

No. 3100—Syrup Pitchers, Nickel Plated, ... each, $3.00
No. 3100—Slop Bowls, Nickel Plated, ... " 2.25
No. 3100—Cream Pitchers, Nickel Plated, ... " 2.75

Cream Pitchers.

No. 4000.

Slop Bowls.

No. 4000.

Spoon Holders.

No. 3100.

No. 4000—Cream Pitchers, Nickel Plated, ... each, $2.75
No. 4000—Slop Bowls, Nickel Plated, ... " 2.25
No. 3100—Spoon Holders, Nickel Plated, ... " 2.50

Butter Dishes.

Sugar Bowls.

No. 3100.

No. 3100.

Sugar Bowls.

No. 4000.

No. 3100—Butter Dishes, Nickel Plated, ... each, $4.00
No. 3100—Sugar Bowls, Nickel Plated, ... " 3.50
No. 4000—Sugar Bowls, Nickel Plated, ... " 3.50

GRANITE IRON WARE.

Covered Buckets.

Slop Jars.

Wine Coolers.

Chambers.

Water Carriers.

Pattie Pans.

Seamless Water Buckets.

Cake Moulds.

Dinner Plates.

Seamless Water Buckets.

Miners' Cups.

Tumblers.

Cups and Saucers.

ROUND CAKE GRIDDLES.

PITCHERS AND BOWLS.

BASTING SPOONS.
SOUP LADLES.
DIPPERS.

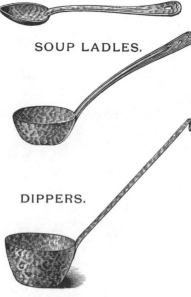

FRY PANS

JELLY CAKE PANS.

EGG FRY PANS.

MILK, RICE AND VEGETABLE BOILERS.

MILK PANS.

MOUNTAIN CAKE PANS.

DISH PANS

DEEP PUDDING PANS.

MILK OR RICE BOILERS.

FUNNELS.

"BELLE" TEA POTS.

LIPPED SAUCE PANS.

WALL SOAP DISHES.

TEA KETTLES.

Pit Bottom.

FOR HEATING STOVES.

TEA KETTLES.

Flat Bottom.

POT.

KETTLE.

TEA KETTLE.

SPIDER.

ROUND BAKE PANS.

PRESERVING KETTLES.

STOVE KETTLES.

With Granite Covers.

LIPPED WATER PAILS.

COFFEE BOILERS.

WASH BOWLS, WITH RINGS.

Wm. Frankfurth Hardware Co.

PERFECTION GRANITE IRON WARE.

TEA.   COFFEE.

No. 5500 Series, Assorted Colors.

No. 8500 Series, Assorted Colors.

Handsomely Mottled Colors, White Porcelain Inside, Nickel Plated Covers.

No. 5500, 3 Pints,	per dozen, $17 00	No. 8500, 3 Pints,	per dozen, $17 00
5510, 4 "	" 19 00	8510, 4 "	" 19 00
5520, 5 "	" 21 00	8520, 5 "	" 21 00
5530, 6 "	" 23 00	8530, 6 "	" 23 00

TEA.   COFFEE.

No. 4100 Series.

No. 5100 Series.

With White Metal Mountings.   With White Metal Mountings.

No. 4100, 3 Pints,	per dozen, $21 00	No. 5100, 3 Pints,	per dozen, $21 00
4110, 4 "	" 24 00	5110, 4 "	" 24 00
4120, 5 "	" 27 00	5120, 5 "	" 27 00
4130, 6 "	" 33 00	5130, 6 "	" 33 00

Milwaukee, Wisconsin.

PERFECTION GRANITE IRON WARE.

No. 4700 Series.

With Deep Infusion Sacks.

	Metal, Per Dozen.	Nickel Plated, Each.
No. 4710, 4 Pints,	$27 00	3 00
4720, 5 "	30 00	3 25
4730, 6 "	33 00	3 50

No. 2000 Series, Decoration Z.

With Deep Infusion Sacks.

No. 2020, 5 Pints, Nickel Plated, - - each, $4 75
2030, 6 "   "   "   - - " 5 25

SOUP LADLES.

No. 113.

No. 113, Nickel Plated Mountings, 3½-inch Bowl, - - - - per dozen, $14 00

No. 114.

No. 114, White Metal Mountings, Enameled Wood Handles, 3½-inch Bowl, - per dozen, $7 00

PERFECTION GRANITE IRON WARE.

FRENCH COFFEE POTS.

With White Metal Filters.

Sectional View of French Coffee Pots.
All Granite Pots with Metal or Planished Filters are now made with this Ball Valve.

No. 3404, 4 Pints, White Metal Mountings and Protection Bands, - - per dozen, $34 00
    3405, 5 " " " " " " " - - - " 38 00
    3406, 6 " " " " " " " - - - - " 42 00

"GOOD MORNING" COFFEE MAKER.

No. 6800 Series. Decoration A.

Decoration A with Patent Combination Textile Filter. Decoration A, Morning Glories, hand-painted in Mineral Colors.

                                        EACH
No. 6820, 5 Pints, Decoration A, Nickel Plated, $5 50
    6830, 6 " " " " " 6 00
    6820, 5 " " Z, " " 4 50
    6830, 6 " " " " " 5 00

No. 3600 Series.

Granite Iron with Planished Covers, with Patent Combination, Textile Filter.

No. 3620, 5 pint, White Metal, - per dozen, $24 00
    3630, 6 " " " - " 27 00

Milwaukee, Wisconsin.

PERFECTION GRANITE IRON WARE.

No. 9200 Series. Tea or Coffee. Decoration P.

Decoration P, Stork in Bayou, hand-painted in Mineral Colors. Cool Porcelain Handle.

No. 9200, 3 Pint, Nickel Plated,	each,	$4 50
9210, 4 " " "	"	5 00
9220, 5 " " "	"	5 50
9200, Sugar Bowl,	"	4 50
9200, Cream Pitcher,	"	3 75
9200, Spoon Holder,	"	3 50

No. 9200 Series. Tea or Coffee. Decoration Z.

Decoration Z is a handsomely mottled Blue and White. Cool Porcelain Handle.

No. 9200, 3 Pint, Nickel Plated,	each,	$3 50
9210, 4 " " "	"	4 00
9220, 5 " " "	"	4 50
9200, Sugar Bowl, "	"	3 50
9200, Cream Pitcher, "	"	3 00
9200, Spoon Holder, "	"	2 75

No. 8000 Series. Decoration L.

Decoration L, Wreath of Violets, hand-painted in Mineral Colors.

Decoration Z is a handsomely mottled Blue and White.

	EACH
No. 8000, 3 Pint, Decoration L, Nickel Plated,	$4 25
8010, 4 " " " " "	4 50
8020, 5 " " " " "	5 00
8000, 3 " " Z " "	3 50
8010, 4 " " " " "	3 75
8020, 5 " " " " "	4 25

No. 4000 Series.

White Metal Mountings and Protection Bands.

No. 4000, 3 Pint, Metal,	per dozen,	$27 00
4010, 4 " "	"	31 00
4020, 5 " "	"	35 00
4030, 6 " "	"	39 00
4000, 3 " Nickel Plated,	each,	3 25
4010, 4 " " "	"	3 50
4020, 5 " " "	"	4 00
4030, 6 " " "	"	4 50

Wm. Frankfurth Hardware Co.

WATER COOLERS.

No. 10.

Gallons,	3	4	6
Each,	$4 50	5 50	6 50

nickel plated brass Faucets

Coffee Urns,
Made from extra heavy copper, heavily tinned pure block tin; outside heavily nickel plate finished.

No. 20.

3	4	6	8
$3 50	4 00	5 00	6 00

walls of galvanized iron.

No. 804, Holds, Ice 5 lbs.,
    Water, 2¼ gals., each, $5 00
No. 806, Holds, Ice 6 lbs.,
    Water, 3¾ gals., each, 6 00

Galvanized iron reservoir and ice cylinder, nickel plated Lever Faucet.

Cooler Drainers.

Gallons,	3	4	6	8	10
Each,	$1 75	2 00	2 30	2 65	3 00

For Nos. 10 and 20 Coolers, and "Clear Water" Filters. Finished in neutral colors to match Coolers of any color.

Cooler Stands.

Nos.	1	2	3
For	3 and 4 Gal.	6 Gal.	8 and 10 Gal.
Each,	$4 00	4 50	4 50

Made of malleable iron and handsomely Japanned. They are shipped "knocked down." Can be screwed to the floor.

STEVENS' FILTERS.

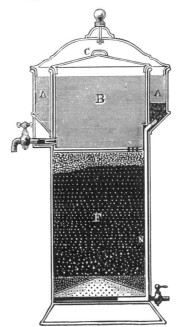

No. 12—15 Gallons per hour, . each, $16.00 | No. 14—18 Gallons per hour, . . each, $20.00
No. 17—30 Gallons per hour, . . " 35.00

Centennial Coolers.

Filters.

Filters and Coolers.

No. 6.   No. 73.

Centennial Coolers.

Gallons, ...	2	3	4	6	8	10
Each, ...	$5.50	7.00	8.00	10.00	12.00	15.00

Filters.

No. 6—Filters, Capacity 6 Gallons, for Rain or Cistern Water, ... each, $ 7.50
No. 73—Filters and Coolers, Reservoir, 11 Quarts, for Rain or Cistern Water, . . . " 11.00

N. B.—The Stevens' Filters will filter perfectly the muddy water from any of the Western Rivers.

REFRIGERATORS.

Solid Walnut Sideboard Refrigerators and Water-Coolers Combined.

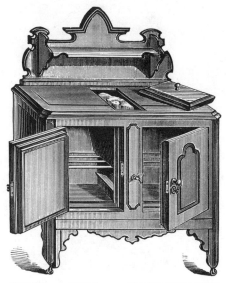

	Length.	Depth.	Height, with Back,	Each.
No. 26.	35 in.	20½ in.	50 in.	$32.00
No. 27.	38 in.	21½ in.	53 in.	38.00

Cedar Chests.

Size, Inside Measurement.

Chests.	Length.	Depth.	Width.	Each.
No. 1,	28 in.	15 in.	15 in.	$ 8.50
No. 2,	34 in.	18 in.	19 in.	10.00
No. 3,	38 in.	19 in.	20 in.	11.00
No. 4,	46 in.	21 in.	22 in.	12.50

Alaska Refrigerators.

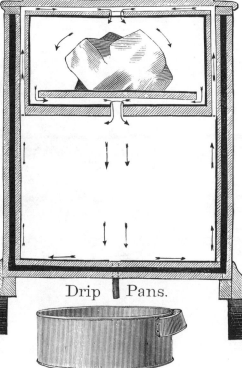

Drip Pans.

Nos. 2, 3, 4, 5.   Nos. 0 and 1.

Beautifully grained in Oak and Black Walnut, Lined with Zinc, with Porcelain Castors and Silver Mounted Trimmings.

	Length.	Depth.	Height.	Each.		Length.	Depth.	Height.	Each.
No. 0.	27 in.	19 in.	42 in.	$13.00	No. 3.	36 in.	24 in.	48 in.	$25.00
No. 1.	31 in.	20 in.	44 in.	17.00	No. 4.	40 in.	24 in.	51 in.	30.00
No. 2.	34 in.	21 in.	46 in.	22.00	No. 5.	43 in.	26 in.	52 in.	35.00

Refrigerator Drip Pans, Galvanized, ...65

CENTENNIAL ICE CHESTS.

Oak Grained, with filled lid. Paneled Front and Ends.

	Length.	Depth.	Height.	Price.
No. 16,	32 in.	21 in.	26 in.	$ 8.00
No. 17,	35 in.	23 in.	28 in.	9.00
No. 18,	38 in.	25 in.	30 in.	11.00
No. 19,	41 in.	27 in.	32 in.	14.00

This Refrigerator, like the Centennial, is guaranteed to be free from Sweat in the Provision Chamber. Beautifully Grained in Oak, with Walnut Panels and on Castors. Panel fronts and sides with double lid.

	Length.	Depth.	Height.	Price.
No. 11,	36 in.	23 in	34 in.	$16.00
No. 12,	41 in.	26 in.	37 in.	20.00
No. 13,	47 in.	29 in.	40 in.	24 00

Centennial Refrigerators.

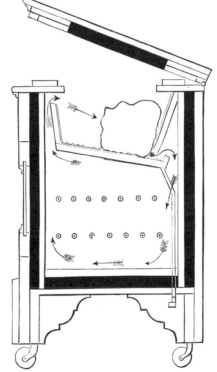

Nos. 0, 1, 2, 3 and 4. Nos. 5, 6, 7, 8 and 9.

Beautifully Grained in Oak and French Burl Panels, with Over-lapping Doors, Porcelain Castors, and Silver Mounted Trimmings.

	Length.	Depth.	Height.	Price.		Length.	Depth.	Height.	Price.
No. 0,	35 in.	22 in.	45 in.	$22.00	No. 5,	26 in.	19 in.	41 in.	$13.50
No. 1,	37 in	24 in.	47 in.	24.00	No. 6,	28 in.	21 in.	41 in.	16.00
No. 2,	39 in.	26 in.	48 in.	26.00	No. 7,	29 in.	22 in.	42 in.	19.00
No. 3,	42 in.	28 in.	49 in.	30.00	No. 8,	32 in.	23 in.	45 in.	21.00
No. 4,	46 in.	32 in.	51 in.	35.00					

SCHALL & CO., # Metal Ice Cream Moulds NEW YORK

No. 589. Chrysanthemum.

No. 365. Daisy Bunch.

No. 310. Sea Shell.

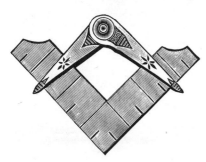

No. 323. Masonic Emblem.

No. 562. Automobile.   Chauffeur.

No. 580. Bunch of Grapes.

Fixtures for Ice Cream Moulds

Coon with Banjo.

Coon with Watermelon.

Cupid with Golden Oars.

Mother Goose.

Female Diver.

Cupid, Bow and Arrow

NO. 37, BATTLESHIP, TWO QUARTS.

Ice Tools.

BUYING A
Peerless Freezer
WITH THE
Vacuum Screw Dasher

Economy Ice Cutters.

each, $ .35

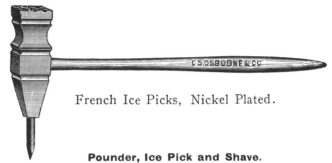

French Ice Picks, Nickel Plated.

$ .50

No. 33,	Tinned	$6.00 per dozen
No. 34,	Nickeled,	15.00 per dozen

Pounder, Ice Pick and Shave.

No handles to get loose as is the case with all wood handled ice picks. The "Star" is the housekeepers' delight; buy no other. Price, per dozen, $2.00.

ICE SAWS, WITH IRON HANDLE.
28-inch,

each, $2 00

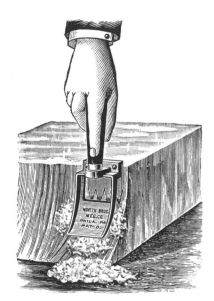

"CROWN" ICE CHIPPER

The only practical tool for chipping ice quickly to small uniform pieces for use in Ice Cream Freezers, in place of the old time bag and hatchet. A 20 lb. block of ice can be reduced to small pieces, size of peanuts, in five minutes. Saves all waste of ice. Useful in Bar Rooms, Restaurants, etc.

"GEM" ICE SHAVE

For shaving ice from blocks in Refrigerator without removal. Shaved ice is desirable for cooling Cantaloupes, Melons, Tomatoes, Fruits, Oysters on half-shell, Wines, Liquors, Lemonades, and other iced drinks, at home or at Soda Fountains, Bar Rooms and Restaurants.

Milwaukee, Wisconsin.

ICE CREAM FREEZERS.

WHITE MOUNTAIN.

Ice Cream Dishing Spoons
WOOD HANDLE, RIVETED

Square End Spoon

Inches....	12	14	16	18
Each.....	25c.	28c.	30c.	35c.

Pointed End Spoon

	12	14	16	18
	25c.	28c.	30c.	35c.

Round Bowl Spoon

............................35c.
Nickel plated..............

Fig. 5327
Seamless, retinned cups,

PEERLESS.

Quarts,	3	4	6	8	10
Each,	$4 50	5 50	7 00	9 00	12 00

The can sets squarely anywhere. The dasher is self-adjusting, and when placed in the can, goes directly to its centre and rest, whether the can is filled with cream or not. The can and the dasher revolve in opposite directions, but the dasher turns with greater speed, and rapidly rubs the cream to perfect fineness. One arm of the dasher detaches the frozen cream from the sides of the can and whirls it to the centre, while the other, following, spreads the liquid cream upon the sides of the can, alternately freezing and removing, which makes the work more rapid and cream even.

When the cream is all evenly frozen and mixed, the dasher can be removed, and the can may be revolved without it, until the cream is sufficiently solid.

The crank may be used on the top or side of the Freezer, as preferred.

For Confectioners and Hotels.

Quarts,	-	-	14	18	21	25
Each,	-	-	$20 00	25 00	30 00	35 00

They have all the improvements of the "Peerless" Freezer, and in addition, are so heavy and strongly made as to be almost everlasting, and scarcely ever require repairs.

American Machine Company's Ice Cream Freezers.

American. Star.

Crown, Single Action. Crown, Double Action. Gem.

Best Quality Cedar Tubs, Galvanized Iron Hoops.

The growing demand for small size Freezers with Fly Wheels has prompted us to add them to our "Wonder" line on all sizes. The labor of freezing the Cream is so greatly lessened by the addition of the Fly Wheel that any one who uses one once will never again be satisfied with a crank Freezer.

CHAMPAGNE TAPS.

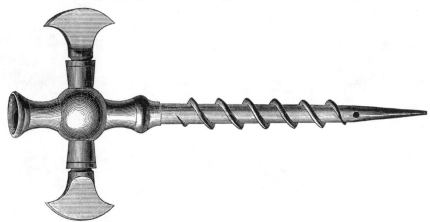

No. 100—Champagne Taps, Silver Plated, ... each $ .85

NUT CRACKS.

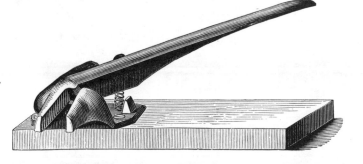

No. 10, japanned, stained wood base, per dozen. $5.40

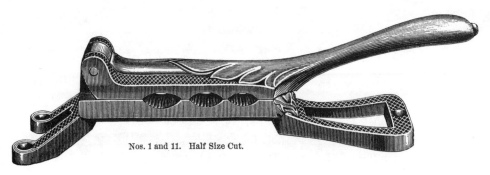

Nos. 1 and 11. Half Size Cut.

No. 1, Japanned, ... per dozen, $2 80
No. 11, Fancy Bronzed . . . . " 3 60

Cork Pressers.

Nos. 2 and 12. Half Size Cut.

No. 2, Japanned, ... per dozen, $2 25
No. 12, Fancy Bronzed, ... " 3 00

CORK SCREWS.

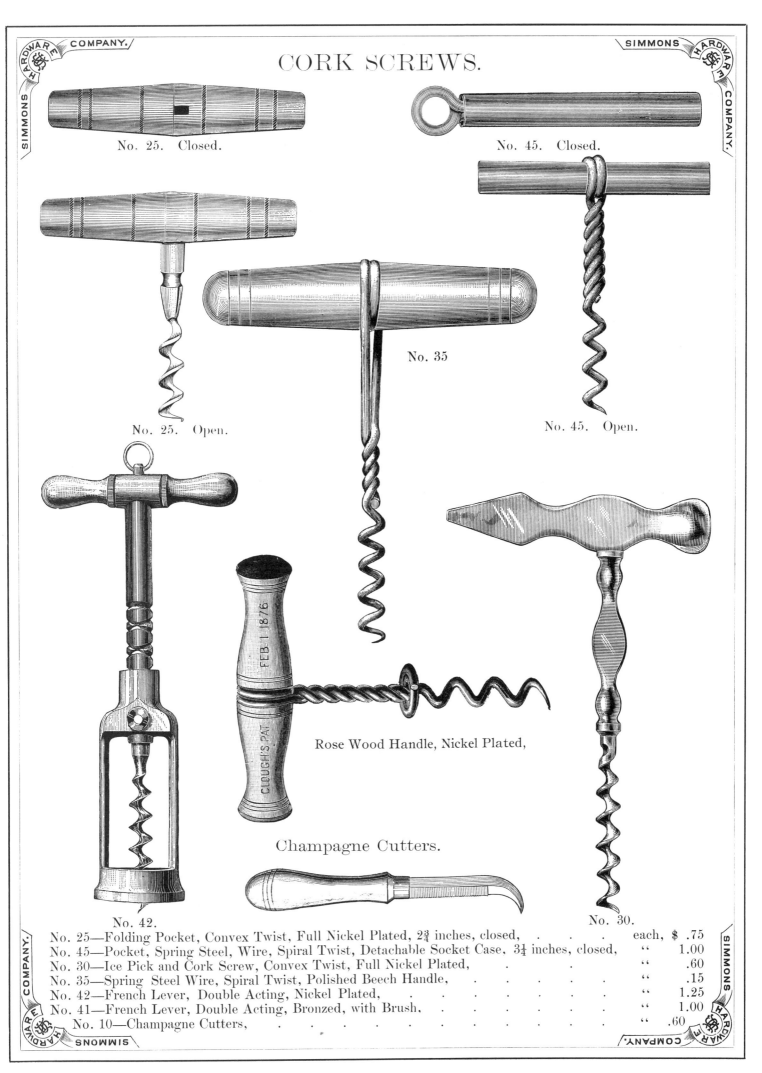

No. 25. Closed.
No. 45. Closed.
No. 25. Open.
No. 35
No. 45. Open.
Rose Wood Handle, Nickel Plated,
Champagne Cutters.
No. 42.
No. 30.

No. 25—Folding Pocket, Convex Twist, Full Nickel Plated, 2¾ inches, closed, . . . each, $ .75
No. 45—Pocket, Spring Steel, Wire, Spiral Twist, Detachable Socket Case, 3¼ inches, closed, " 1.00
No. 30—Ice Pick and Cork Screw, Convex Twist, Full Nickel Plated, . . . " .60
No. 35—Spring Steel Wire, Spiral Twist, Polished Beech Handle, . . . " .15
No. 42—French Lever, Double Acting, Nickel Plated, . . . " 1.25
No. 41—French Lever, Double Acting, Bronzed, with Brush, . . . " 1.00
No. 10—Champagne Cutters, . . . " .60

CORK SCREWS.

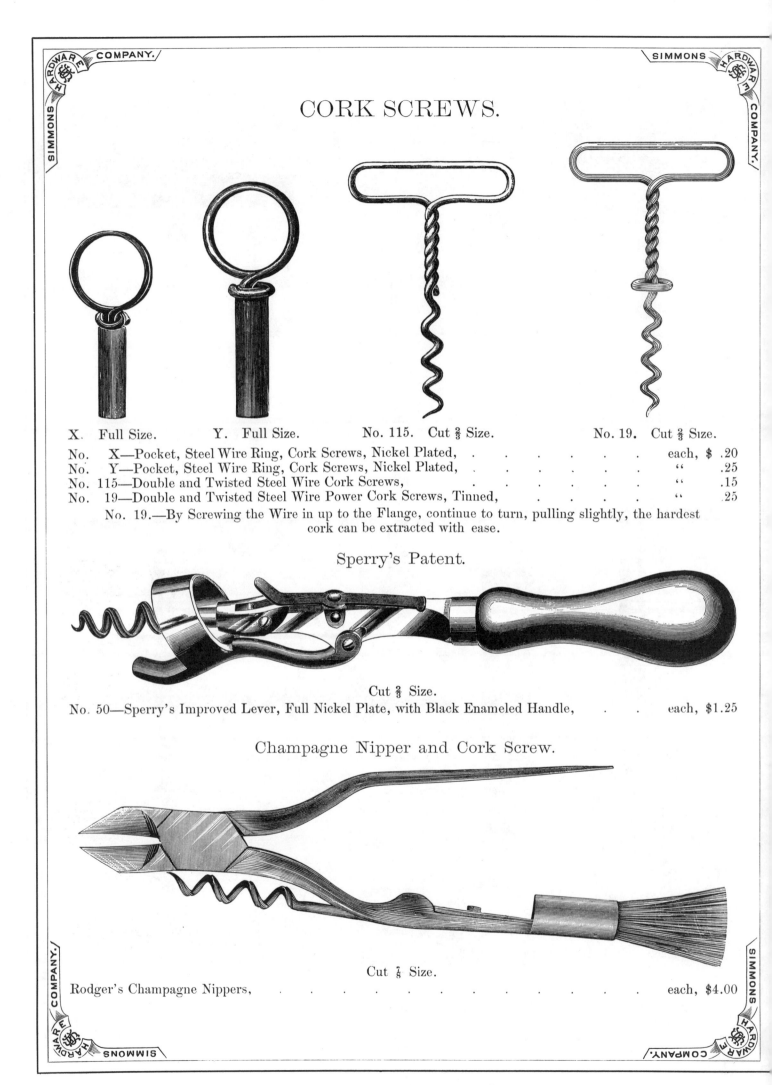

| X. Full Size. | Y. Full Size. | No. 115. Cut ⅔ Size. | No. 19. Cut ⅔ Size. |

No.   X—Pocket, Steel Wire Ring, Cork Screws, Nickel Plated, ... each, $ .20
No.   Y—Pocket, Steel Wire Ring, Cork Screws, Nickel Plated, ... "   .25
No. 115—Double and Twisted Steel Wire Cork Screws, ... "   .15
No. 19—Double and Twisted Steel Wire Power Cork Screws, Tinned, ... "   .25

No. 19.—By Screwing the Wire in up to the Flange, continue to turn, pulling slightly, the hardest cork can be extracted with ease.

Sperry's Patent.

Cut ⅔ Size.

No. 50—Sperry's Improved Lever, Full Nickel Plate, with Black Enameled Handle, . . each, $1.25

Champagne Nipper and Cork Screw.

Cut ⅞ Size.

Rodger's Champagne Nippers, ... each, $4.00

WILLIAM MARPLES & SONS LTD., SHEFFIELD.

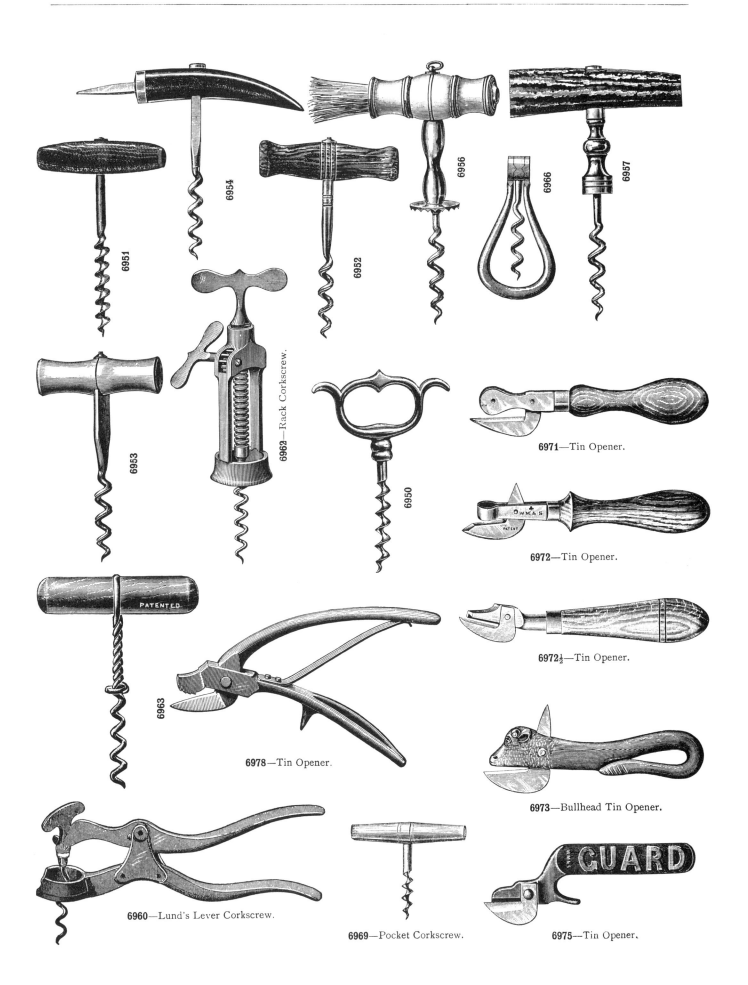

6971—Tin Opener.
6972—Tin Opener.
6972½—Tin Opener.
6978—Tin Opener.
6973—Bullhead Tin Opener.
6960—Lund's Lever Corkscrew.
6969—Pocket Corkscrew.
6975—Tin Opener.

CORK SCREWS.

Half Size Cut of Nos. 45, 47 and 48.

MAGIC.
Half Size Cut.

GLOBE.
Half Size Cut.

No. 45, Cast Steel, Japanned Malleable Loop Handle,	per dozen,	$3 50
47, " Copper Bronzed " " "	"	5 00
48, " Nickel Plated " " "	"	6 00
Magic, " Self-raising Cork Screw, Wood Handle,	"	6 00
No. 1, Globe " " " " "	"	8 00

HERCULES.
Half Size Cut of No. 1717.

NICKEL PLATED.
Self puller, seal lifter, stag handle, nickel finish, No. 450.

POWER.
One-third Size Cut of No. 950.

No. 1717, Hercules, Cast Steel, Coppered Steel Spring, Bronzed Frame,	per dozen,	$8 00
950, Power, " Nickel Plated,	"	10 00

One-half Dozen in a Box.

Milwaukee, Wisconsin.

CORK SCREWS.

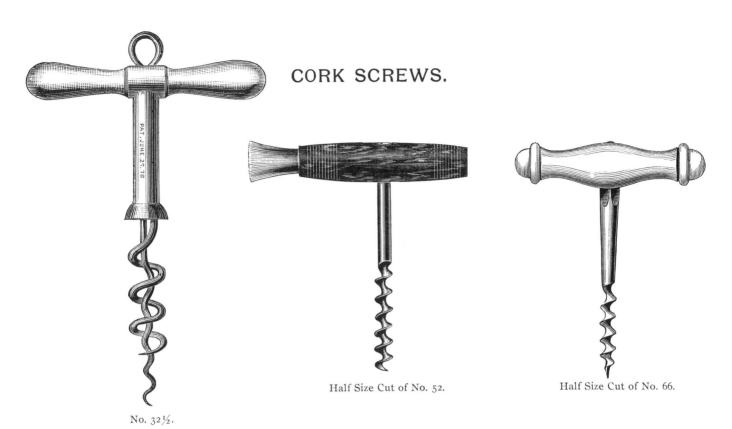

No. 32½. Half Size Cut of No. 52. Half Size Cut of No. 66.

No. 51, Cast and Steel Hand Forged, Rosewood Handle, - - - per dozen, $2 50
    52, " " " " " with Brush, - - " 3 20
    66, " Heavy Hand Forged, Rosewood Handle, Riveted Shank, - " 4 50
    32½, " " Duplex, Solid Metal Handle and Shank, Nickel Plated, - - " 2 87

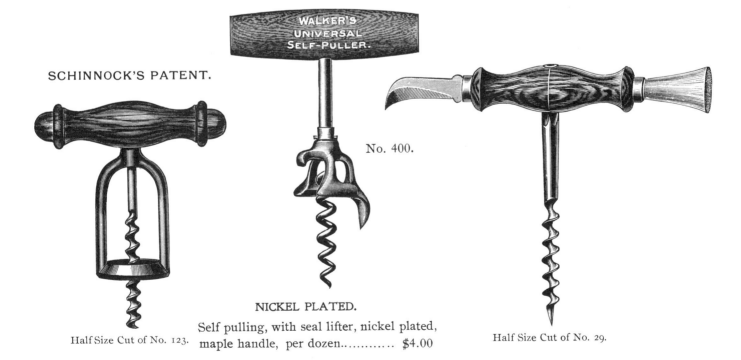

SCHINNOCK'S PATENT.

No. 400.

Half Size Cut of No. 123. NICKEL PLATED. Self pulling, with seal lifter, nickel plated, maple handle, per dozen............ $4.00 Half Size Cut of No. 29.

No. 123, Cast Steel Hand Forged, Bronzed Frame, Rosewood Handle, - - per dozen, $4 50
    29, " " with Champagne Opener and Brush, Rosewood Handle, - - " 8 50
    6, " " Extra Heavy, Champagne Opener and Brush, Rosewood Handle, " 12 00

All One Dozen in a Box.

Etiquette of the Table.

THE TABLE--HOW TO SET AND ARRANGE IT.

THE dinner-hour will completely test the refinement, the culture and good breeding which the individual may possess. To appear advantageously at the table, the person must not only understand the laws of etiquette, but he must have had the advantage of polite society. It is the province of this chapter to show what the laws of the table are. It will be the duty of the reader, in the varied relations of life, to make such use of them as circumstances shall permit.

Rules to be Observed.

Sit upright, neither too close nor too far away from the table.

Open and spread upon your lap or breast a napkin, if one is provided—otherwise a handkerchief.

Do not be in haste; compose yourself; put your mind into a pleasant condition, and resolve to eat slowly.

Keep the hands from the table until your time comes to be served. It is rude to take knife and fork in hand and commence drumming on the table while you are waiting.

Possibly grace will be said by some one present, and the most respectful attention and quietude should be observed until the exercise is passed.

It is the most appropriate time, while you wait to be served, for you to put into practice your knowledge of small talk and pleasant words with those whom you are sitting near. By interchange of thought, much valuable information may be acquired at the table.

Do not be impatient to be served. With social chit-chat and eating, the meal-time should always be prolonged from thirty minutes to an hour.

Taking ample time in eating will give you better health, greater wealth, longer life and more happiness. These are what we may obtain by eating slowly in a pleasant frame of mind, thoroughly masticating the food.

If soup comes first, and you do not desire it, you will simply say, "No, I thank you," but make no comment; or you may take it and eat as little as you choose. The other course will be along soon. In receiving it you do not break the order of serving; it looks odd to see you waiting while all the rest are partaking of the first course. Eccentricity should be avoided as much as possible at the table.

The soup should be eaten with a medium-sized spoon, so slowly and carefully that you will drop none upon your person or the table-cloth. Making an effort to get the last drop, and all unusual noise when eating, should be avoided.

If asked at the next course what you desire, you will quietly state, and upon its reception you will, without display, proceed to put your food in order for eating. If furnished with potatoes in small dishes, you will put the skins back into the dish again; and thus where there are side-dishes all refuse should be placed in them—otherwise potato-skins will be placed upon the table-cloth, and bones upon the side of the plate. If possible, avoid putting waste matter upon the cloth. Especial pains should always be taken to keep the table-cover as clean as may be.

Eating with the Fork.

Fashions continually change. It does not follow, because he does not keep up with them, that a man lacks brains; still to keep somewhere near the prevailing style, in habit, costume and general deportment, is to avoid attracting unpleasant attention.

Fashions change in modes of eating. Unquestionably primitive man conveyed food to his mouth with his fingers. In process of time he cut it with a sharpened instrument, and held it, while he did so, with something pointed. In due time, with the advancement of civilization, there came the two-tined fork for holding and the broad-bladed knife for cutting the food and conveying it to the mouth. As years have passed on, bringing their changes, the three and four-tined forks have come into use,

Fig. 9 The general arrangement of the table set for a party of twelve persons. The plates are often left off, and furnished by the waiter afterwards.

and the habit of conveying food with them to the mouth; the advantage being that there is less danger to the mouth from using the fork, and food is less liable to drop from it when being conveyed from the plate. Thus the knife, which is now only used for cutting meat, mashing potatoes, and for a few other purposes at the table, is no longer placed to the mouth by those who give attention to the etiquette of the table.

Set the table as beautifully as possible. Use only the snowiest of linen, the brightest of cutlery, and the cleanest of china. The setting of the table (Fig. 9) will have fruit-plates, castors and other dishes for general use, conveniently placed near the center. The specific arrangement (Fig. 10) of plate, knife, fork, napkin, goblet and salt-cup, is shown in the accompanying illustration.

Fig. 10. Relative position of plate, napkin, goblet, salt-cup, knife and fork, when the table is set.

It is customary for the gentleman who is the head of the household, in the ordinary family circle, to sit at the side of the table, in the center, having plates at his right hand, with food near by. When all the family are seated, and all in readiness, he will serve the guests who may be present; he will next serve the eldest lady of the household, then the ladies and gentlemen as they come in order. The hostess will sit opposite her husband, and preside over the tea, sauces, etc.

TRADE PRICE LIST.—BENJ. ALLEN & CO., CHICAGO.

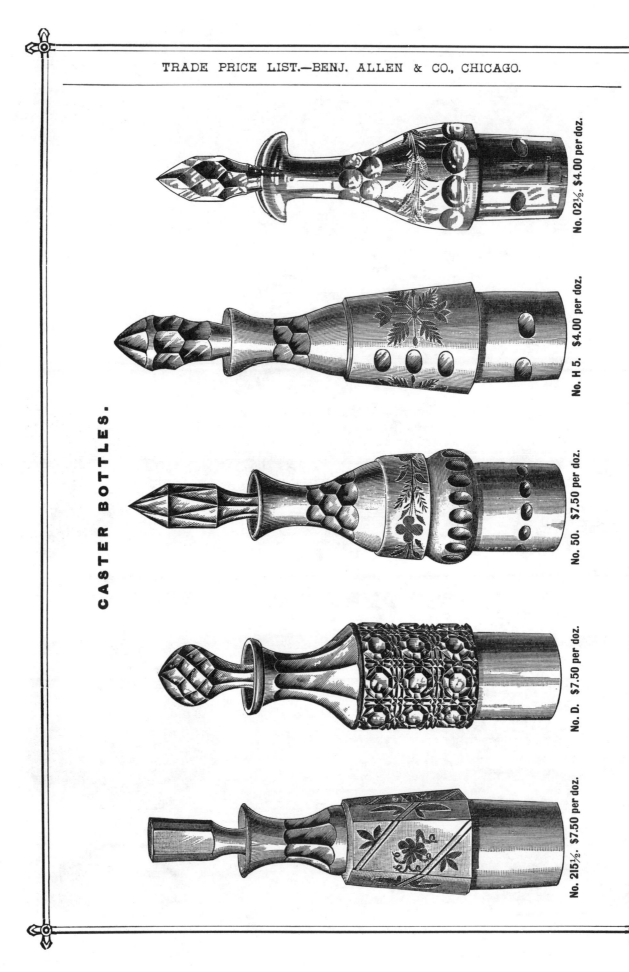

CASTER BOTTLES.

No. 02½. $4.00 per doz.
No. H 5. $4.00 per doz.
No. 50. $7.50 per doz.
No. D. $7.50 per doz.
No. 2155½. $7.50 per doz.

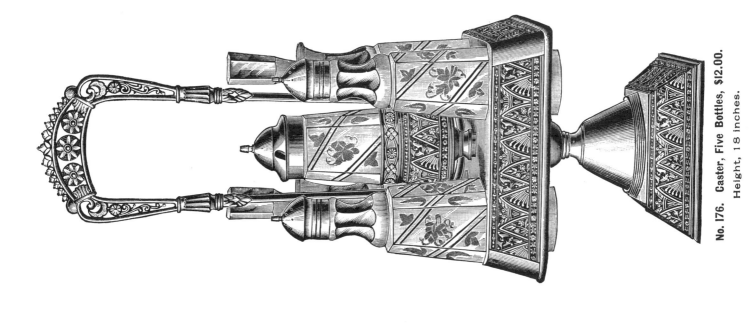

No. 176. Caster, Five Bottles, $12.00.
Height, 18 inches.

No. 183. Caster. 5 Bottles. Fluted, $7.25.
Height, 17 Inches.

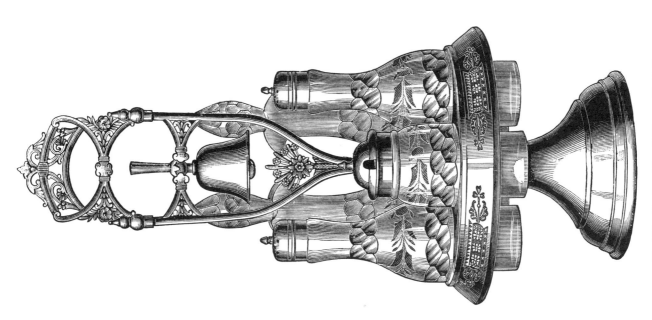

No. 110. Caster, Five Bottles, with Bell.
Plain, $5 75. Chased $6 25.

No. 157. Tilting Water Set.
Chased, with Gold Lined Slop and Goblets, $42.00.
Height, 22 inches.

ELECTRO SILVER PLATE.

No. 821—Chased Tilting Pitcher Sets, $19.00.

No. 88—Double Walls, Triple Plate, $8.00.

No. 99—Double Walls, Triple Plate, $9.00.

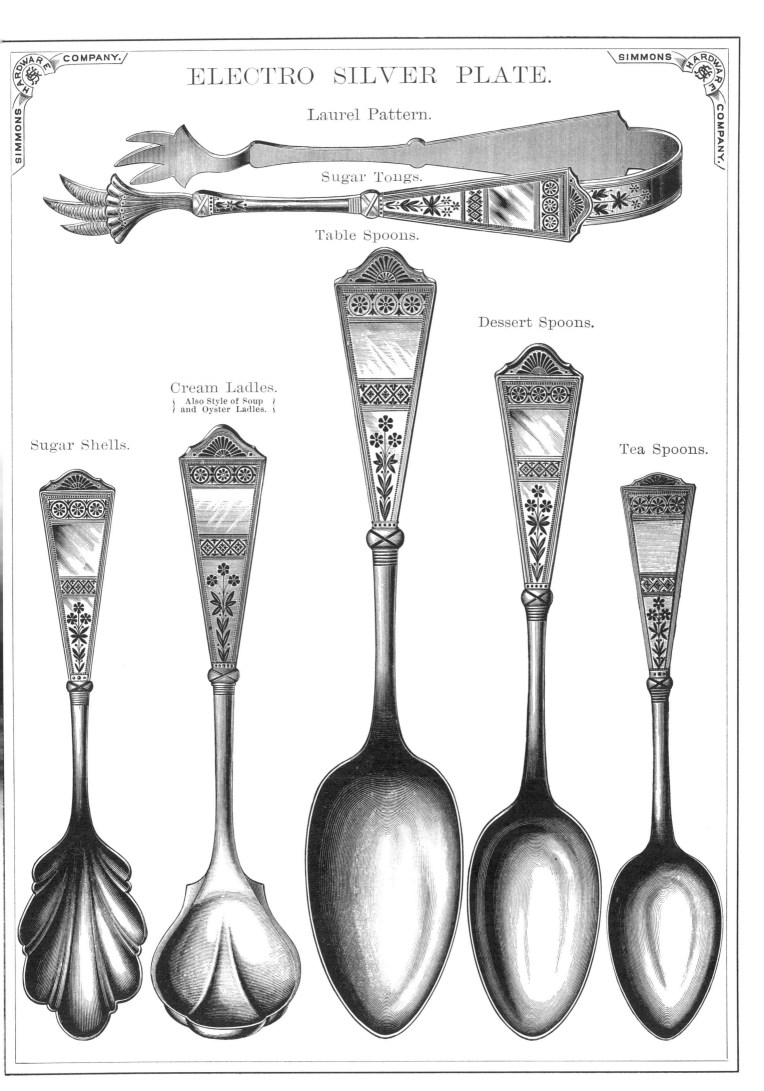

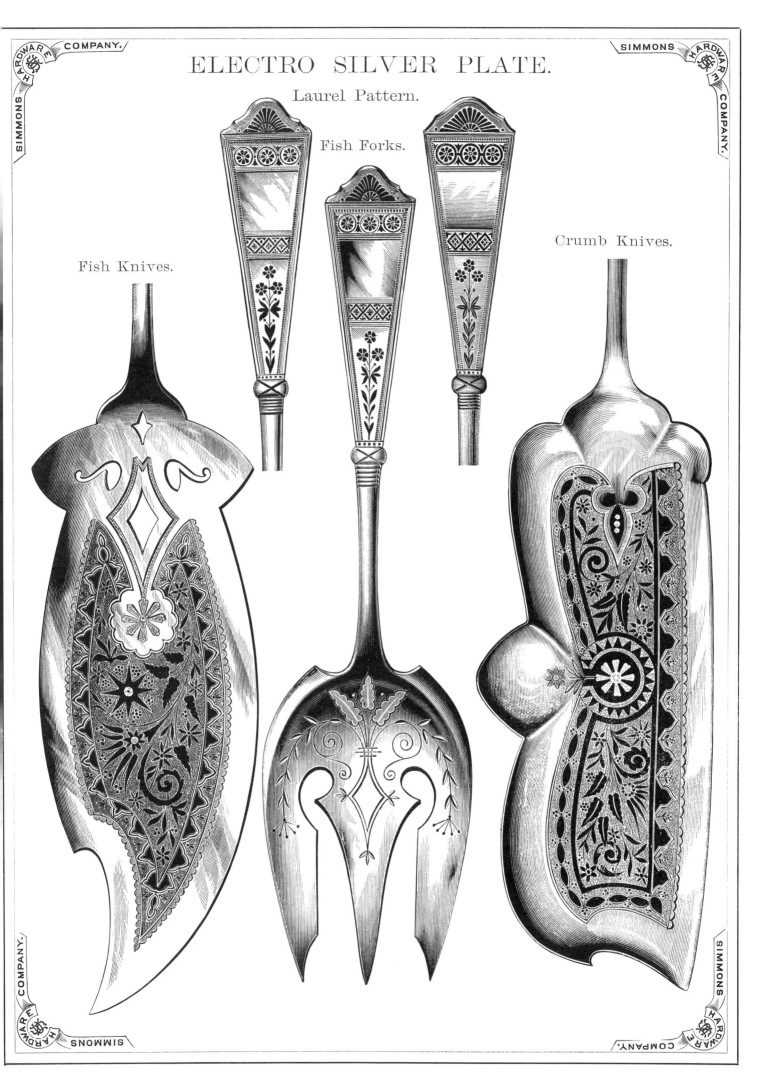

TABLE KNIVES AND FORKS.

No. 1953 C.

No. 1953 C, Cocoa Handle, Double Bolster, 5¼-inch Cimeter Blade,   - -   per gross, $17 00

No. 1946 C.

No. 1946 C, Cocoa Handle, Double Bolster, 5¼-inch Cimeter Blade,   - -   per gross, $21 00
    1746 C, Ebony   "   "   "   5¼ "   "   " - - -   "   22 00

No. 1947 C.

No. 1947 C, Cocoa Handle, Double Bolster, 5¼-inch Cimeter Blade,   - -   per gross, $24 00

No. 1943 C.

No. 1943 C, Cocoa Handle, Double Bolster, 5¼-inch Cimeter Blade,   - -   per gross, $22 00
    1743 C, Ebony   "   "   "   5¼ "   "   " - - -   "   23 00

No. 1952 C.

No. 1952 C, Cocoa Handle, Double Bolster, 5¼-inch Cimeter Blade,   - -   per gross, $23 00
    1752 C, Ebony   "   "   "   5¼ "   "   " - - -   "   24 00

TABLE CUTLERY.

Solid Steel, Silver Plated. (Knives only.)

Rogers Bros.' 1847, 12 ounce or Triple Plate, Medium, ... per dozen, $4.00
Rogers Bros.' 1847, 12 ounce or Triple Plate, Medium, Satin Handle, ... " 4.50
Rogers Bros.' 1847, 12 ounce or Triple Plate, Dessert, ... " 4.00
Rogers Bros.' 1847, 12 ounce or Triple Plate, Dessert, Satin Handle, ... " 4.50

Solid Steel, Silver Plated. (Knives only.)

Royal.

Royal—Medium Knives, Quadruple Plate, ... per dozen, $8.00
Royal—Dessert Knives, Quadruple Plate, ... " 7.00

Forged from one piece of Steel.

One Arm Man Knives.

Cut full size.

One Arm Man Knives, White Bone Handle, Lap Bolster, Capped End, ... each, $1.25
One Arm Man Knives, Ivory Handle, ... " 2.50

Butter Knives.

No. 66. Cut full size.

No. 66—Butter Knives, Solid Nickel Handle, Ivory Inlaid, Quadruple Silver Plated Blade, . each, $1.00
No. 460—Butter Knives, Solid White Bone Handle, Plain Blade, ... " .75
No. 8619—Butter Knives, Rubber Handle, Fancy Bolster, ... " .75

Fruit Knives.

Cut full size.

Fruit Knives, Solid Steel, Silver Plated, ... per dozen, $4.00

BREAD KNIVES.

No. 44.

No. 44, Goodell's, 8-inch Blade, Lap Bolster, Solid Cocoa Handle, - - - per dozen, $5 00

No. 3599.

No. 3599, J. Rodgers & Sons', 6½-inch Blade, Fancy Carved Beech Handle, - - per dozen, $6 00

No. 244.

No. 244, Royal Slicers, 8½-inch Blade, Star, Bolster, Cocoa Handle, - - per dozen, $7 50

8½ INCH CAST STEEL BLADE.

No. 900.   Cocobolo handle.   Per dozen $3.50

No. 14, Tubb's Bread Knives, 8-inch Blade, Solid Beechwood Handle, Brass Ferrule, - per dozen, $2 50

8¼ INCH CAST STEEL BLADE.

No. 972.   Cocobolo handle, plain bolster.   Per dozen $5.50

CARVING KNIVES AND FORKS.

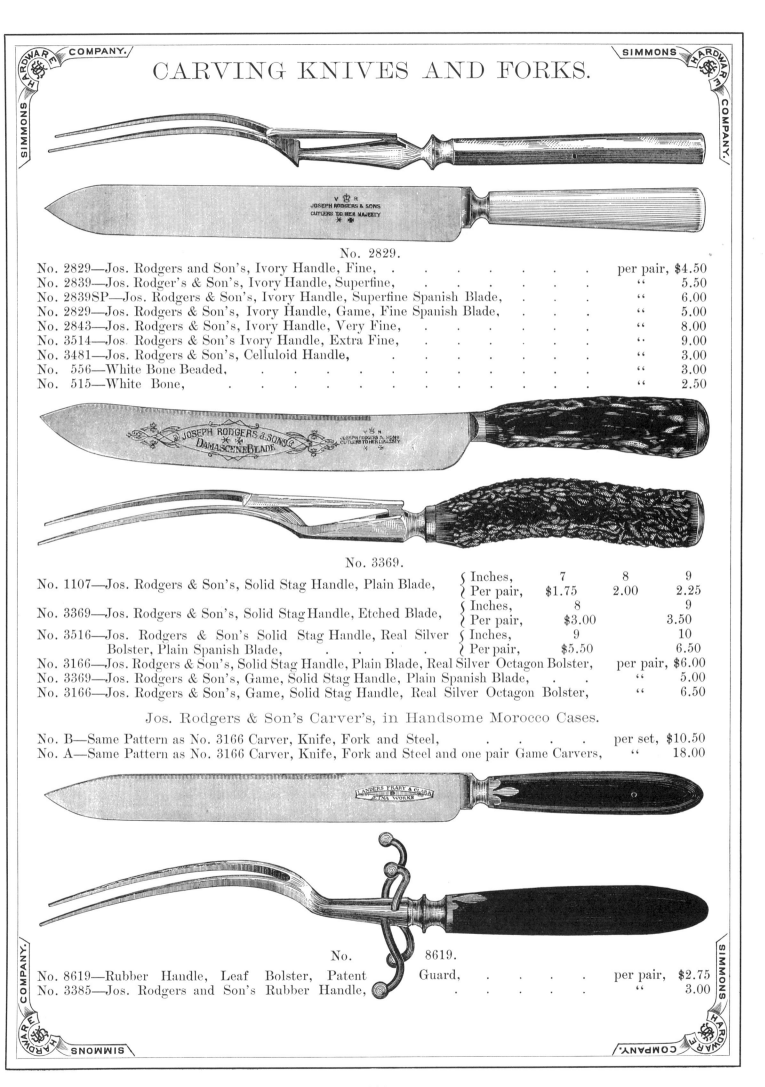

No. 2829.

No. 2829—Jos. Rodgers and Son's, Ivory Handle, Fine,	per pair,	$4.50
No. 2839—Jos. Rodger's & Son's, Ivory Handle, Superfine,	"	5.50
No. 2839SP—Jos. Rodgers & Son's, Ivory Handle, Superfine Spanish Blade,	"	6.00
No. 2829—Jos. Rodgers & Son's, Ivory Handle, Game, Fine Spanish Blade,	"	5.00
No. 2843—Jos. Rodgers & Son's, Ivory Handle, Very Fine,	"	8.00
No. 3514—Jos. Rodgers & Son's Ivory Handle, Extra Fine,	"	9.00
No. 3481—Jos. Rodgers & Son's, Celluloid Handle,	"	3.00
No. 556—White Bone Beaded,	"	3.00
No. 515—White Bone,	"	2.50

No. 3369.

No. 1107—Jos. Rodgers & Son's, Solid Stag Handle, Plain Blade,	Inches,	7	8	9
	Per pair,	$1.75	2.00	2.25
No. 3369—Jos. Rodgers & Son's, Solid Stag Handle, Etched Blade,	Inches,		8	9
	Per pair,		$3.00	3.50
No. 3516—Jos. Rodgers & Son's Solid Stag Handle, Real Silver Bolster, Plain Spanish Blade,	Inches,		9	10
	Per pair,		$5.50	6.50

No. 3166—Jos. Rodgers & Son's, Solid Stag Handle, Plain Blade, Real Silver Octagon Bolster,	per pair,	$6.00
No. 3369—Jos. Rodgers & Son's, Game, Solid Stag Handle, Plain Spanish Blade,	"	5.00
No. 3166—Jos. Rodgers & Son's, Game, Solid Stag Handle, Real Silver Octagon Bolster,	"	6.50

Jos. Rodgers & Son's Carver's, in Handsome Morocco Cases.

| No. B—Same Pattern as No. 3166 Carver, Knife, Fork and Steel, | per set, | $10.50 |
| No. A—Same Pattern as No. 3166 Carver, Knife, Fork and Steel and one pair Game Carvers, | " | 18.00 |

No. 8619.

| No. 8619—Rubber Handle, Leaf Bolster, Patent Guard, | per pair, | $2.75 |
| No. 3385—Jos. Rodgers and Son's Rubber Handle, | " | 3.00 |

WM. ROGERS MFG. CO.'S SILVER PLATED WARE.

BUTTER KNIFE—BENT.

Tipped Pattern.

AA or Extra Plate, Tipped, Flat Handle, Bent, - - - - - - per dozen, $8 50

BUTTER KNIFE—TWIST.

Saratoga Pattern.

AA or Extra Plate, Tipped, Twist Handle, - - - - - per dozen, $9 50
AA or "   "   Saratoga, "   "   - - - - - - " 10 50

SARATOGA CAKE KNIFE.

AA or Extra Plate, Saratoga Cake Knife, - - - - - - each, $2 00

SARATOGA PIE KNIFE.

AA or Extra Plate, Saratoga Pie Knife. - - - each $2 00

WM. ROGERS MFG. CO.'S SILVER PLATED WARE.

DESSERT FORKS.

Tipped Pattern.

AA	or Extra Plate, Tipped, Dessert Forks,		per dozen,	$7 50
AA	or " " " Medium "		"	8 50
12 oz. or Tripple " " Dessert "			"	11 50
12 oz. or " " " Medium "			"	13 50

Saratoga Pattern.

AA	or Extra Plate, Saratoga Dessert Forks,		per dozen,	$8 50
AA	or " " " Medium "		"	9 50
12 oz. or Tripple " " Dessert "			"	12 50
12 oz. or " " " Medium "			"	14 50

SUGAR SHELLS.

Saratoga Pattern.

AA or Extra Plate, Tipped, Sugar Shells,	per dozen,	$6 50
AA or " " Saratoga " "	"	9 00

Assyrian Pattern.

AA or Extra Plate, Assyrian Sugar Shells,	per dozen,	$9 00

C. Rogers & Bros.' Silver-Plated Flat-Ware.

Child's Sets, Assorted Fancy Patterns, put up on Beautifully Colored Cards ....$ 1 75
Child's Sets, Assyrian Pattern, old Silver, on Fine Colored Cards 2 00
Child's Set, Tipped Pattern, on Fancy Colored Card, (Plated Steel Knife)................................Each 1 38

Pie Knife. $3.50 each.   Siren Hollow Handle Cheese Scoop. $4.00 each.   Assyrian Pattern. Medium Fork.   Newport Pattern. Table Spoon.

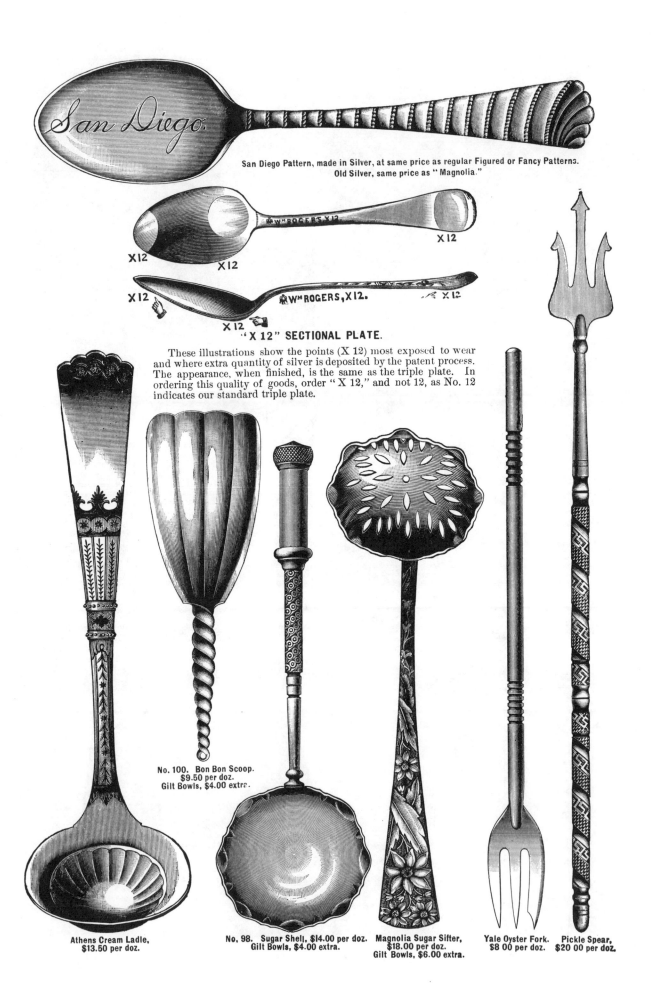

San Diego Pattern, made in Silver, at same price as regular Figured or Fancy Patterns. Old Silver, same price as "Magnolia."

"X 12" SECTIONAL PLATE.

These illustrations show the points (X 12) most exposed to wear and where extra quantity of silver is deposited by the patent process. The appearance, when finished, is the same as the triple plate. In ordering this quality of goods, order "X 12," and not 12, as No. 12 indicates our standard triple plate.

Athens Cream Ladle, $13.50 per doz.

No. 100. Bon Bon Scoop. $9.50 per doz. Gilt Bowls, $4.00 extra.

No. 98. Sugar Shell, $14.00 per doz. Gilt Bowls, $4.00 extra.

Magnolia Sugar Sifter, $18.00 per doz. Gilt Bowls, $6.00 extra.

Yale Oyster Fork. $8 00 per doz.

Pickle Spear, $20 00 per doz.

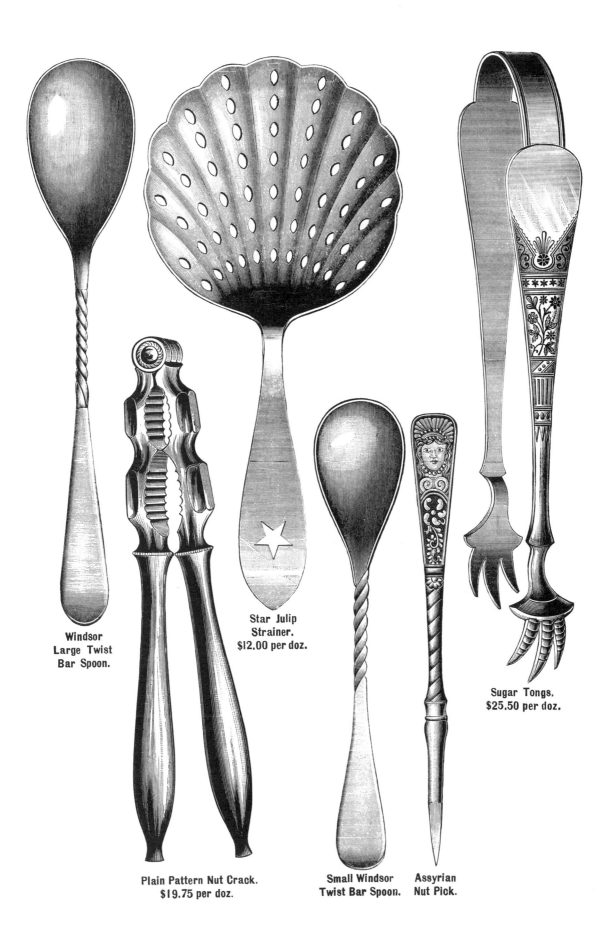

ELECTRO SILVER PLATE.

Embossed Chased Tea Sets.

No. 91912.

Set of Six Pieces.	Coffee.	Tea, Six Half Pints.	Tea, Five Half Pints.	Sugar.	Cream, Gold Lined.	Slop, Gold Lined.
$68.00.	$16.50.	$13.25.	$12.25.	$8.50.	$9.25.	$8.25.

No. 545—26 inch Waiter, Chased to match the Set, $60.00.

A Great Variety of Tea Sets of the Latest Designs Always in Stock.

170

ELECTRO SILVER PLATE.

Syrups.

No. 0812. Engine Syrup, Patent Cut-off, $5.50.

No. 39812½. Chased Syrup, with Plate, $7.50.

No. 91912. Embossed Syrup, with Plate, $8.00

Butter Dishes.

No. 16945—Chased Butter Dishes, $5.50.

No. 62912. Chased Butter Dishes, $8.25.

No. 64945. Butter Dishes, $4.00.

TRADE PRICE LIST.—BENJ. ALLEN & CO., CHICAGO.

No. 1872. Combination Sugar Bowl and Spoon Holder, Chased Satin, 9.50.

No. 30. Spoon Rack, $6.50.
With 12 Gilt Spoons, $7.75 extra.

No. 75. Celery Stand, $5.75.

No. 66. Celery Stand, $6.00.

No. 79. Celery Stand.
Venetian Glass, Rose Tinted, $10.50

No. 20. Spoon Holder.
Plain, $3 50. Chased, $4 00.
Chased and Gilt, $4 50

Height, 5 inches.

No. 3001. Spoon Holder, $7.50.
Gold Lined, $1.00 extra.

No. 28   Spoon Tray. $5.50
Length inside, 6 inches.

Gilt _____ $5.00
not Gilt _____ 4.25

No. 2093. Spoon Holder.
Plain or Satin $5 00
"   "   Gilt 6 00
Engraved Bright Cut 6 00
"   "   "   Gilt 7 00

No. 23. Spoon Holder.
Plain, $4 00. Chased, $4 50.
Chased Gilt, $5 00.

No. 9. Spoon Holder.
Plain Gilt, $3 75; Chased Gilt, $4 50.
Plain, $3 00; Chased, $3 75.

ADAMS, QUINCY, FRANKLIN, FIFTH AVE., CHICAGO.

Quadruple Plate Silver-Ware.

No. 32—Bonbon.
Gold-Lined.....................$6 00

No. 31—Bonbon.
Gold-Lined..............$3 50

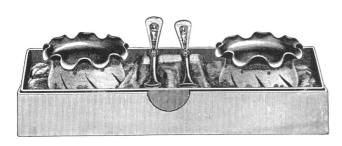

No. 176—Set Two Salts (with Spoons).
In Satin-Lined Box..................................$3 00

No. 191—Table Salt.
Gold-Lined.....................$2 75

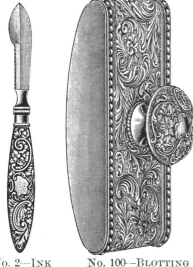

No. 27—Bonbon Tray.
Bright or Old Silver, Gold-Lined, $4 50

No. 2—Ink Eraser. $2 00   No. 100—Blotting Pad. $3 50

No. 2542—Ink. $5 00   No. 5—Trinket Tray. $2 00

No. 1783—Card Tray.
Satin Shield, 6-inch, as shown, $3 50

TRADE PRICE LIST.—BENJ. ALLEN & CO., CHICAGO.

No. 042. Cup, "Hammered" Finish.
Plain, $2 50. Chased, $3 00.
Chased Gilt, $3 50.

No. 057. Cup. Basket Finish, $4.00.
Gold Lined, $4.50.
Old Silver, Gold Lined, $5.00.

No. 045. Cup. "Hammered" Finish.
Plain, $2 75. Chased, $3 25.

No. 64. Cup. Plain, $1 75. Chased, $2 25.
Chased Gilt, $3 00.

No. 047. Cup. Plain, $2 25. Plain Gilt, $2 75.

No. 18 Cup.
Chased Satin, $2 25. Chased Gilt $3 00.

Cuts one-half size.

TRADE PRICE LIST.—BENJ. ALLEN & CO., CHICAGO.

No. 5. Salt Celler, 75c each.

No. 30. Pepper, $1 50.

No. 28. Pepper, $1 50.

No. 8. Owl Pepper,
$2 00 each.

No. 31. Pepper, $1 75

No. 15. Pepper Box,
Silver Finish, $1 50 each.

No. 14. Knife Rest, per dozen.
Cuts full size.

No. 125. Napkin Ring,
Satin, Engraved.

FINE QUADRUPLE SILVER PLATED WARE.

2095. Sugar Bowl, Bright Cut.
2095. Slop, Bright Cut.
2095. Creamer, Bright Cut.
2095. No. 5. Tea Pot, Bright Cut.
No. 2095. Coffee Pot, Bright Cut.
2095. No. 6 Tea Pot, Bright Cut.

No. 2095. Tea Set, Chased or Bright Cut, 6 pieces		$54 00	$58 00
Coffee Pot, Chased or Bright Cut $13 50		Sugar Bowl, Chased or Bright Cut.. 8 00	
No. 6. Tea Pot, Chased or Bright Cut........ 11 00		Creamer, " " " 6 50	8 50
No 5. " " " " 9 50		Slop, " " " .. 5 50	7 50

No. 400. 26 Inch Square Waiter, Chased.

MARSHALL FIELD & CO.,

Quadruple Plate Silver-Ware.

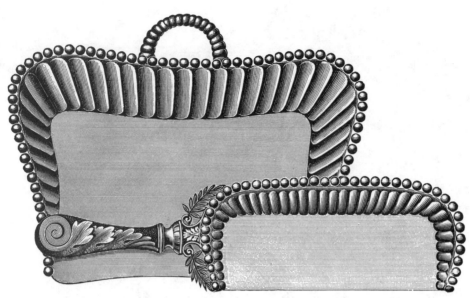

No. 2555—CRUMB TRAY. Bright Silver......................$7 50
No.   9—CRUMB SCRAPER. Bright Silver.................. 4 25

No. 2951—BRANDY FLASK.

No. 0252—NAPKIN RING.
Engraved Satin, Per Doz., $6 50

050—STAMP BOX
$3 75

No. 2560—CRUMB TRAY. Satin Shield.....................$6 25
No. 2561—CRUMB SCRAPER. Satin Shield.................. 4 00

No. 78.

No. 51—MATCH HOLDER.
Satin; Gold-Lined........$1 75

No. 079—TOOTHPICK HOLDER.
Acorn, Old Silver, Gold-Lined, $2 00

No. 0265—NAPKIN RING.
Bright Silver, Gold-Lined, Per Doz., $12 50

ADAMS, QUINCY, FRANKLIN, FIFTH AVE., CHICAGO.

Quadruple Plate Pickle Casters.

No. 663.
Rose or Ruby Glass, $6 00   Etched, $6 50

No. 638.
Single Plate, Crystal, $1 75

No. 659.
$3 50

No. 661.
Small, Crystal Glass, $2 75

No. 637.
$3 00

No. 652.
Pink Rose, Velvet Finish Glass, $6 00

No. 660.   Cut, half size.
$2 75

No. 4100—Mucilage Bottle.
Satin, $3 50

No. 1994.
Plain, complete $7 00

No. 445.
Satin....................$1 50
Satin, Gold-Lined...... 2 00

No. 0803—Cup.
As shown............$3 00
Gold-Lined, extra.... 50

No. 0805—Cup.
Embossed, as shown...$2 75
Gold-Lined, extra..... 50

No. 0807—Cup.
Chas'd, Emb. Sat., as shown, $3 50
Gold-Lined, extra............ 50

No. 01311—Syrup.
$6 50

No. 0804—Cup.
As shown...............$2 50
Gold-Lined, extra 50

No. 1825—Gravy Boat.
Height, 4 inches............$ 8 50
With Plate.................. 12 25

No. 01309—Syrup.
With Plate..........$6 00

Quadruple Plate Silver

No. 1151—Dessert Set.

Repoussé Chased, Gold-Lined, Per Set.................................$24 25

No. 1150—Dessert Set.

Repoussé Chased, Gold-Lined, Per Set.................................$24 25

No. 1134—Cream.	No. 1134—Sugar.	No. 1134—Spoon Holder.
Bright Silver, Gold-Lined....$6 75	Bright Silver, Gold-Lined........$7 75	Bright Silver, Gold-Lined...$6 75

Solid Sterling Silver

Gorham Manufacturing Co.

REPOUSSE CHASED.

KETTLE—Height, 12½ in. Capacity, 3½ pints. COFFEE—Height, 7½ in. Capacity, 2½ pints.
TEA—Height, 5½ in. Capacity, 2½ pints.

KETTLE　　COFFEE　　TEA POT　　SUGAR BOWL　　CREAMER　　SLOP BOWL

Quadruple Plate Napkin Rings.

No. 77.
Satin, Per Dozen............$10 00

No. 47.

No. 335—Napkin Ring.
Engraved Satin, Per Dozen, $18 00

No. 68.

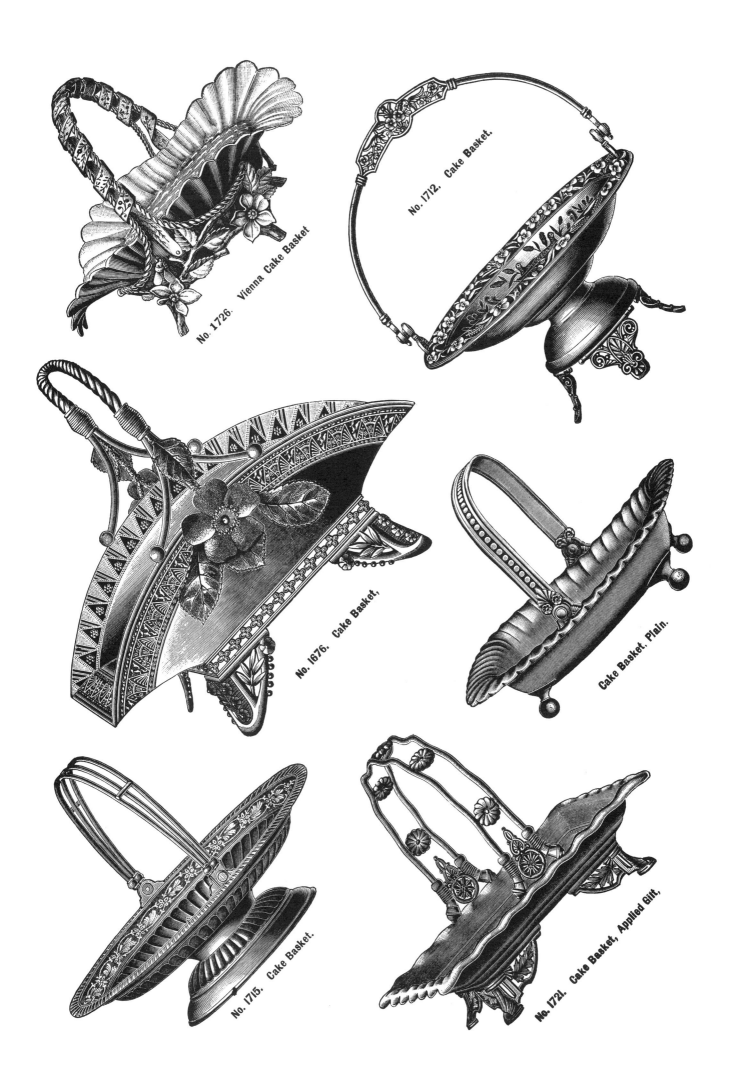

Quadruple Plate Berry Dishes.

No. 311. Fruit Dish.
Rich Etched Glass, Pink or Turkish Blue, $14.50.

No. 307. Berry Dish, Pink Tinted Glass.........$8.00

No. 142. Berry Dish............$8.50.
Height, 12½ inches.

No. 170. Berry Dish, $8.00. Amberetta Glass.
Height, 10¾ inches.

No. 147. Berry Dish, $4.50.

No. 312. Berry Dish, Crystal or Canary Glass, $6.50.

No. 300. Berry Dish, Crystal Glass, $5.25 each.

No. 2199.
Ivory Glass, Turquoise or Yellow Edge............$9 75

Milwaukee, Wisconsin.

JAPANNED WARE.

BOXES.

FORK AND SPOON.

No. 4.

COMBINATION KNIFE, FORK AND SPOON.

No. 4, Perforated Fork and Spoon Boxes, 14x8 inches, - - - - per dozen, $25 00
Combination Knife, Fork and Spoon Boxes, 13½x10x2⅜ inches, - - - " 20 00

BILL HEAD.

POST OFFICE.

Parts,	1	2	3	Parts,	1	2	3
Per Dozen,	$12 00	17 50	25 00	Per Dozen,	$15 00	18 75	25 00

SPICE BOXES.

CEYLON.

NOVELTY.

No.	8		
Inches,	5½x8	Inches,	6x8½
Per Dozen,	$10 00	Per Dozen,	$15 00

SQUARE.

No. 5.

ROUND.

No. 7.

DESK TOP.

No. 9.

No. 5, 9½-inch Square Spice Boxes, with Round Inside Boxes, - - - per dozen, $20 00
7, 8½ " Round " " " " " - - - " 20 00
9, 9½ " Desk Top Spice Boxes, - - - - - " 22 50

JAPANNED WARE.

Spoon and Fork Boxes.

Spoon and Fork Boxes, Flannel Lined, No. 1, 9½ inches,	each, $1.25
Spoon and Fork Boxes, Flannel Lined, No. 2, 12¾ inches,	" 1.50
Round Cornered Knife Trays,	" .85

Fancy Flour Boxes.

Cash Boxes.

Cash Boxes, Sunk Handles.

Inches,	8½	9½	10½	11½	13
Each,	$1.00	1.25	1.50	1.65	1.75

Box Graters.

Box Graters, ... each, $ .05

Sprinklers.

Common Tea Canisters.   Green.   French.

Green Sprinklers.

Quarts,	1	2	4	6	8	10	12	16	20
Each,	$ .45	.60	.70	.85	1.00	1.15	1.50	2.00	2.50

Square Cake Boxes.   Round Sugar Boxes.   Round Cake Boxes.

JAPANNED WARE.

Cuspadores.

Nos. 11, 13 and 15.

Assorted Colors.

3, 70 and 100.

No. 11—Light Blue, Bronzed Band, Screw Top, $ .85	No. 3—Cuspadores, Light,	.35
No. 13—Maroon, Bronzed Band, Screw Top, .85	No. 70—Cuspadores, Light, Weighted Bottom,	.50
No. 15—Lavender, Bronzed Band, Screw Top, .85	No. 100—Cuspadores, Heavy, Weighted Bottom,	.90

Cuspadores.

Nos. 2 and 2½.

Spittoons.

No. 7.

Spit Cups.

No. 2.   Fancy.

No. 2—Solid Brass, ... each, $1.25
No. 2½—Solid Brass, Nickel Plated, ... " 1.50
No. 7—Spittoons, Cast Iron, Painted, ... " 1.75
No. 2—Fancy Japanned, ... " .35

GRANITE IRON WARE.

Cuspadores.

Cuspadores.
6¼×8 inches, each, $1.35

Cuspadores.

Cuspadores.
5×7½ inches, each, $1.00

Spittoons.

3½×10 inches, each, $1.00

Japanned Dust Pans.

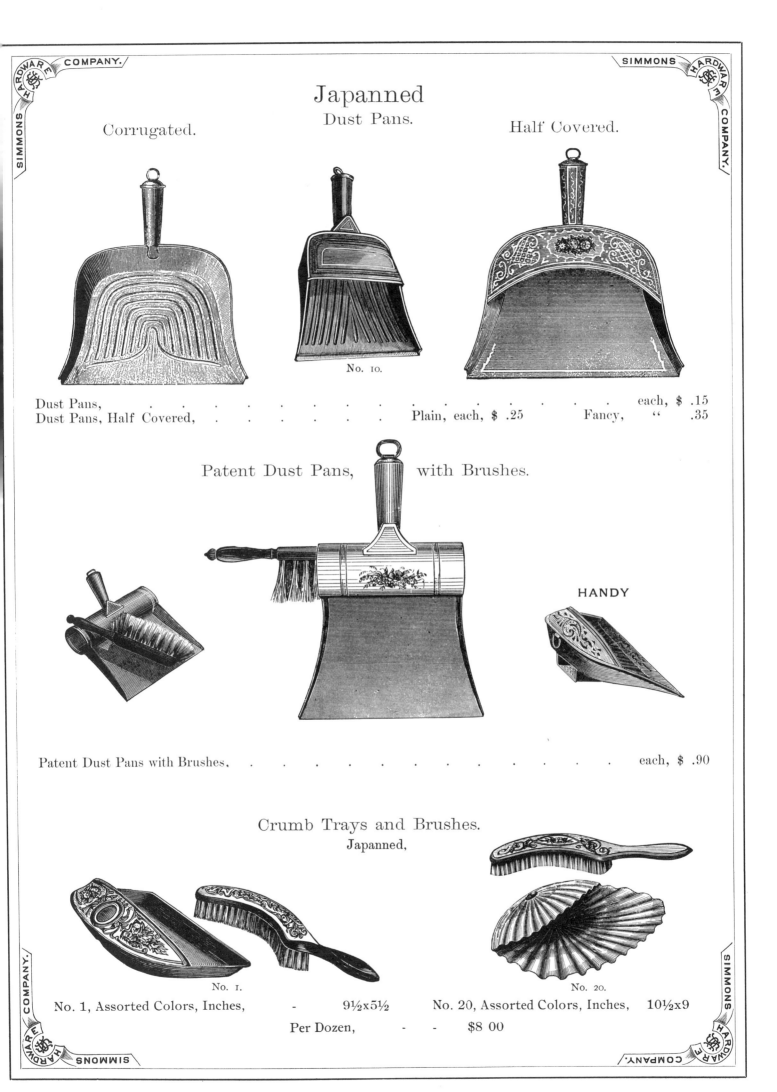

Corrugated.   No. 10.   Half Covered.

Dust Pans, ... each, $ .15
Dust Pans, Half Covered, ... Plain, each, $ .25   Fancy, " .35

Patent Dust Pans, with Brushes.

HANDY

Patent Dust Pans with Brushes, ... each, $ .90

Crumb Trays and Brushes.
Japanned,

No. 1.   No. 20.

No. 1, Assorted Colors, Inches,   -   9½x5½   No. 20, Assorted Colors, Inches,   10½x9

Per Dozen,   -   - $8 00

CARPET SWEEPERS.

Ladies' Friend.

Ladies' Friend, Carpet Sweepers, End Dump, Oiled Walnut Case, ... each, 2.50

Bissell's.

Top View.

Bottom View.

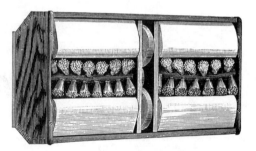

Bissell's Carpet Sweepers, Oiled Walnut Case, Handsomely Finished, ... each, $2.50

PORTABLE, SELF-ADJUSTING, AUTOMATIC CARPET STRETCHER.

Wm. Frankfurth Hardware Co.

Bullard's Patent Carpet Stretchers.

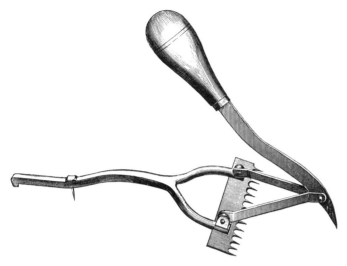

Bullard's Carpet Stretchers, per dozen, $6 00

It holds the carpet in position after it is drawn to its proper place, thus giving the operator the free use of both hands with which to do the nailing.

Simple, powerful, and warranted not to injure the finest carpet.

No. 1.

Bullard's.

Excelsior Carpet Stretcher and Tack Hammer   dozen, $ 7 50
Noyes'   "   "   "   Tacker   15 00

No. 2—Socket Carpet Stretchers, ... $ .50
No. 1—Steel Blade, Malleable Frame, Carpet Stretchers,   .25
Bullard's Patent (it holds the Carpet in position, each,   .75

193

Wm. Frankfurth Hardware Co.

BIRD CAGE TRIMMINGS.

OPAL FOUNTS.

No. 22, For Brass Cages,
Per Dozen, - - $0 80

FLINT FOUNTS.

No. 1, For Japanned Cages,
Per Dozen, - - $0 60

Hooks

No. 5, Green, Red and Blue, Fancy Bronzed, Per dozen, $2 20

One-third Size

WIRE BIRD NESTS.

per dozen, $0 90

OPAL MOCKING BIRD BATHS.

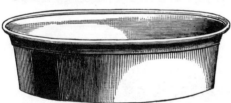

No. 9, Opal Mocking Bird Baths,
8, " " " Two Sizes Nested,

USE THE EXCELSIOR PATENT **Waterproof Cage Mat**
Greatest invention of the age. Will keep the cage *clean and free from vermin*. Highly recommended by all who have used them as an article of great value to all having pet birds. 5 pkgs., to any address, $1.00; 2 pkgs., 50c. (12 mats in a pkge.) Send inside dimensions.

OPAL BIRD BATHS.

- - - - per dozen, $2 00
- - - - " 80

BRASS BIRD CAGE SPRINGS.

EUREKA.

No. 1, Eureka, - - - - - - - - - per dozen, $0 75

NOVELTY.

No. 2, Novelty with Safety Hook, - - - - - - - per dozen, $1 25

No. 3, Handy with Chain 3½ Feet Extension, - - - per dozen, $1 75

JEWETT'S REVOLVING PERCH BIRD CAGES.

No. 602.

16 inches Long, 10 inches Wide, 16½ inches High.
Per Dozen, - - - - $40 00

No. 606.

16 inches Long, 10 inches Wide, 19 inches High.
Per Dozen, - - - $50 00

The Revolving Perch is the most entertaining novelty in the cage line ever invented.

SQUIRREL CAGES.

No. 6.

For Red or Chip Squirrels, - each, $3 75

Size over all, 18 inches Long, 13 inches Wide, 16 inches High. Size of body, 9½ inches Wide, 16 inches High. Size of wheel, 6½ inches Long, 9½ inches in Diameter.

Top is hinged, giving access to the nest to clean.

No. 7.

For Red, Gray or Black Squirrels, - each, $6 00

Size over all, 23½ inches Long, 16 inches Wide, 19 inches High. Size of body, 12 inches Wide, 19 inches High. Size of wheel, 9 inches Long, 12 inches in Diameter.

Top is hinged, giving access to the nest to clean.

Wm. Frankfurth Hardware Co.

JEWETT'S JAPANNED BIRD CAGES.

ARTISTIC IN SHAPE, COLOR AND DECORATION.

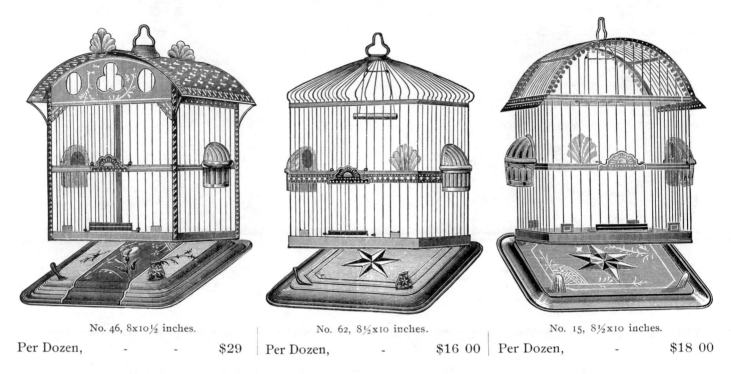

No. 46, 8x10½ inches.	No. 62, 8½x10 inches.	No. 15, 8½x10 inches.
Per Dozen, - - $29	Per Dozen, - $16 00	Per Dozen, - $18 00

The Jewett Bird Savings and Patent Seed and Water Fountains, also Patent Self Locking Hooks, are important features in these Cages.

No. 41, 8¾x8¾ inches.	No. 83, 8½x10 inches.	No. 21, 8¼x11½ inches.
Per Dozen, - $17 50	Per Dozen, - $26 00	Per Dozen, - $27 00

No. 99, 9x10 inches.

Per Dozen, - $18 00

Trimmings for each Cage are packed separately in paste board Boxes.

The Delusion Mouse Trap.

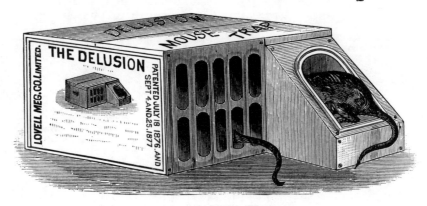

Self-Setting, Perfect in Operation.

"The Mouse goes in to get the bait,
And shuts the door by his own weight,
And then he jumps right through a hole,
And thinks he's out; but, bless his soul,
He's in a cage, somehow or other,
And sets the trap to catch another."

Delusion Mouse Traps, ... per dozen, $1 75

Full cases contain one gross.

Wood Mouse Traps.

Holes,	1	2	3	4	5	6
Per dozen Traps,	$0 30	60	90	1 20	1 50	1 80

One dozen in a package.

Lane's Carriage Jacks.

Patent Iron Jacks.

No. 2, Patent All Iron Jacks, . . . each, $1 50
No. 3, " " " " . . " 2 75

Champion Jacks.

No. 1, Wrought Iron and Wood, . . each, $2 50
No. 2, " " " " . . " 3 50

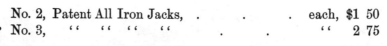

Patent Iron Jack.   Champion.

Wm. Frankfurth Hardware Co.

RAT TRAPS.

Lovet's.

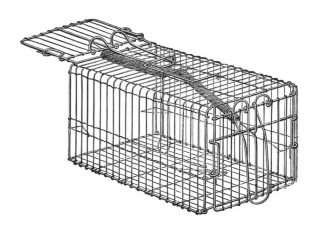

Wood Bottom, Patent Drop Lock, 6 x 11 inches Square, Wire Rat Traps, - - per dozen, $6 00
Wire " " " " 6 x 11 " " " " " - - " 7 00

One-half Dozen in a Crate.

ROUND
GLOBE.

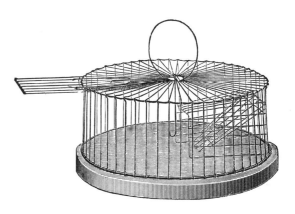

12-inch Wood Bottom, Round Wire Rat Traps, - - - - - per dozen, $5 50
12 " " " " " Globe " - - - - - " 7 50

Round, All Wire.

SLAYER RAT TRAP.

Slayer " " Coppered Steel Spring, - - - - - " 3 00

MOUSE TRAPS.

Wood Chokers.

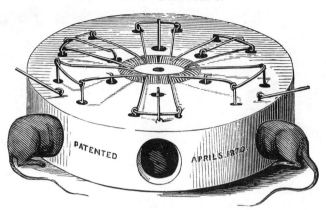

Wood Choker Mouse Traps,
Number of Holes,	1	2	3	4	5	6
Each,	$ .05	.10	.15	.20	25	.30

Revolving Wire.   Perpetual.

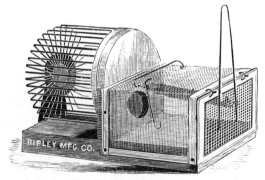

Revolving Wire Mouse Traps, ... each, $ .30
Perpetual Mouse Traps, Self Setting, ... " .25

CATCH-'EM-ALIVE.

Tin Mouse Trap, Self-setting, ... per dozen, $3 50

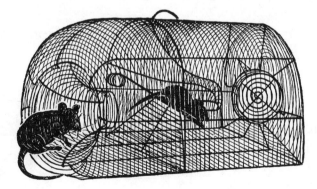

No. 3, Marty rat trap, family size, all wire, self-setting, weight per dozen, 30 pounds,   per dozen.... $9.00

FLY TRAPS

Roach Traps.

Roach Traps, Perforated Tin .35

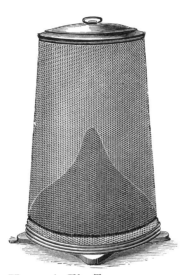

Harper's Fly Traps, - - - per dozen, $3 75
Balloon " " - - - - " 3 50

Window Screen Cloth.

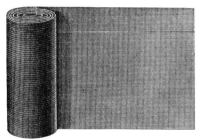
Plain.

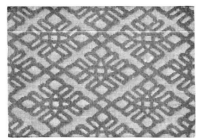
Figured.

Landscape.

Plain Colors—Green, Drab, or Black, ... per 100 square feet.
Figured, ... " "
Landscape, ... " "

FLY FANS.

FLY KILLERS.

LITTLE PEACH.
Wire screen, plain handle, per dozen......... $1.00

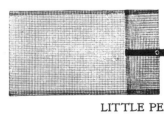

BIGELOW.
Bright steel wire, enameled handle finely finished, per dozen... $2.00

Per dozen, $30 00

Fly Fans,
 It is self-acting; winds up like a stem-winding watch; complete in itself; requires no key; will run an hour and a half at each winding, and can be rewound at any time by simply using the thumb screw at bottom, or turning base or stem.

Japanned Foot Scrapers.

One-third Size Cut of No. 20.   per dozen, $8 00

Half Size Cut of No. 12.   per dozen, $3 50

One-third Size Cut of No. 25.   per dozen, $20 00

Boot Jacks.

One-third Size Cut of Nos. 3 and 13.

American Bull-dog, Lacquered,

One-third Size Cut of Nos. 2 and 12.

No. 2, Japanned, . per dozen, $2 75
No. 12, Bronzed, . . " 3 25

One-third Size Cut of No. 25.

Double, Two Sizes in One.
                                Per dozen.
No. 25, Berlin Bronzed, . . $2 75

One-third Size Cut of No. 15. Patented.

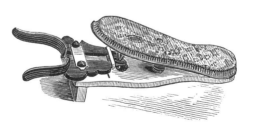

Ripley's Automatic Plain Finish,

One-third Size Cut of Nos. 4 and 14.

Beetle Pattern.
No. 4, Japanned, . per dozen, $3 25
No. 14, Bronzed, . . " 3 75

JAPANNED WARE.

Toilet Stands.

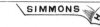
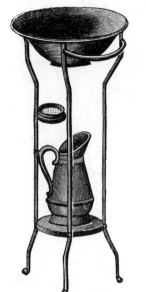

Umbrella Stands.

Towel Stands.

No. 6—Umbrella Stands, Height, 25 inches, Width, 19 inches, ... each, $2.25
No. 2—Towel Stands, Height, 31 inches, Length, 28 inches, ... " 2.00
No. 3—Toilet Stands, Complete, Assorted Colors, ... " 2.75

Soap Brackets.

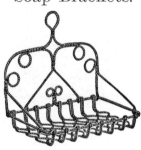

Toilet Stands.

Brush and Soap Holders.

Tooth Brush Stands.

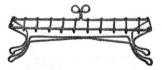

WIRE WARE.

Closet Paper Racks.

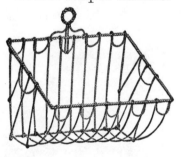

Sponge Baskets.

Complete—Assorted Colors. 32½ inches High. 17½ inches Diameter at Base.
No.   5—Blue Curtain, Gold Trimmings, ... each, $7.50
No. 155—Drab, with Light Green Band, Deep Green Vine, Gold Trimmings, ... " 6.75

TOILET WARE.

In Sets Only.

No. 321.

In Sets Only.

No. 310.

In Sets Only.

No. 316.

No. 321—Pea Green Body, Cherry Red Trimmings, Japanese Ornaments; 3 Pieces in Set, . per set, $3.50
No. 310—Black Body, Red Trimmings, Japanese; 3 Pieces in Set, ... " 3.50
No. 316—Green Body, Brown Trimmings, Deer Heads and Artists' Work; 3 Pieces in Set . " 3.50

In Sets Only.

In Sets or Separate.

No. 303.

No. 103.

In Sets Only.

No. 104.

No. 303—Oak Grained, Gilt Hoops; 3 Pieces in Set, ... per set, $3.00
   (Water Carriers, each, $1.00; Slop Jars, each, $1.15; Foot Baths, each, $1.00.)
No. 103—Red and Pea Green, Deer and Artists Work; 3 Pieces in Set, ... " 4.00
No. 104—Dark green and Salmon, Birds, &c., 3 Pieces in Set, ... " 4.00

In Sets Only.

In Sets Only.

No. 145.

In Sets or Separate.

Nos. 325, 326, 327.

Wash Bowls and Pitchers.
Wash Bowls and Pitchers, Assorted Colors, ... per set, $1.50
No. 145—Blue Ground, Brown Band, Japanese and Artists' Work; 3 Pieces in Set, . . " 4.00
Nos. 325 (Green), 326 (Scarlet), 327 (Blue); 3 Pieces in Set, ... " 2.80
   Nos. 325, 326, 327, Water Carriers, each, $1.00; Slop Jars, each, $1.00; Foot Baths, each, $1.00

Most folks cooked up their own lye soap in a kettle on the back porch. But for those who could afford store-bought goods, Pears' soap was available in every city in the world.

The bath water recipes are from a popular book of household hints published in 1908.

People still believed a good hot bath could cure anything.

PEARS' SOAP
a Specialty for Children

Salt-water Bath. — Add 4 or 5 pounds of sea salt, which can be purchased of any druggist, to a full bath at the temperature of 65° F. The patient should remain in this bath from 10 to 20 minutes, and afterwards should rest for half an hour in a recumbent position.

Mustard Bath. — The addition of 3 or 4 tablespoonfuls of powdered mustard to a hot footbath in cases of chill is a preventive against taking cold, and is also useful in the early stages of colds to induce perspiration. The feet should be taken out of this bath as soon as the skin reddens and begins to smart. The parts bathed should be carefully cleansed, rinsed, and wiped dry. Great care should be exercised in giving mustard baths to children, else the skin may become badly blistered.

The Bran Bath. — Make a decoction of wheat bran by boiling 4 or 5 pounds of wheat bran in a linen bag. The juice extracted, and also the bran itself, should be put into the water. This is for a full bath at a temperature of about 90° F. This bath is of service in all skin affections accompanied by itching.

Benjamin Franklin brought home his own shoe-shaped bathtub from Paris in 1778. He was suffering from a skin ailment at the time and took hot baths often. He sat erect in the heel with his legs extending into the vamp of this giant shoe. There was a device in the heel to keep the water warm and the toe contained a spigot drain for emptying. A book rack was built into the tub at eye level for leisurely reading. The entire contraption was made of copper for easy maintenance.

Seventy-two years elapsed from the time of Ben Franklin's pioneering efforts until a bathtub was actually installed in the White House, by President Millard Fillmore, in 1850. Sometime earlier the state of Virginia had placed an annual tax of thirty dollars on every bathtub brought into the state, and Boston had made bathing unlawful except on medical advice. President Buchanan even lacked a tub at his private residence in the late 1860's.

In early America the bathroom was of course nonexistent. Bathless homes were a normal state-of-being. Washing your face, neck, arms and feet once a week was considered adequate for general social acceptance. Special containers were kept handy for this chore in upper class bedrooms, where bathing customarily took place. These furnishings consisted of a washstand, a bowl, a pitcher, a slop jar (for used water), a foot bath, and a towel horse. Warm water was carried upstairs by servants; plumbing in bedrooms was a late nineteenth century invention.

During the 1870's, when all plumbing began to move indoors, cast iron bathtubs gradually replaced the earlier portable tin variety which were dragged out for use in front of the fireplace or woodstove on Saturday night. It had taken four thousand years to come full circle. Bathers were back in the tub where they belonged!

JAPANNED WARE.

Hip Bath Tubs.

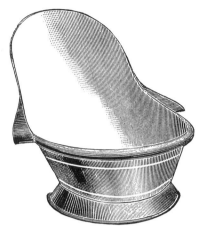

Combination Bath Tubs.

Hip Bath Tubs,	each, $5.00
Combination Bath Tubs,	" 7.00
Dr. McPheeters' Shallow Hip Bath,	" 6.00

Plunge Bath Tubs.

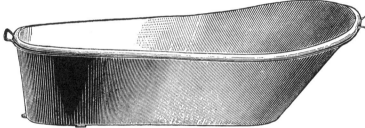

Infant Bath Tubs.

Plunge Bath Tubs, ... each, $12.00

Sponge Bath Tubs.

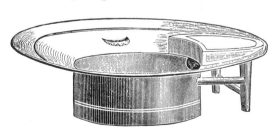

Foot Tubs.

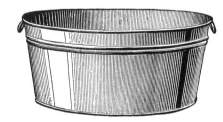

Sponge Bath Tubs, ... each, $6.00

Infant Bath Tubs.

		Nos.	1	2	3		Nos.	1	2	3
Japanned,	{	Inches,	30	38	41	Galvanized, {	Inches,	30	38	41
		Each,	$2.50	3.20	4.00		Each,	$3.00	4.00	5.00

TURNBULL'S FAMILY SCALES.

No. 56.

No. 53,— 8 lbs. Scale with Platform,	per dozen,	$30 00
No. 54,— 8 " " 15 inch Tin Scoop,	"	33 00
No. 55,—16 " " with Platform,	"	33 00
No. 56,—16 " " 18 inch Tin Scoop,	"	36 00

TURNBULL'S MARKET SCALES.

No. 33.

No. 26,—16 lb., x ½ oz., Marble Slab, Glass Sash,	each,	$13 50
No. 33,—32 " x 1 " " "	"	15 00
No. 43,—64 " x 2 " " "	"	16 50

SPRING BALANCES.

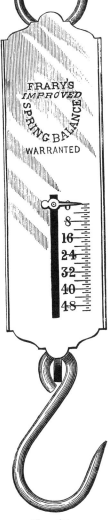

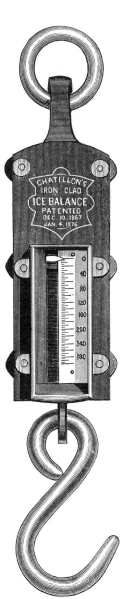

	German Crab Ice Scales.		
No. 70.		Nos. 50, 51, and 81.	No. 84.	No. 160

Nos.	SHAPE.	CAPACITY.	KIND OF DISH.	PRICE. EACH.
70	Round,	24 pounds × ½ pounds,	Hook only,	$ .40
81	Straight,	24 pounds × ½ pounds,	Hook only,	.40
84	Straight,	50 pounds × pounds,	Hook only,	.85
94	Straight,	50 pounds × pounds,	Round Tin Dish,	2.00
50	Straight,	24 pounds × ½ pounds,	Hook only,	.20
51	Straight,	48 pounds × pounds,	Hook only,	.35
52	Straight,	24 pounds × ½ pounds,	Round Tin Dish,	.40

LETTER.	NEW IMPROVED PACKAGE.	TOBACCO.

No. 601, Improved Letter Balances,	Weighs ½ oz. to 8 oz.,	-	each, $3 00
603, " " "	" ½ oz. to 10 oz.,	- -	" 4 00
604, " " "	" ½ oz. to 34 oz.,	-	" 6 00
605, " Package "	" ½ oz. to 4 lbs.,	- -	" 8 00
495, Tobacco Scale with Brass Scoop,	" ¼ oz. to 1 lb.,	-	" 6 50

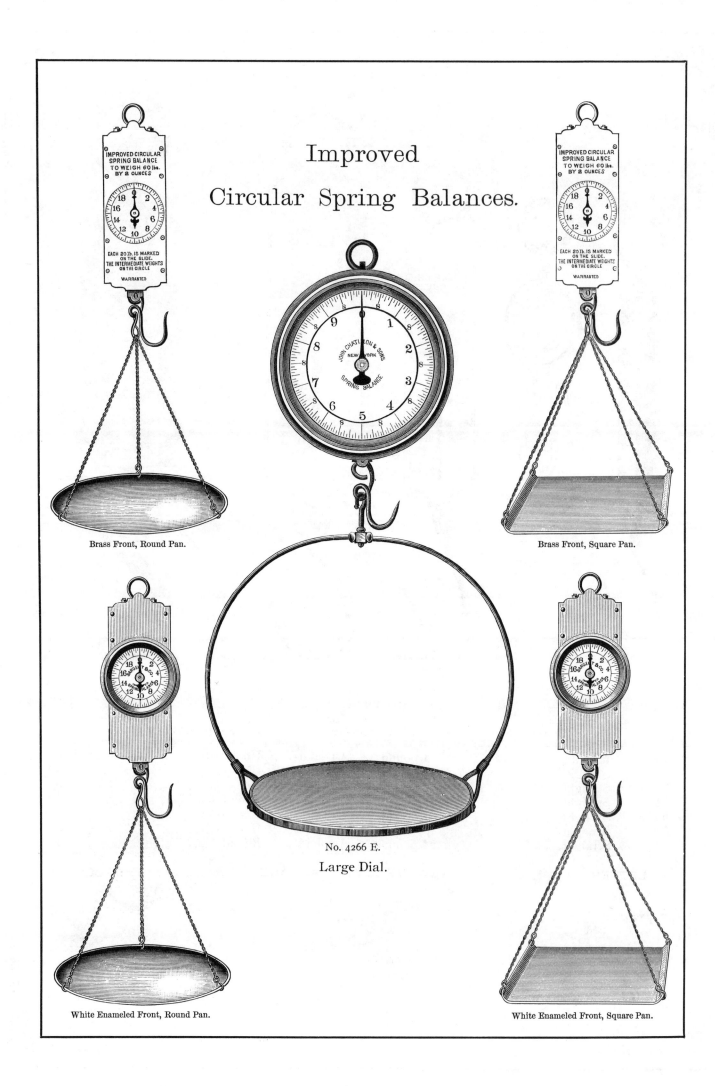

Improved Circular Spring Balances.

Brass Front, Round Pan.

Brass Front, Square Pan.

No. 4266 E.
Large Dial.

White Enameled Front, Round Pan.

White Enameled Front, Square Pan.

STANDARD FAMILY SCALES.

No. 170.

No. 192.
Turnbull's Patent, each, $4.00

Novelty.

No. 198.

Windsor.

No. 198—12 pound Scale, to weigh by ounces, with Platform, ... each, $2.00

Windsor Family Scale; weighing 15 pounds by ounces, ... " 3.00

GROCERS' EVEN BALANCE TRIP SCALES WITH FUNNEL SCOOP.

With Iron Weights.

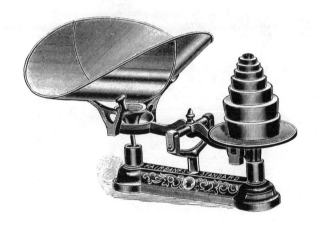

With Side Beam.

No. 661, Even Balance, ½ oz. to 10 lbs., Tin Scoop, - - - - - - each, $7 00
    673,   "   "   ½ oz. to 6 lbs., "   " - - - - - - - "   6 00

COUNTER.
BUTTER.

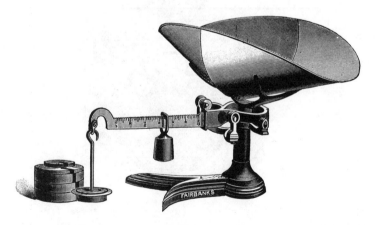
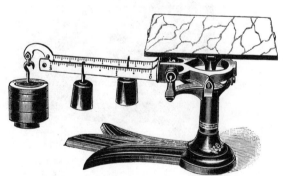

With Square Marble Plate 16x12 Inches.
Double Beam.

No. 554, Counter, with Tin Scoop, Weighs ½ oz. to 36 lbs., - - - - each, $10 00
    556,   "   " Brass "   "   ½ oz. to 36 lbs., - - - - "   11 50
    547, Butter without "   Marble Plate, Weighs ½ oz. to 62 lbs., - - - "   13 75
    551,   "   with Tin "   "   "   "   ½ oz. to 62 lbs., - - - "   15 00
    553,   "   "   Brass "   "   "   "   ½ oz. to 62 lbs., - - - "   16 50

Grocers' Scales.

Polished Brass Beam, Tin Scoop.

No. 132, To weigh 36 lbs.,	each, $6 00
No. 133, " " 26 "	" 5 00

Hatch's Tea and Counter Scales.

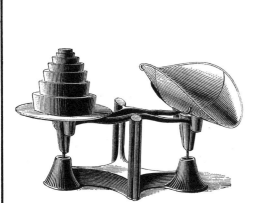

Tea Scale, No. 161.

Counter Scale, No 171.

Tea Scales.

To weigh ½ ounce to 4 lbs.

No. 161, Tea Scales, . . per dozen, $15 00

Counter Scales.

To weigh ½ ounce to 8 lbs.

No. 171, Counter Scales, per dozen, $42 00

No. 44.

Platform Scales.

Nos.		Each.
40, To weigh 240 lbs., Japanned, ...		$6 00
41, " " 240 " " Striped in Bronze,		6 25
44, " " 244 " " " " "		
Steel Bearings, ...		6 60
51, To weigh 240 lbs., Painted Red, Striped in Bronze,		6 75
54, " " 244 " " " " " "		
Steel Bearings, ...		7 10
64, To weigh 244 lbs., Same as No. 44 with Dbl. Beam,		9 00

UNION HARDWARE AND METAL COMPANY.

SPRING SCALES.

COLUMBIA FAMILY.

WHITE ENAMELED DIAL, SQUARE STEEL TOP.
No. 102 weighs 24 pounds by ounces. Without scoop, per dozen.. $16.00
No. 1102 weighs 24 pounds by ounces. With scoop, per dozen.. 17.50

UNIVERSAL FAMILY.

TIN SCOOP, ALUMINUM DIAL, RUBBER FEET.
12 pound, by ounces, per dozen................................. $39.00
24 pound, by ounces, per dozen................................. 45.00
48 pound, by ounces, per dozen................................. 51.00

EVEN BALANCES.

TEA.

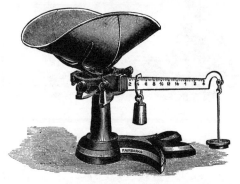

No. 580, Tea Scale, with Tin Scoop, Weighs ¼ oz. to 8 lbs.,
582, " " " Brass " " ¼ oz. to 8 lbs.,

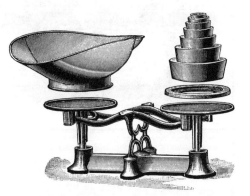

No. 260, with scoop, will weigh to 2 pounds.
Per dozen.................................. $17.00

No. 261, with scoop, will weigh to 4 pounds.

MARKET SCALES.

NO. 32 WITH SCOOP.
No. 32, counter scale, brass scoop, 32 pounds by ounces, 7-inch dial, red finish, each $16.00
No. 33, counter scale, without scoop, 32 pounds by ounces, 7-inch dial, marble platform, red finish, each............ 16.00

CHATILLON'S MARKET SCALES.

No. 69.

With Marble Slab.

No. 69, Weighs from 1 oz. to 32 lbs.,	each,	$14 00
79, " " 2 oz. to 64 lbs.,	"	16 00

FAIRBANK'S STANDARD SCALES.

THE NEW IMPROVED GROCER SCALE.

Has patent balance, whereby Scale may be instantly balanced, with or without scoop. Capacity, 1 ounce to 50 pounds on the beam, with weights to make total capacity 250 pounds.

Platform, 10½ x 13½ inches.

The Double Beam is Nickel-plated and may be placed at eight different Angles with the Platform.

No. 518, With Tin Scoop,	each,	$24 00
522, " " Seamless Scoop,	"	25 00
520, " Brass Scoop,	"	25 00
524, " " Seamless Scoop,	"	27 00

STEELYARDS.

With Balanced Heads and Stops.
American Pattern.

50 Pounds Steel Bar, Weigh by Ounces,	-	-	-	-	per dozen,	$5 50
100 " " " " " 2 "	-	-	-	-	"	7 00
150 " " " " " 4 "	-	-	-	-	"	9 00
200 " " " " " 4 "	-	-	-	-	"	10 00
250 " " " " " 4 "	-	-	-	-	"	11 00

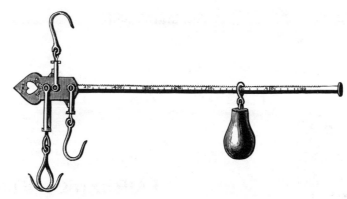

Balanced Heads and Stops. Warranted Correct. Heads and Poises Painted Red.
Farmers' Pattern.

50 Pounds Steel Bar, Polished Malleable Hooks, Weight by Ounces,	-	per dozen,	$ 9 00
100 " " " " " " " " "	-	"	10 00
150 " " " " " " " " "	-	"	12 00
200 " " " " " " " " "	-	"	15 00

SCALE BEAMS.

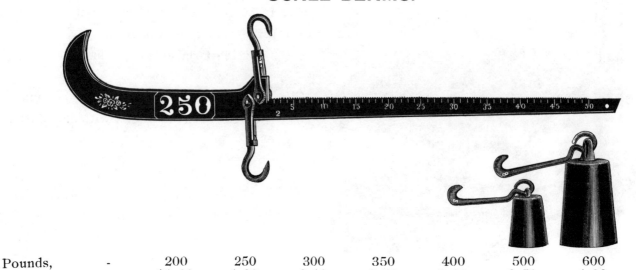

Pounds,	-	200	250	300	350	400	500	600	700
Each,	-	$1 60	1 90	2 10	2 50	2 90	3 50	4 00	4 66

FAIRBANK'S STANDARD SCALES.

DROP LEVER.

With Heavy Wheels and Drop Lever.

No. 1166,	Capacity	2500	pounds,	Platform	23 x 30	inches,	each,	$94 00
1168,	"	2000	"	"	23 x 30	"	"	82 00
1170,	"	1500	"	"	21 x 28	"	"	70 00
1172,	"	1200	"	"	20 x 28	"	"	59 00
1174,	"	1000	"	"	17 x 26	"	"	51 00

PORTABLE PLATFORM, VIBRATORY AXLE.

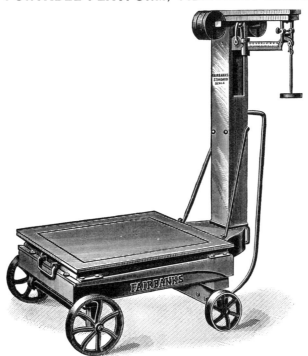

With Wheels and Vibratory Axle, Sliding Poise.

No. 1182,	Capacity	3000	pounds,	Platform	30 x 39	inches,	each,	$135 00
1184,	"	2500	"	"	23 x 30	"	"	102 00
1186,	"	2000	"	"	23 x 30	"	"	90 00
1188,	"	1500	"	"	21 x 28	"	"	77 00
1190,	"	1200	"	"	20 x 28	"	"	65 00
1192,	"	1000	"	"	17 x 26	"	"	57 00

Wm. Frankfurth Hardware Co.

CHAMPION ENTERPRISE BEEF SHAVERS.

ENTERPRISE CHEESE KNIVES.

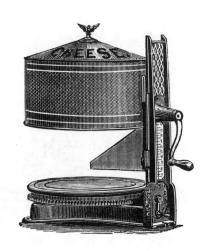

Champion Smoked Beef Shaver, - - - - - - - each, $7 50
Self-Gauging Cheese Knife, Tinned Blade, with Cover, - - - - - " 13 00

CHERRY STONERS.

MILLS.

ENTERPRISE BONE, SHELL AND CORN.

No. 750.
Height, 20 inches; Weight, 60 Pounds.
No. 750, Enterprise Bone Mill, Grinders Warranted Equal to Steel. - - each, $7 50

No. 1, Enterprise Cherry Stoners, .50

SAUSAGE STUFFERS.

PERRY'S PATENT.

No. 1, Family Size, - - - - - - - - - per dozen, $15 00

RAILROAD.

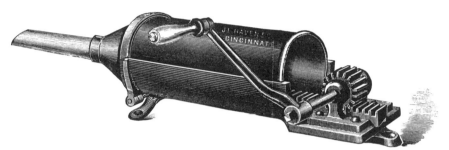

Railroad, Tin Cylinder, Family Size, Weight, Each, 8 Pounds, - - - - per dozen, $15 00

SILVER'S PATENT.

No. 2, Single Geared.

No. 2, Capacity, 8 Pounds, Weight When Crated, 35 Pounds, - - - - each, $9 00

No. 3, Double Geared.

No. 3, Capacity 12 Pounds, Weight, When Crated, 60 Pounds, - - - - each, $16 00
" 4, " 20 " " " " 100 " - - - " 22 00

Wm. Frankfurth Hardware Co.

SAUSAGE STUFFERS.
WAGNER'S.

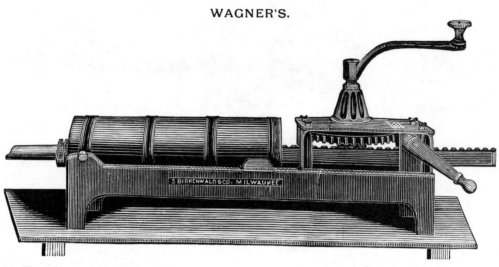

No. 1, Wagner Stuffer, Capacity 38 Pounds, - - - - - - each, $28 00
   2,   "   "   "   20   " - - - - - - "   18 00

BRECHT.

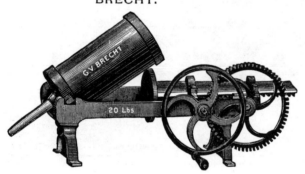

C, Holding 20 Pounds of Sausage Meat, with 5 Tubes, - - each, $20 00

LARD PRESS.

A, Capacity 35 Pounds of Sausage Meat, with 5 Tubes, Size 8x19 Inches, - - - - - each, $35 00

No. 1, Cylinder 12-in. Diam., $10 00
   2,   "   14-in.   "   17 00
   3,   "   17-in.   "   26 00

The weight on all machines does not include packing boxes. The Cylinders are of Heavy Galvanized Iron. These machines are triple geared, very powerful, with quick return motion. All wearing parts can easily be oiled and will not need repairs for a life-time, if properly handled.

The curb is double the height of that in the ordinary press. The whole is in every way better and more strongly made and handsomely finished.

Sausage Stuffers.

No. 22, Sausage Stuffers, dozen, $24 00
No. 24,   "   "   "   32 00

dozen, $15 00

WOODRUFF'S PATENT.

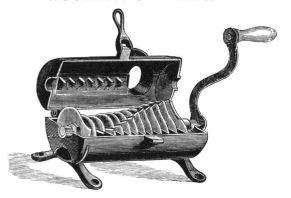

Nos. 100 and 150.

No. 100, Family Size, per dozen, $15 00

THE CHAMPION.

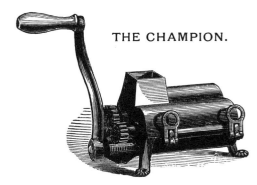

Nos. 200 to 400.

No. 200, Family Size, per dozen, $22 00
    300, Medium,   "   27 00
    400, Large,   -   "   40 00

Meat Cutters.

Japanned.

No. 32, Meat Cutters, . per dozen, $44 00
No. 33,   "   "   . .   "   54 00

Nos. 32 and 33. Patented.

GEM FOOD CHOPPER.

TINNED, SELF CLEANING, SELF SHARPENING.

No. 20, small size, 3 inch hopper, with four steel cutters for cutting coarse, medium, fine and pulverizing, also one nut butter cutter, per dozen........ $33.00

No. 22, medium, 3¾ inch hopper, four cutters and nut butter cutter, per dozen.. 37.00

No. 24, large, 4⅝ inch hopper, four cutters, per dozen............................... 50.00

One in a box.

PULVERIZING.   FINE.   MEDIUM.   COARSE.   NUT BUTTER.

STUFFER ATTACHMENTS.
FOR GEM FOOD CHOPPER.

No. 22, attachment for No. 22 chopper, per dozen..................................... $5.00
No. 24, attachment for No. 24 chopper, per dozen..................................... 5.00

 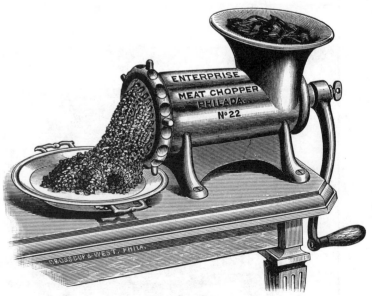

Nos. 5 to 10.   Nos. 12 and 32.

No.		Weight		Chops					
5,	Galvanized,	4½ pounds,		½ pound per minute,	-	-	each,	$2 00
10,	"	8 "		1 " "	-	-	"	3 00
12,	"	7 "		1 " "	-	-	"	2 50
22,	"	12 "		2 " "	-	-	"	4 00
32,	"	18 "		3 " "	-	-	"	6 00

Wm. Frankfurth Hardware Co.

STARRETT'S PATENT
AMERICAN MEAT AND VEGETABLE CHOPPERS.

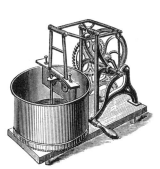
No. 1.

No. 2.

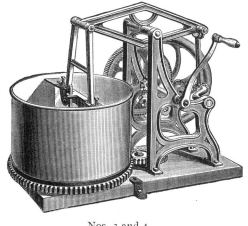
Nos. 3 and 4.

No.		Cylinder	Weight		Cuts						each	
1,	8-inch	"	"	14 pounds,	"	3 pounds in	3 minutes,	-	each,	$ 5	00
2,	10	"	"	21	"	"	5 to 6	"	"	3	"	-	-	"	7	00
3,	12	"	"	37	"	"	8 " 10	"	"	3 or 4	"	-	"	10	00
4,	15	"	"	120	"	"	60 " 70	"	"	1 hour,	-	-	"	25	00

SALISBURY STEAK

Put two pounds of tenderloin of beef in the chopping machine (Fig. 1); this machine is far superior to any other, for in chopping the meats the sinews and other hard parts collect at the bottom of the machine, on the shelf; the meat arising to the surface is the best part; take this out, leaving the hard, fibrous pieces at the bottom. Mold the Salisbury steak in a ring three-quarters of an inch high by three inches in diameter or else in a small empty goose-liver terrine (No. 10). These raw steaks are frequently served without any seasoning or else seasoned and broiled very rare.

MINCE MEAT

Suppress all fibers and skin from half a pound of beef kidney suet, chop it up very finely; have also chopped half a pound of cooked ox heart; seed and pick half a pound of Malaga raisins, half a pound of Smyrna raisins, half a pound of currants, chop up three ounces of citron, cut three ounces of candied orange peel into three-sixteenth of an inch squares, peel and chop finely two pounds of apples. Have two ounces of brown sugar, half an ounce of ground cinnamon, a quarter of an ounce of grated nutmeg, a quarter of an ounce of allspice and ground ginger, and a quarter of an ounce of powdered coriander seeds, one pint of cider, one gill of rum, quarter of a gill of brandy and the peels and juice of two lemons. Mix all the ingredients together and put them into a stone crock leaving it in a cool place for at least fifteen days before using.

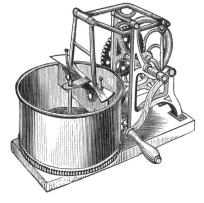

For cutting Pie Meat, Hash, Sausage Meat, Fruit, Vegetables, &., &c. These Choppers have been awarded the Highest Premium over all competitors, both for Butchers' and Family use, wherever exhibited—have taken over 100 First Premiums, including 4 Silver Medals from the principal Fairs in the country within the last two years.

One of the most useful and greatest labor-saving inventions of the day. Every Housewife knows that it requires from 1 to 3 hours tedious labor with the common hand-knife and tray, to cut the meat and apple for an ordinary "batch" of mince pies. With the AMERICAN CHOPPER a child 6 years of age can do the same work in from 5 to 15 minutes with the greatest ease. No Housekeeper can afford to be without one, while for Hotels, Restaurants, Boarding Houses and Bakeries they are absolutely indispensable.

As a SAUSAGE CUTTER for BUTCHERS, they have no equal, and are fast superseding the Rocker Grinder and other cutters. Operating upon the same principle as the old Chopping Knife and Tray, they do not *grind* or *tear* the meat, leaving it in *strings*, but *cut* it evenly and to any required degree of fineness, with astonishing rapidity. At the same time they require *less power than any other cutter whatever* to do the same work.

    Price of Family Sizes, from .... $6 to $15
    " " Butchers' " .... $30 to $75

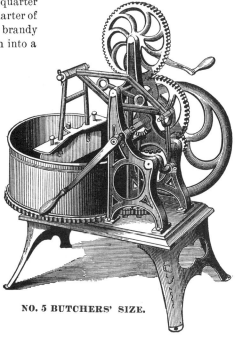

NO. 5 BUTCHERS' SIZE.

Excelsior Lawn Mowers.

Excelsior "Side Wheel Mowers."

Inch Cut,	10	12	14	16	18	20
Each,	$11 00	15 00	17 00	19 00	21 00	23 00

EXCELSIOR

New Pattern, 1874. Hand Mowers.
"Clipper Mowers."

Front View.

Inch Cut,	12	14	16
Each,	$13 00	15 00	17 00

NEW EXCELSIOR HORSE LAWN MOWER

Showing Side-Draft Attachment in Proper Position.

Full directions for Setting-up, Adjusting, and Using, accompany each Mower.

Wm. Frankfurth Hardware Co.

PUMPS.

PITCHER SPOUT.

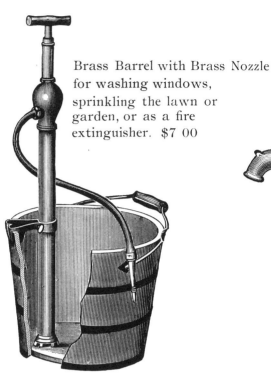

Brass Barrel with Brass Nozzle for washing windows, sprinkling the lawn or garden, or as a fire extinguisher. $7 00

CISTERN.

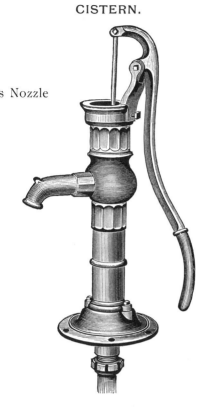

No. 1, Close Top Pitcher Spout, Diameter of Cylinder, 2½-inches,	each, $4 25
2, " " " " " " 3 "	" 4 75
3, " " " " " " 3½ "	" 5 25
4, " " " " " " 4 "	" 5 75
1, " " Cistern, Diameter of Cylinder, 2¼-inches,	" 4 00

Buckets, Well,

CHAIN PUMP FIXTURES.

WELL WHEELS.

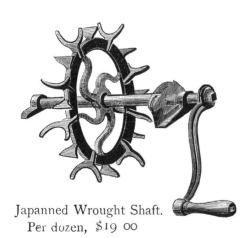

Japanned Wrought Shaft.
Per dozen, $19 00

Inches,	8	10
Per dozen,	$11 00	14 50

Top Ear, Twist Bail,	per dozen, $4 50
" " Swivel "	" 5 50
Strap" Twist "	" 4 50
" " Swivel "	" 5 50
"Crown," Patent Strap Ear, Twist Bail, with Malleable Swivel,	" 6 00

The "Crown" is one of the best Buckets made. Staves are Tongued and Grooved, turned Smooth inside and out and is warranted not to leak. Capacity same as other Buckets.

J. H. SUTCLIFFE & CO., LOUISVILLE, KY.

THE BEST PUMP IN THE WORLD.
J. H. Sutcliffe & Co.'s Sanitary Pump and Water Purifier.

Our 1896 Improved Galvanized Iron Curb,

PURE SWEET WATER, NO FREEZING, NO BUGS.

Purifies the Foulest Well or Cistern by Aeaation, has no Valves, Suckers, or other devices to get out of order.

Our Improved Galvanized Iron Curb Sanitary Pump is now ready for market. It is the most valuable and substantial water drawer in use. The curb instead of being made of wood as heretofore is now made of selected heavy galvanized iron, so constructed as to combine extreme lightness with durability. This perfected pump is endorsed by the medical profession as being the most healthful water drawer in use. There is nothing better made for use in wells or cisterns ranging from 10 to 40 feet. Simple, durable, clean. Each revolution of the wheel carries air into the well, thoroughly aerating the water.

No. 4500 J. H. S. & Co.'s Galvanized Iron Curb Sanitary Pump, for cisterns or wells 10 to 12 feet deep. Price complete $7 55
No. 4501 Same for wells or cisterns 13 to 15 feet deep 8 00
No. 4502 Same for wells or cisterns 17 to 20 feet deep 8 50
No. 4503 Same for wells or cisterns 23 to 25 feet deep, using double gear 9 50

In ordering always give depth of well. We will quote special prices on pumps for wells ranging from 27 to 40 feet.

I. X. L. Steel Chain Pump

O. P. Schriver & Co., Cincinnati, O., manufacture the I. X. L. Patent Galvanized Steel Pump, Curb and Galvanized Steel Tubing, herewith illustrated. The makers in placing this apparatus before the trade, call attention to its superiority over the old wooden suction and iron pitcher pumps, as there are no valves to get out of order or dry out; it will always throw a large stream of water.

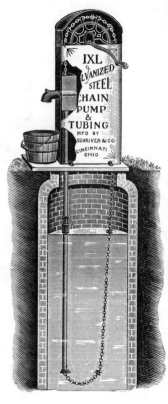

IRON PUMPS.

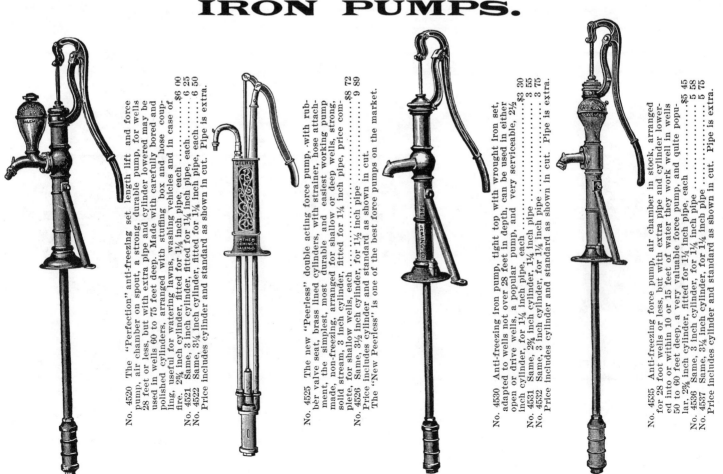

No. 4520 The "Perfection" anti-freezing set length lift and force pump, air chamber on spout, a strong, durable pump, for wells 28 feet or less, but with extra pipe and cylinder lowered may be used in wells 60 to 75 feet deep. Made with carefully bored and polished cylinders, arranged with stuffing box and hose coupling, useful for watering lawns, washing vehicles and in case of fire. 2¾ inch cylinder, fitted for 1¼ inch pipe, each $6 00
No. 4521 Same, 3 inch cylinder, fitted for 1¼ inch pipe, each...... 6 25
No. 4522 Same, 3¼ inch cylinder, fitted for 1¼ inch pipe, each..... 6 50
Price includes cylinder and standard as shown in cut. Pipe is extra.

No. 4525 The new "Peerless" double acting force pump, with rubber valve seat, brass lined cylinders, with strainer, hose attachment, the simplest, most durable and easiest working pump made, non-freezing, arranged for shallow or deep wells, strong, solid stream, 3 inch cylinder, fitted for 1¼ inch pipe, price complete, for shallow wells, each $8 72
No. 4526 Same, 3½ inch cylinder, for 1½ inch pipe 9 89
Price includes cylinder and standard as shown in cut.
The "New Peerless" is one of the best force pumps on the market.

No. 4530 Anti-freezing iron pump, tight top with wrought iron set, adapted to wells not over 28 feet in depth, can be used in either open or drive wells, a popular pump, and very serviceable, 2½ inch cylinder, for 1¼ inch pipe, each $3 30
No. 4531 Same, 2¾ inch cylinder, fitted for 1¼ inch pipe, each.... 3 55
No. 4532 Same, 3 inch cylinder, for 1¼ inch pipe 3 75
Price includes cylinder and standard as shown in cut. Pipe is extra.

No. 4535 Anti-freezing force pump, air chamber in stock, arranged for 28 foot wells or less, but with extra pipe and cylinder lowered into or within 10 or 15 feet of water they work well in wells 50 to 60 feet deep, a very valuable force pump, and quite popular, 2¾ inch cylinder, fitted for 1¼ inch pipe, each $5 45
No. 4536 Same, 3 inch cylinder, for 1¼ inch pipe 5 58
No. 4537 Same, 3¼ inch cylinder, for 1¼ inch pipe 5 75
Price includes cylinder and standard as shown in cut. Pipe is extra.

POST HOLE DIGGERS.

EUREKA.   SAMSON.

Eureka Post Hole Digger, Steel Blades, Split Handle, - - - - - per dozen, $27 00
Samson   "   "   "   "   " Solid   " - - - - - "   36 00

LITTLE GIANT POST HOLE DIGGER.

Little Giant Post Hole Digger, Detachable Joint, - - - - - per dozen, $30 00

Wm. Frankfurth Hardware Co.

MOUNTED GRINDSTONES.

MOUNTED WITH SPECIALLY SELECTED STONE.

No. 1, Stone Weighs 100 to 110 lbs.,		each, $6 00
2, " " 70 to 80 "		" 5 00
3, " " 40 to 50 "		" 4 50

The usual thickness of stone mounted is from 1¾ to 2¼ Inches. Every frame is provided with Treadle, Crank and Bucket Holder, and the Crank can be used as a Wrench to tighten any nut on the frame.

The Douglas Axe Manufacturing Company.
Hunt's and Sharp's Hatchets and Edge Tools.

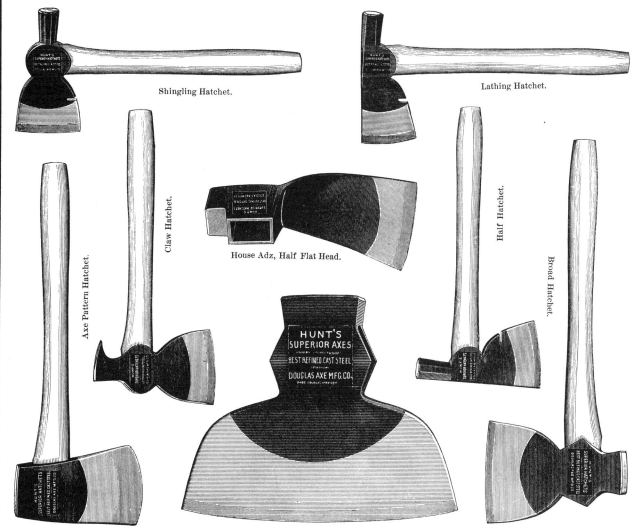

BROAD AXES.

Western Pattern.

Pennsylvania Pattern.

Assorted from 11 to 13 inch Cut, and from 5½ to 8 lbs., . per dozen, $32 00

Assorted from 11 to 13 inch Cut, and from 5½ to 7½ lbs., . per dozen, $32 00

HAY KNIVES.

HEATHS.

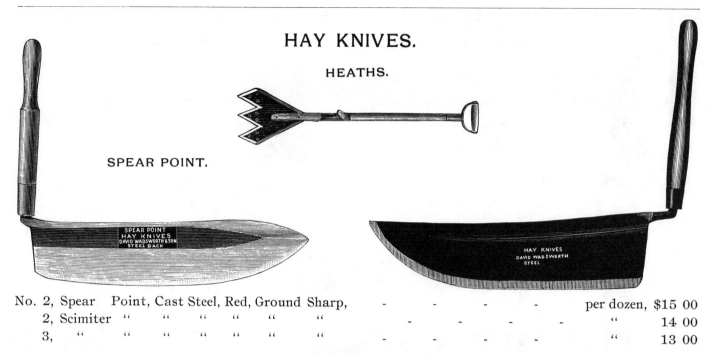

SPEAR POINT.

No. 2, Spear Point, Cast Steel, Red, Ground Sharp,	per dozen, $15 00
2, Scimiter " " " " "	" 14 00
3, " " " " " " "	" 13 00

One-half Dozen in a Bundle.

LIGHTNING.

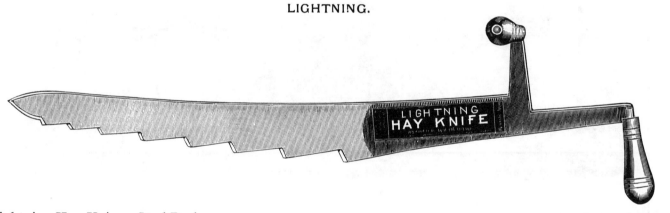

Lightning Hay Knives, Steel Back, — per dozen, $18 00

GEM.

Gem Patent Hay Knives, Steel Back,	per dozen, $17 00
Electric " " " " "	" 17 00

One Dozen in a Box,

HAY RAKES.

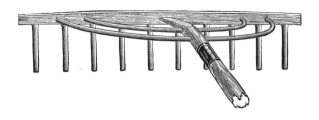

No. 2, Ten Teeth,	Mortise Head,	Common Timber, Bent Handles,	per dozen,	$2 00
1, " "	" "	Selected " " "	"	2 25
1, " " Geneva	" "	" " " " Oiled,	"	2 50

GRAIN CRADLES.

GENUINE MORGAN.

With Scythes.

Genuine Russell Morgan Cradles, Wood Brace, — — — — — per dozen, $33 00

COKE AND COAL FORKS.

POTATO SCOOP.

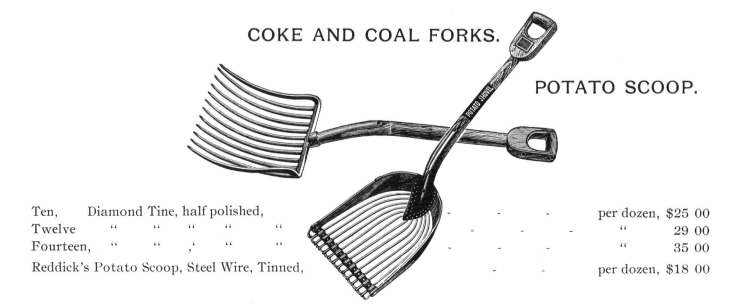

Ten, Diamond Tine, half polished,	per dozen, $25 00
Twelve " " " " "	" 29 00
Fourteen, " " , " "	" 35 00
Reddick's Potato Scoop, Steel Wire, Tinned,	per dozen, $18 00

CORN SHELLERS.

ONE HOLE.

IMPROVED HOCKING VALLEY TWO HOLE.

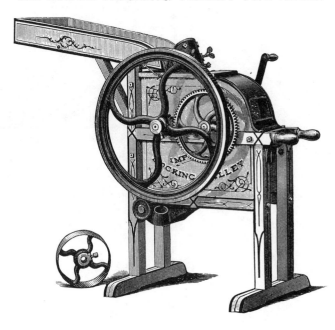

One Hole Sheller, for Hand, - - - - - - - - each, $10 00
Improved Two Hole Corn Sheller, with double its capacity. - - - - each, $20 00

BURRALL CORN SHELLER.

HAND CORN SHELLER

No. 3515 The improved "C-C" Corn Sheller. One of the best portable shellers in the market; cleans all size ears, easily adjusted, shells almost as fast as the large rotary machines, simple and durable. Retails everywhere at $3.00. Our price, each, $1.90
Per dozen 21 50

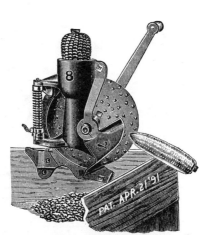

Burrall's Right Hand Corn Shellers, $6 00

A good salesman can make a neat profit handling these shellers, they retail everywhere at $3.00 each and being the best device of the kind on the market they sell quickly and give excellent satisfaction.

SEED STRIPPER.

No. 3510 Seed Gatherer or Stripper, especially suited for all kinds of grass seed. Will gather about 20 bushels a day. Wood handle and receiver with 9 inch steel teeth. Price each $1 75

No. 3503 No. 1 Poulterers' Mill, weight 35 pounds, each $3 89

EAGLE CIDER MILLS

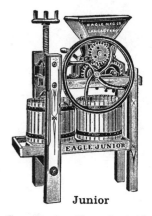
Junior

Senior

—Capacity 2 to 4 barrels of cider per day; 43 in. high. Each $18.00

—Capacity 6 to 10 barrels of cider per day; 48 in. high $32.00

CORN HUSKERS

No. 60.

HALL'S PATENT.

No. 70.

BRINKERHOFF'S PATENT.

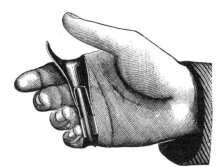

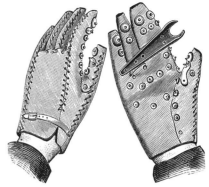
No. 50.

HUSKING PINS.   HUSKING GLOVES.

No. 60, Single Point, with Straps, Adjustable, - - - - per gross, $ 7 50
    70, "   "   " Finger Cot and Thumb Attachment, - - - " 27 00
Brinkerhoff's, Brass, Universal Extension, - - - - - per gross, $24 00
No. 50, Full Gloves, Laced, Heavy Oil Tanned Leather, Full Metal Plates, dozen pairs, $12 00

BUCKEYE.

No. 1, Buckeye Combined Husker and Binder, Tinned, - - - - per gross, $5 50

TRIUMPH.

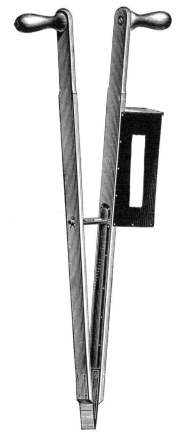

CHAMPION.

CORN PLANTERS.

Batcheller's

TRIUMPH,

With Pumpkin
Seed Attachment.

Champion Planters, with Pumpkin Seed Attachment, - - - - per dozen, $20 00
Triumph   " - - - - - - - " 12 00
   "   " with Pumpkin Seed Attachment, - - - " 18 00

Bull Snaps.

Length 7⅛ Inches
Malleable Iron.

No. 500, Tinned

Showing the Bull Snap attached to Handle

SANFORD'S PATENT OX SHOES.

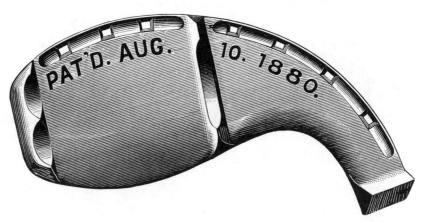

Malleable Iron.

These Shoes are Perfectly Fitting, being level, and made to cover the ball of the foot, without pressure, and entirely protecting the hoof and foot from injury.

These Shoes will cling to the foot much longer than an ordinary hand-made shoe. They are made from bar iron, of superior, refined quality, and withal are cheaper than they can be produced by hand.

Nos. 11 to 25

Bull Rings.

Nos. 31, 61 and 72
Patented July 2, 1889 and June 9, 1891

FARMER'S OX YOKES.

				Each.
Large	Size Farmers' Ox Yokes,	-	$5 00
Medium	" " "	-	4 50
Small	" " "	-	-	4 00

CATTLE LEADER

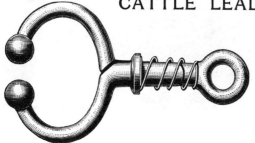

Nos. 51 and 52.

Ox Bow Pins.

Half Size of No. 72
Malleable Iron.

No. 72, Steel Spring, 2 Inch, Japanned   .   . per hundred, $32 00

Wm. Frankfurth Hardware Co.

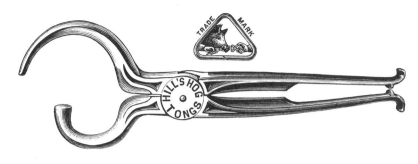

Hill's Improved Hog Tongs, - - - - - - - per dozen, $5 00

CHAMPION.

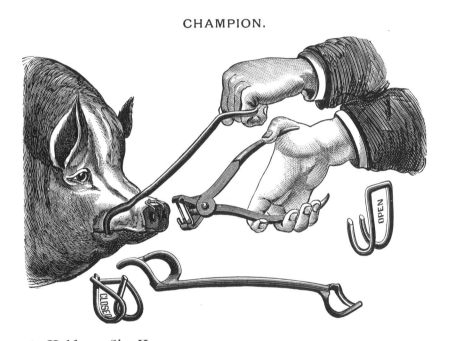

Champion Hog Tongs, to Hold any Size Hog, - - - - - per dozen, 4 50

ANIMAL CATCHERS.

Improved Animal Catchers, - - - - - - - per dozen, $12 00

BARROWS.

Garden Barrows.

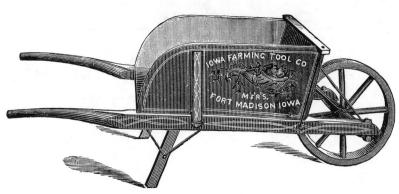

Medium Size, with Wood Wheel,	each, $5.00
Medium Size, with Iron Wheel,	" 5.75
Large Size, with Wood Wheel,	" 6.00
Large Size, with Iron Wheel,	" 6.75

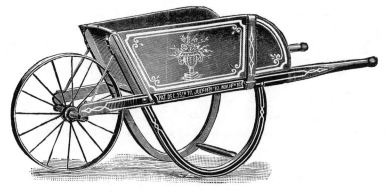

No. 6, Garden Barrow, Wood Wheel,	each, $3 50
6, " Steel "	" 4 00

It has Jointless Bent Legs, as shown in Cut, which have no equal for Lightness and Strength. Can be slid on the lawn, if necessary, without tearing up the sod or punching holes in the same, as other garden barrows do. Has a 20-inch Steel or Wood Wheel, with a 1½-inch tread. Neatly painted, striped and varnished. Has Chilled Cap Boxes.

Common Railroad Barrow.

Common Railroad Barrows, Set Up,	each, $2.50

Wm. Frankfurth Hardware Co.

BARLEY FORKS.

STEEL.   WOOD.

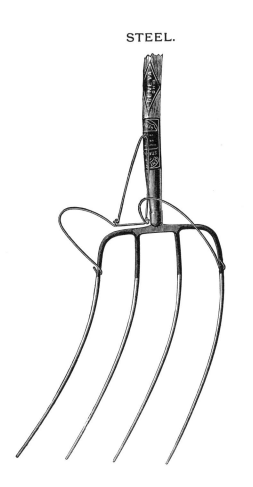

No. 12, Four Oval Tine Steel, 18-inch, Plain Ferrule,	per dozen,	$22 50
012, " " " " 18 " Strap "	"	24 00
Four Tined, Wood, Wire Braced,	"	7 00
Six " " " "	"	8 00

One-half Dozen in a Bundle.

WOODEN FIELD AND STABLE FORKS.

Wooden Field and Stable Forks, — per dozen, $6 00

This is a nice tool for stall use, as the tines being of wood, will not injure an animal. The lightness of the tool commends it for field use. These forks are made of a single bar of second growth White Ash, finished by hand and varnished, making them strong and neat.

One-half Dozen in a Bundle.

SHELF BRACKETS.

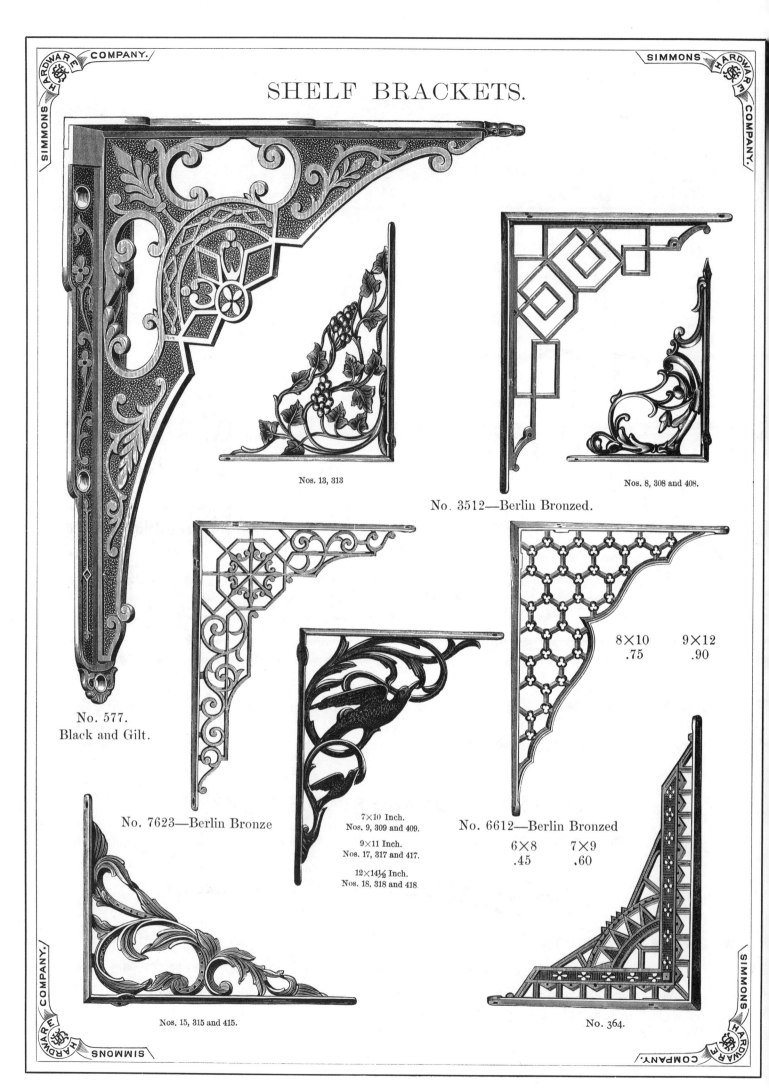

Coat and Hat Hooks.

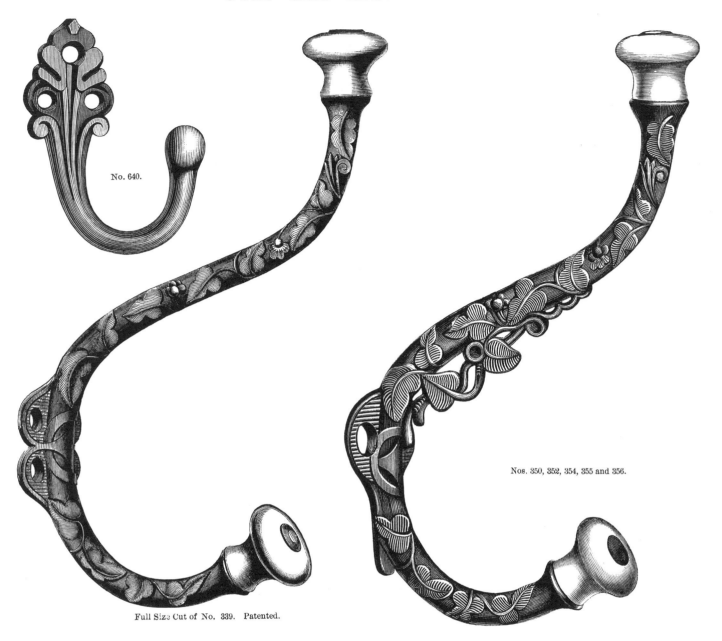

No. 640.

Nos. 350, 352, 354, 355 and 356.

Full Size Cut of No. 339. Patented.

	Per gross.		Per gross
No. 350, Japanned,	$24 00	No. 360, Japanned,	$30 00
No. 352, Berlin Bronzed, with Screws,	27 00	No. 362, Berlin Bronzed, with Screws,	36 00
No. 354, Gold Bronzed,	30 00	No. 364, Gold Bronzed,	39 00
No. 355, Cast Brass,	84 00	No. 365, Cast Brass,	132 00
No. 356, Bronze Metal, with Screws,	108 00	No. 366, Bronze Metal, with Screws,	159 00

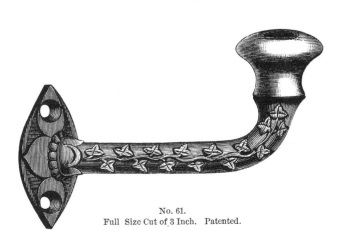

No. 61.
Full Size Cut of 3 Inch. Patented.

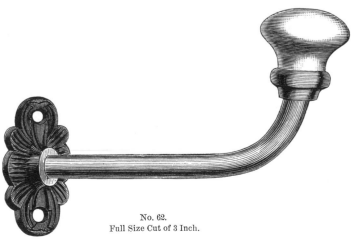

No. 62.
Full Size Cut of 3 Inch.

Full Size Cut of No. 57.

Full Size Cut of No. 93.

Full Size Cut of No. 7½.

No. 3½.

No. 490, Berlin Bronzed, per gross, $22 60

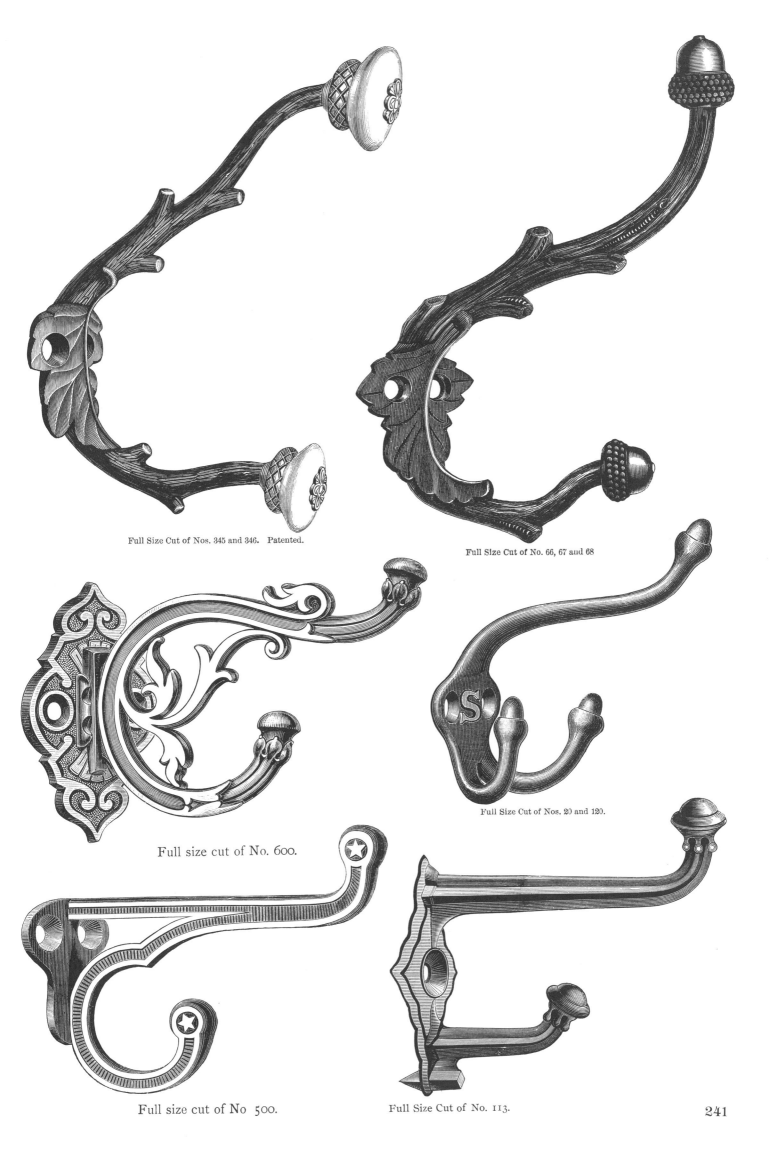

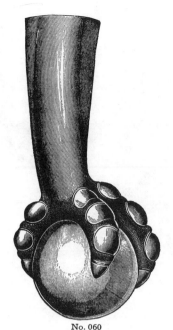

No. 060

Dozen $3.75

No. 77

Cast iron, brass-plated, dozen $2.50

No. 050

Cast iron, glass ball, dozen $1.44

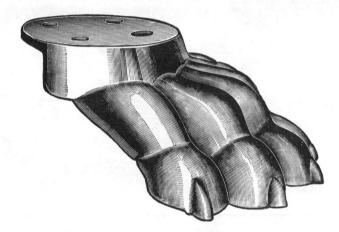

No. 4395

Solid brass, polished, dozen $2.40

No. 4395½

Iron, brass-plated, polished, dozen $1.35

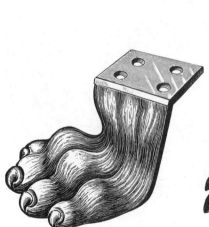

No. 6190

Solid brass, polished, dozen $1.60

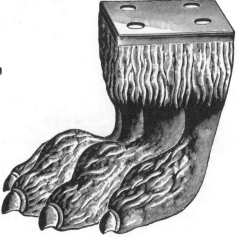

No. 4586 Solid brass, polished, dozen. $4.50

No. 4586½ Iron, brass-plated, polished, dozen 1.60

No. 34

Solid brass, polished, dozen $4.80

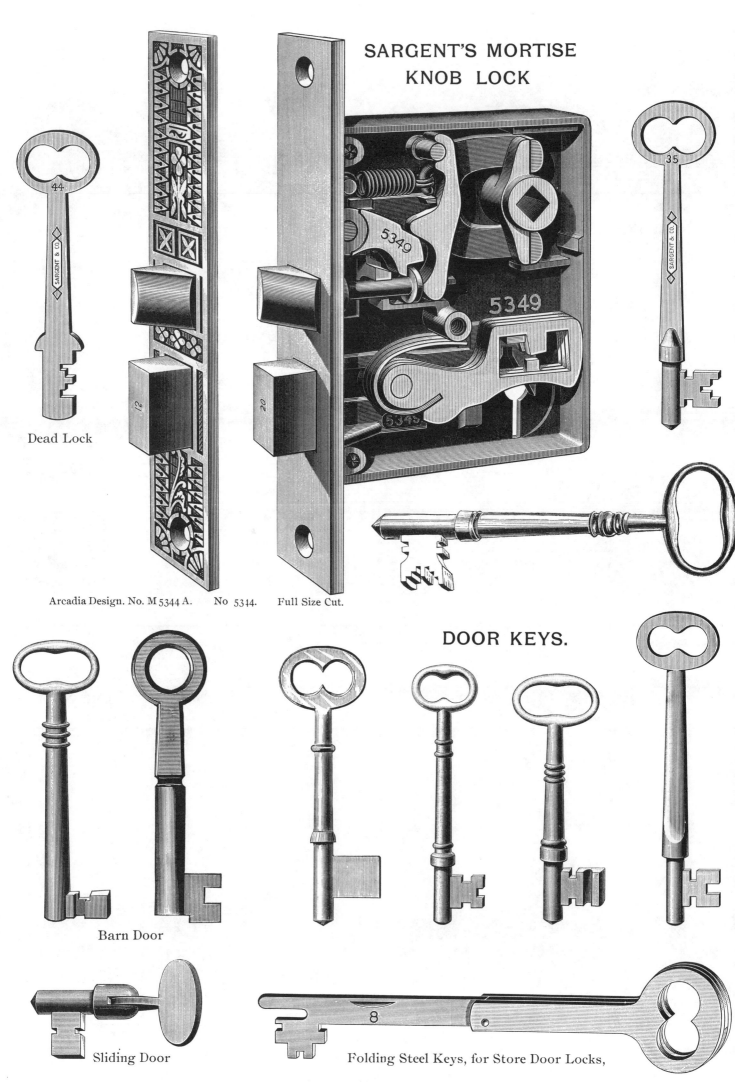

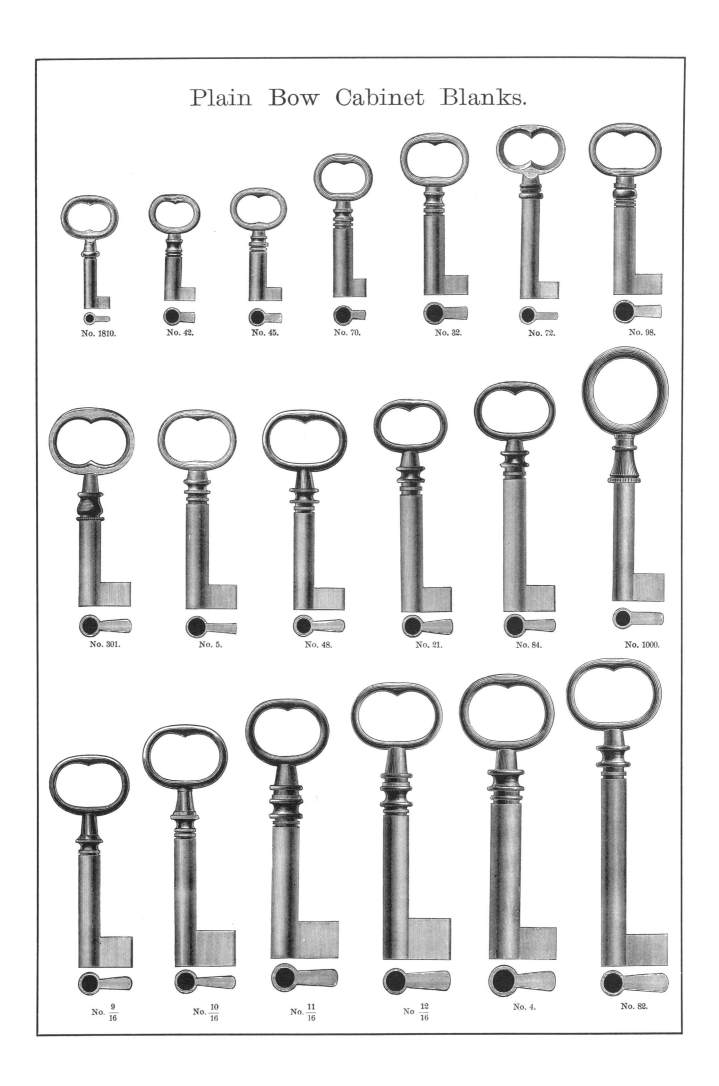

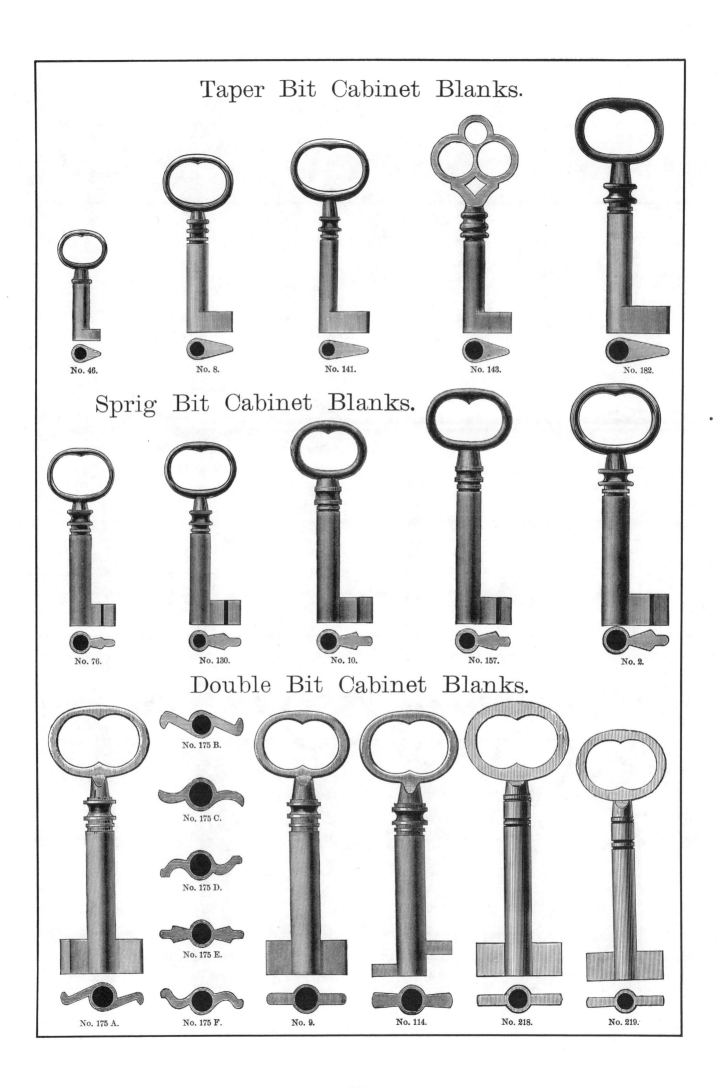

Fancy Bow Cabinet Blanks—Iron.

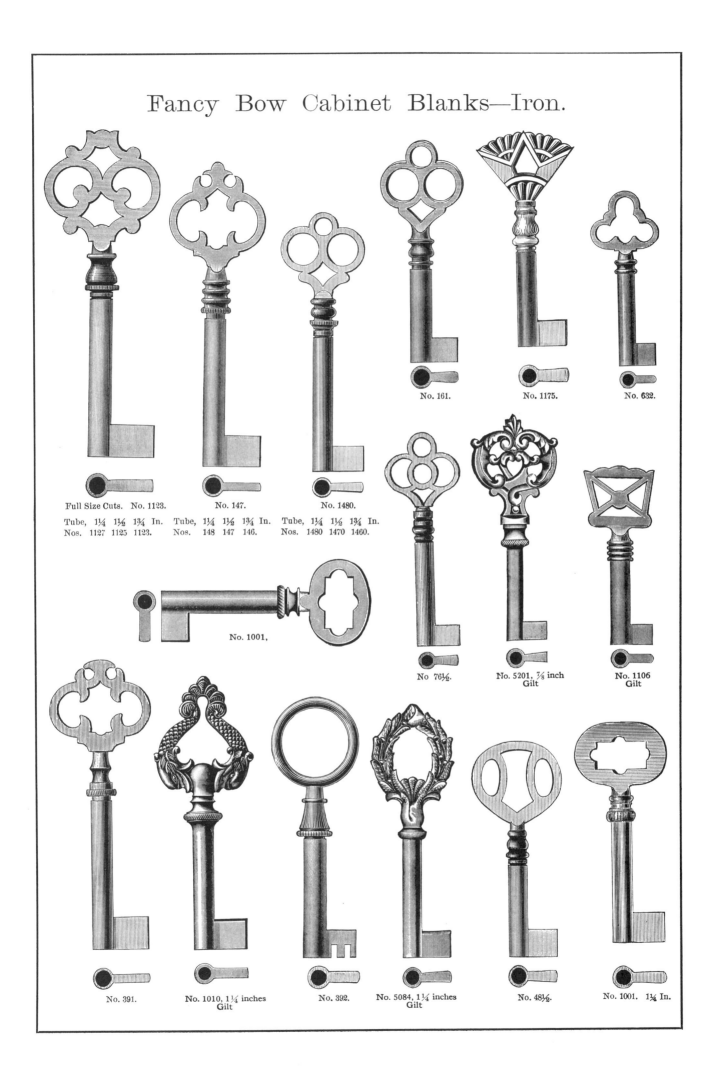

ESCUTCHEONS.
ORNAMENTAL PLATE.

Nos. 423. K 423. Full Size Cut of No. M 870. Full Size Cut of No. M 890.

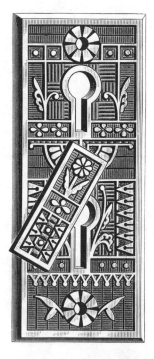

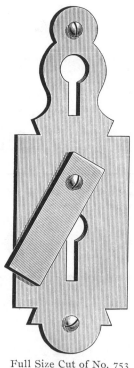

Full Size Cut of No. 131. Full Size Cut of No. 753.

No. 533. No. 506. No. 508. No. 515.

248

ROSE AND ESCUTCHEON COMBINED.

Full Size Cut of No. M 865 A.

For Inside Doors, 1 Key Hole.

Full Size Cut of M 876 A.

No. 2603½.

No. 2634½ and 3634½.

No. 2601.

ORNAMENTAL FLUSH ESCUTCHEONS.

PACKED WITH SLIDING DOOR LOCKS

Full Size Cut of No. M 563 A.

Full Size Cut of Nos. M 843 A and K 443 A.

Full Size Cut of Nos. M 883 A and K 483 A. Arcadia Design.

No. M 563 A, Genuine Bronze Metal, Arcadia Design, - - - - per pair, $3 30
  M 843 A, " " " " " - - - - " 1 85

SUN RIM LATCHES.

Bronze Store Door Handles and Lock

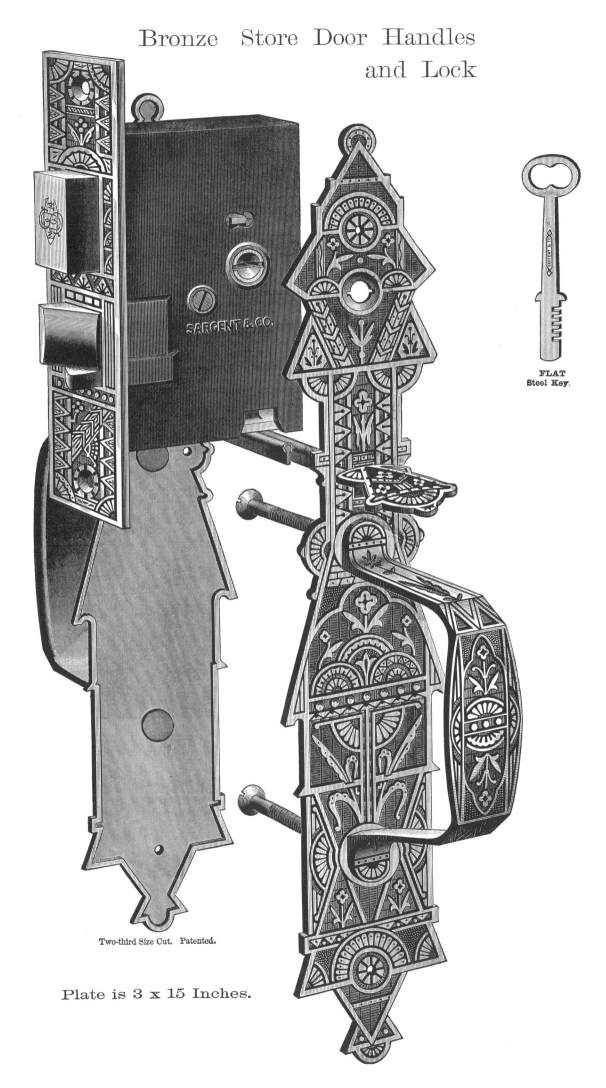

Two-third Size Cut. Patented.

Plate is 3 x 15 Inches.

FLAT Steel Key.

Bank Door Pulls.

EXTRA HEAVY, BRONZE METAL.

Half Size Cut of No. 587. Patented.

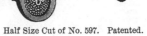

Half Size Cut of No. 597. Patented.

Half Size Cut of Nos. 598 and 599.

Bronze Metal.

Packed with Bronze Metal Screws.

| No. 587, | . per dozen, $28 00 | No. 597, | . per dozen, $40 00 | No. 598, | . per dozen, $50 00 |

Imperial Bronze.

No. 599, Imperial Bronze, Packed with Bronze Metal Screws, . . . per dozen, $50 00

Push Plates.

No. 421, Berlin Bronzed.   No. 821, Bronze Metal.   No. 422, Berlin Bronzed.   No. 822, Bronze Metal.

Packed with Round Head Screws.

Whole Length, 11 Inches.	Whole Length, 12 Inches.
No. 421, Berlin Bronzed, . per dozen, $3 75	No. 422, Berlin Bronzed, . per dozen, $4 75
No. 821, Bronze Metal, . . " 17 00	No. 822, Bronze Metal, . . " 25 50

Loose Joint.
Ball Tips.

DOUBLE STEEL BUSHED.

No. 898, Imperial Bronze.

4×4½	4×5	4½×4½	5×5	5×6	5×7	5½×5½
6 20	6 90	7 30	8 30	10 75	11 20	9 50

No. 1894, Bronze Metal, Hand-Cut Lines.

Inch,	3×3	3½×3½	4×4	4×4½	4×5	4½×4½	5×5	5×6	5×7
Per pair,	$4 75	5 75	7 50	9 00	9 25	9 50	11 25	13 50	16 00

Imperial Bronze Butts.

No. 899, Imperial Bronze.

4×4	4×4½	4×5	4½×4½	5×5	5×6	5×7
5 10	5 40	6 00	6 37	7 20	9 37	9 75

No. 333, BRONZE PLATED.

3½×3½	3½×4	4×4	4×4½	4×5	4½×4½	5×5	5×6
2 20	2 30	2 40	2 60	2 70	2 70	3 20	4 10

DOOR KNOBS.

MINERAL.

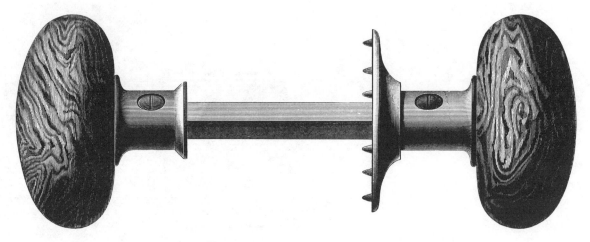

PORCELAIN.

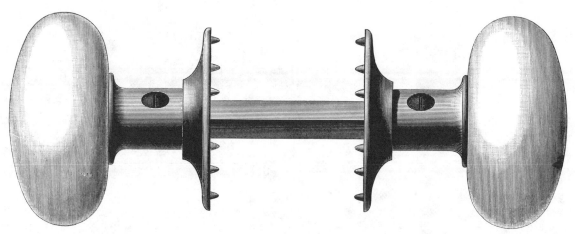

2¼-inch Mortise Knobs. Full Size Cut of Porcelain Knobs Nos. 1212 and 1714.

No. 1212, 2¼-inch, Porcelain Mortise, Japanned Mountings, ⁵⁄₁₆-inch Spindle, - per dozen, $2 35

    1213, 2¼ "   " Rim, " " ⁵⁄₁₆ " " - - " 2 35

    1714, 2¼ "   " Mortise, Nickel Plated " ⁵⁄₁₆ " " - " 7 00

    1715, 2¼ "   " Rim, " " " ⁵⁄₁₆ " " - - " 7 00

    1724, 2¼ "   " Mortise, Porcelain Roses Nickel Mountings, ⁵⁄₁₆-inch Spindle, " 7 80

Full Size Cut of 2¼-inch Knob. No. M 1542 A.

Full Size Cut of 2¼-inch Knob. No. 1872 P. Style also of 2½-inch, No. 1874 P.

No. M 1542 A, 2¼-inch, Bronze Metal, Mortise, M Finish, ⁵⁄₁₆-inch Spindle, - per dozen, $16 80

    1872 P, 2¼ " " " " Light Bronze, ⁵⁄₁₆ " " - - " 29 40

    1874 P, 2½ x 2¼ " " " " " ⁵⁄₁₆ " " - " 32 40

ORNAMENTAL DOOR KNOBS.

Full Size Cut of No. M 1814 A.

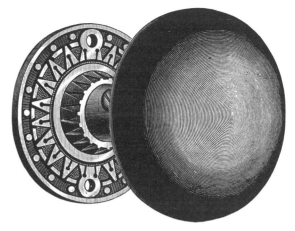

Full Size Cut of No. K 1434 A.

No. M 1814 A, 2¼-inch, Porcelain Mortise, Bronze Metal Mountings, M Finish, ⁵⁄₁₆ Spindle, per dozen, $8 20
K 1434 A, 2¼-inch, Ebony   "   Tokio Bronze Plated Mountings, ⁵⁄₁₆   "   "   6 00

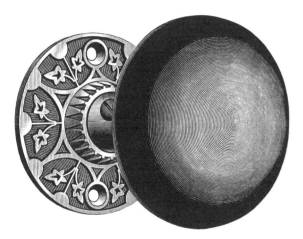

Full Size Cut of No. M 1832 A.

Full Size Cut of Nos. 1452 A, K 1452 A.

No. M 1832 A, 2¼-in., Ebony Mortise, Bronze Metal Mountings, M Finish,   ⁵⁄₁₆ Spindle, per dozen, $7 30
1452 A, 2¼-in., Berlin Bronzed Mortise, Berlin Bronzed Mountings,   ⁵⁄₁₆   "   "   7 70

Full Size Cut of 2¼-inch Knob.   No. M 1872 A.

Full Size Cut of 2¼-inch Knob.

Nos. M 1882 A and M 1886 A, 5-16-inch Straight Spindle, } Style also of
No. M 1887 A, 3-8 " Swivel " } 2½-inch Knobs.

No. M 1872 A, 2¼-inch, Bronze Metal Mortise, M Finish, ⁵⁄₁₆-inch Spindle,   -   per dozen, $22 80

Rural Cupboard Catches.

Full Size Cut of Nos. 507 and 607. Patented.

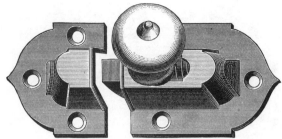

Full Size Cut of No. 305.

Full Size Cut of No. 318.

Full Size Cut of No. 311.

Brass and Plated Cupboard Catches.

Full Size Cut of No. 1260. Patented.

Full Size Cut of Nos. 2412 and 2812. Patented.

Full Size Cut of Nos. K 1492 A. and M 1892 A.

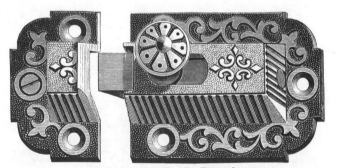

Full Size Cut of No. 262. Patented.

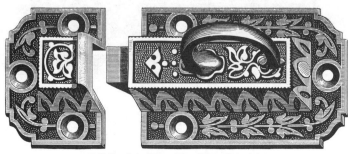

Full Size Cut of Nos. 493 and 893. Patented.

Full Size Cut of No. 1267. Patented.

Cupboard Turns—Beveled Edges.

PATENT TRIANGULAR BOLT.

Full Size Cut of No. 4230. Patented.

Full Size Cut of No. 4240. Patented.

Full Size Cut of Nos. 4231, 4430 and 4830. Patented.

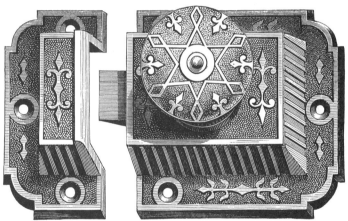

Full Size Cut of Nos. 4241, 4440 and 4840. Patented.

Full Size Cut of Nos. 4432 and 4832. Patented.

Full Size Cut of Nos. 4442 and 4842. Patented.

Tuscan and Berlin Bronzed, one dozen in a box; Bronze Metal, half dozen.

ORNAMENTAL CORNER BRACKETS.

ORNAMENTAL WALL BRACKETS.

Nos 501 and 502

7 inches Wide, 10 inches High.

Per dozen.
No. 501, Maroon Finish, . . $5 20
No. 502, " " and Bronze, 6 00

11 inches Wide, 9 inches High.

Per dozen.
No. 510, Maroon Finish, Plain, . $9 18
No. 511, " " Ornam'ted, 10 00
No. 511½, " " Fancy, . 10 28

14 inches Wide, 10 inches High.

21 inches High, 9 and 11 inches Front.

Per dozen.
No. 630, Maroon Finish, Plain, . $18 60
No. 631, " " Ornamented, . 20 22
No. 631½, " " " Fancy, 20 62

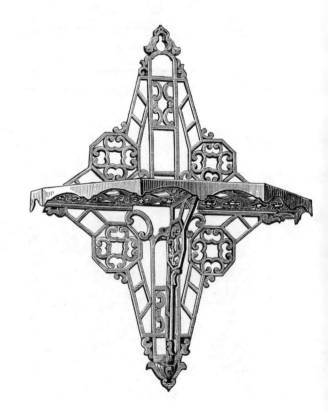

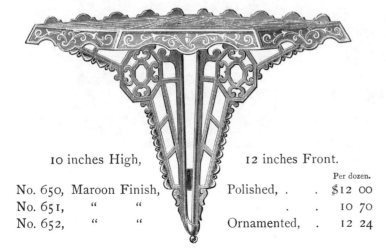

10 inches High, 12 inches Front.

Per dozen.
No. 650, Maroon Finish, Polished, . $12 00
No. 651, " " . . 10 70
No. 652, " " Ornamented, . 12 24

9 inches Wide, 12 inches High.

Per dozen.
No. 522, Maroon Finish, Polished, $7 88
No. 523, " " "

260

FLOWER POT BRACKETS.

12-inch Arm, Diameter of Shelves 5½-inch and 4-inch,

No. 2089. Black and Gilt,

No. 2261, Ebony and Gold.

Wire Stands.

No. 5.
36 inches long, 18 inches deep,

No. 3.
36 inches long, 26 inches deep,

One-quarter Size Cut

Wm. Frankfurth Hardware Co.

DOOR BELLS.

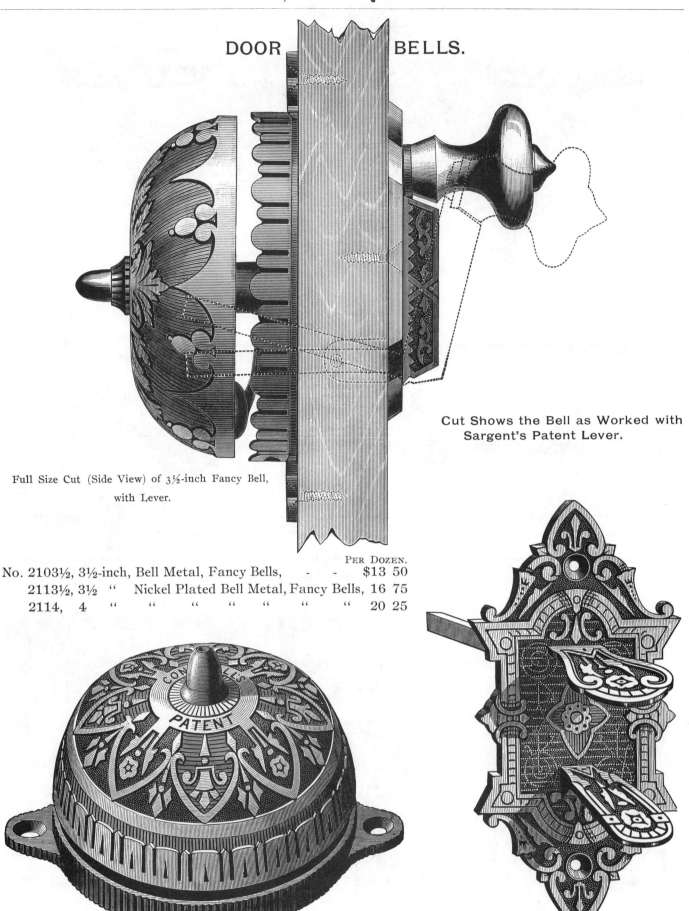

Full Size Cut (Side View) of 3½-inch Fancy Bell, with Lever.

Cut Shows the Bell as Worked with Sargent's Patent Lever.

		Per Dozen.
No. 2103½, 3½-inch, Bell Metal, Fancy Bells,		$13 50
2113½, 3½ " Nickel Plated Bell Metal, Fancy Bells,		16 75
2114, 4 " " " " " " "		20 25

No. 820, 4-inch, Nickel Plated Bell, with Engraved Thumb Latch, - - per dozen, $22 50

Milwaukee, Wisconsin.

BELL LEVERS.

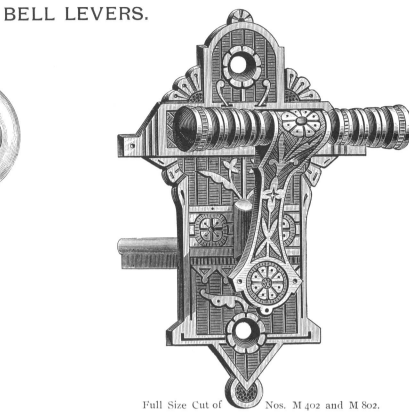

Full Size Cut of Nos. 401 and 601.   Full Size Cut of Nos. M 402 and M 802.

No. 401,	Berlin Bronzed, Porcelain Knob, with Screws,	per dozen,	$4 40
601,	Silver Plated on Cast Brass, Porcelain Knob, with Screws,	"	11 25
M 402,	Berlin Bronzed, Packed with Screws,	"	3 00
M 802,	Imperial Bronze, " " "	"	10 00

Full Size Cut of No. 807.   Full Size Cut of No. 809.

No. 807,	Bronze Metal, Packed with Screws,	per dozen,	$14 00
809,	" " " " "	"	16 00

Letter Box Plates.

Nos. 126 and 129. Three-quarter Size Cut. Patented.

Cover on Outside, Packed with Screws.

No. 126, Berlin Bronzed, Packed with Screws, ... per dozen,	$5 00
No. 129, Bronze Metal, Packed with Bronze Metal Screws, ... "	18 00

Nos. 326, 328 and 329 without Inside Plate.   Nos. 336, 338 and 339 with Inside Plate.   Three-quarter Size Cut.

With Machine Screws.

No. 326, Berlin Bronzed, Packed with Machine Screws, ... per dozen,	$5 00
No. 328, Imperial Bronze, " " " " . . . . "	18 00
No. 329, Bronze Metal, " " " " . . . . "	18 00

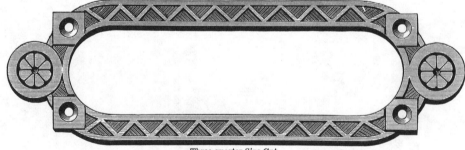

Three-quarter Size Cut.

With Plate for inside of Door.

No. 336, Berlin Bronzed Outside Plate (same as No. 326) and Berlin Bronzed Plate for Inside of Door, Packed with Screws, ... per dozen, $6 00

No. 338, Imperial Bronze Outside Plate (same as No. 328) and Imperial Bronze Plate for Inside of Door, Packed with Screws, ... per dozen, 22 00

Letter Box Plates.

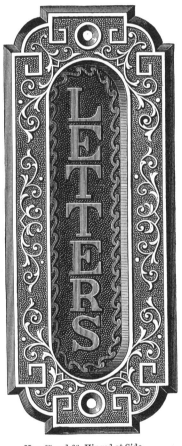

Nos. 65 and 66, Hinged at Side.
No. 166, Hinged at Top.
Three-quarter Size Cut. Patented.

No. 86, Hinged at Side.
Three-quarter Size Cut. Patented.

Packed with Screws,	With Machine Screws.
Per dozen.	Per dozen.
No. 65, Tuscan Bronzed, Hinged at Side, . $3 75	No. 86, Berlin Bronzed, Hinged at Side, . $5 50
No. 66, Berlin " " " . 4 20	

RUSSELL & ERWIN MANUFACTURING CO.

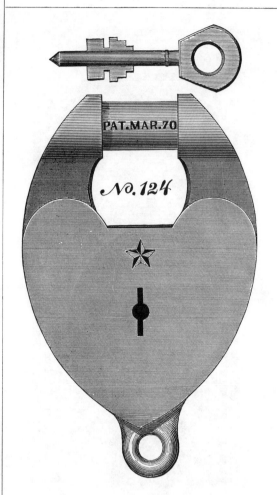

STAR PADLOCKS.

Malleable Iron, Brown Japanned.
2 Keys each.

No. 124, 3¾ inch Long, per dozen, $12 00
No. 125, 3   "   "   10 00
No. 126, 2½   "   "   8 00
With Chains add   "   1 00

Small Size for Watch Chains.

Per dozen.
No. 128, ¾ in. Long, Nickel Plated, $10 00
No. 128, ¾   "   Fine Gilt,   11 00

The cut is full size of No. 124.

MILLER'S PATENT BRASS PADLOCKS.

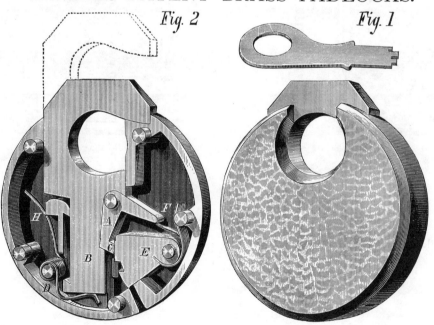

No. 1, ... per dozen, $20 00

Scandinavian Padlocks.

Malleable Iron, Two Keys Each, Loose and Fast Shackle.

Style of Key of Nos. 104 to 114.

Loose Shackle, Red Finish.

Nos.	104	106	108	110	112	114
Per dozen,	$7 25	9 00	11 00	13 00	16 00	24 00

Style of Key of Nos. 304 to 314.

Fast Shackle, Red Finish.

Nos.	304	306	308	310	312	314
Per dozen,	$8 25	10 00	12 00	14 00	17 50	26 00

With Chain, Nos. 307   309   311
Per dozen, $16 50   19 50   22 00

Style of Key of Nos. 406 to 411.

Fast Shackle, Flat Key, Japanned.

Nos.	406	408	410
Per dozen,	$13 50	16 00	18 50

With Chain, Nos. 407   409   411
Per dozen, $20 00   23 00   26 00

Horse Shoe Padlocks.

No. 610.

HORSE SHOE PADLOCKS.

Cast Bronze, 2 Flat Steel Keys.

No. 610, Cast Bronze, 2¼ Inch,	per dozen,	$9 00
Extra Keys, with Locks or to Sample,	"	1 80
Key Blanks,	"	90

Half dozen in a box.

Brown's Patent Padlocks.

2¼ Inch. Full Size Cut of No. 42.

Nos.	Per dozen. 1 Key.	Per dozen. 2 Keys.	Size.	All Brass. Spring Shackle.	
32	$3 25	$4 15	1⅛ Inch.		
34	3 75	4 75	1¼ "		
36	5 00	6 00	1½ "		Compound Tumbler. All over 1½ Inch have both Compound and Ratchet Tumblers, and twelve changes to the dozen.
38	6 25	7 25	1⅝ "		
40	6 75	8 00	1⅞ "		
40 A	7 50	8 75	1⅞ "	With Drop.	
40 B	8 25	9 50	1⅞ "	Spring Drop.	
42	7 50	8 75	2¼ "		
42 A	8 25	9 50	2¼ "	With Drop.	
42 B	9 00	10 25	2¼ "	Spring Drop.	
44	9 00	10 50	2½ "		
44 A	9 75	11 25	2½ "	With Drop.	
44 B	10 50	12 00	2½ "	Spring Drop.	

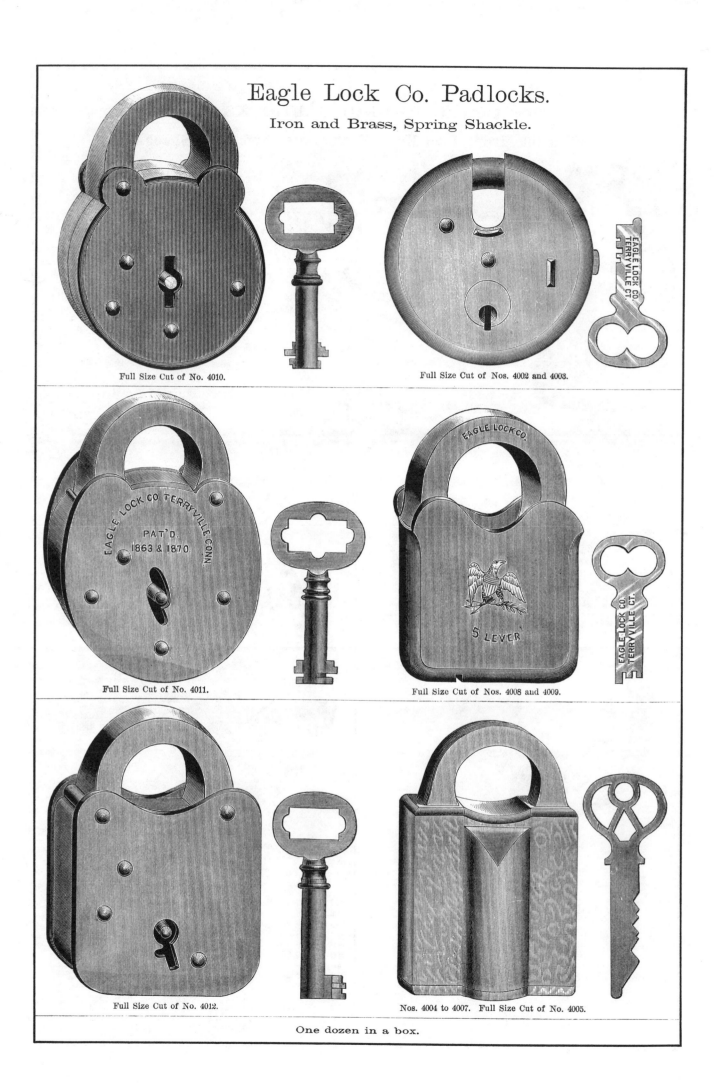

Hammacher Schlemmer & Co.
SINCE 1848 — NEW YORK

Padlocks
Full Size Cuts

Corbin
2⅛ inches. Heavy spring shackle. Three wards. Two levers. Two nickel-plated malleable iron keys to each lock, all different in a dozen. 12 changes.

No. 02865½ Bronze case, polished, bronze shackle, dozen $7.80

1⅞ inches. Self-locking. Spring shackle. Steel, ivory black case. Iron spring shackle, self-locking. Six changes with two steel keys.

No. 4448 Dozen $2.00

Key cannot be removed until it is locked, for railroads, express and telegraph companies, brass inside works, heavy cast bronze metal case.

Small Padlocks.
For Bags, Small Boxes, Dog Collars, etc.
Self-Locking Spring Shackle.

Nos. 30 and 70

Nos. 31 and 71

Nos. 32 and 72

Leather, German Silver Plate, to Lock, Length 14 to 20 inches,

Wrought Iron Tumbler Padlocks.

No. 200

Two Keys.

No. **200**, 2¼ Inch, Japanned . . . . . per hundred, $13 35

No. 402

Two Wide-Bit Keys.

Brass Bushing, Wheel, Side and Bridge Wards.

No. **402**, 2½ Inch, Japanned, 12 Changes . . per hundred, $74 45

No. 201

With Brass Drop.
Two Keys.

No. **201**, 2¼ Inch, Japanned . . . . Void

No. 403

Two Wide-Bit Flat Steel Keys.

Brass Bushing, Wheel, Side and Bridge Wards.
Revolving Key Pin, Monogram Brass Bushing on Back.

No. **403**, 2½ Inch, Japanned, 12 Changes . . per hundred, $77 35

No. 342

Two Double-Bitted Keys.

Brass Bushing, Wheel, Side and Bridge Wards, Double Chamber.

No. **342**, 2½ Inch, Japanned, 6 Changes . . . per hundred, $71 70

No. 404

Two Double-Bitted Flat Steel Keys.

Brass Bushing, One Wheel and Bridge Ward, Double Chamber.
Revolving Key Pin, Monogram Brass Bushing on Back.

No. **404**, 2½ Inch, Japanned, 6 Changes . . per hundred, $80 65

Eagle Lock Co.'s Tumbler Trunk Locks.

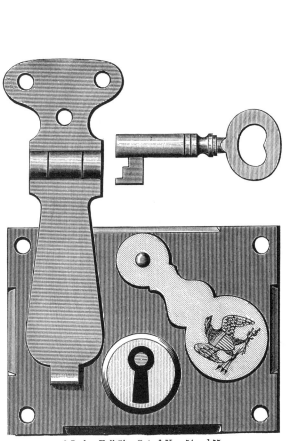

3 Inch. Full Size Cut of Nos. 54 and 55.

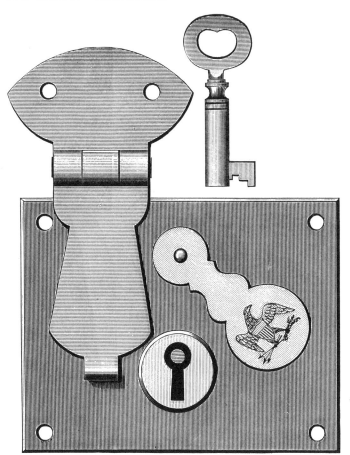

3½ Inch. Full Size Cut of No. 61.

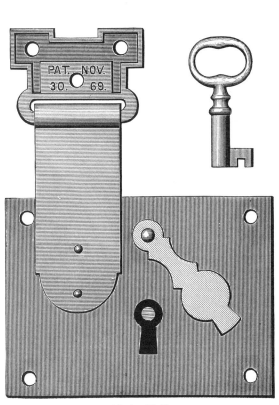

3 Inch. Tin Drop.
Full Size Cut of Nos. 8 and 26.

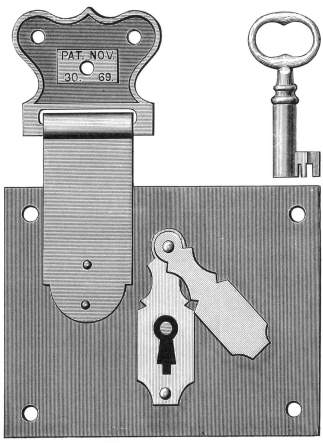

3½ Inch. Brass Bush and Drop.
Full Size Cut of Nos. 15 and 27.

RUSSELL & ERWIN MANUFACTURING CO.

PENDANT PATTERN DRAWER PULLS.

Full size cut of Nos. 801 and 802.   Full size cut of Nos. 851 and 852.

No. 801, Light Antique Finish,	per dozen,	$1 50
No. 802, Dark " "	"	1 50
No. 851, Light " "	"	1 74
No. 852, Dark " "	"	1 74

Packed with Screws.

Full size cut of No. 1005.   Full size cut of No. 1015.

No. 1005, Imperial Bronze, Star Medium,	per gross,	$36 00
No. 1006, " " " Large,	"	36 00
No. 1015, " " Loop Medium,	"	36 00
No. 1016, " " " Large,	"	36 00

RUSSELL & ERWIN MANUFACTURING CO.

BRONZED IRON DRAWER PULLS.

Full size cut of No. 61.

No.	41	51	61	71	81
Inches,	$2\frac{7}{8}$	$3\frac{1}{4}$	$3\frac{7}{8}$	$4\frac{1}{8}$	$5\frac{1}{2}$
Per gross,	$5 75	6 00	7 50	9 00	10 50

Full size cut of No. 52.

No.	52	62	72
Inches,	4	$4\frac{1}{4}$	$4\frac{5}{8}$
Per gross,	$5 75	7 25	8 75

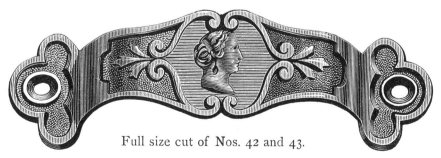

Full size cut of Nos. 42 and 43.

No. 42, Bronzed, 4½ inches, ... per gross, $5 00
No. 43, Copper Bronzed, 4½ inches, ... " 4 50

Above packed complete with Screws.

Drawer Pulls.

Full Size Cut of No. 516.

Full Size Cut of Nos. 1903, 1905, 1906, 1907 and 1908.   Patented.

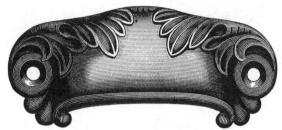

Full Size Cut of Nos. 420½, 421 and 422.

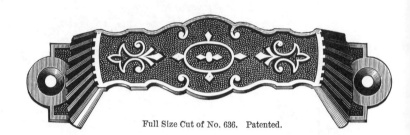

Full Size Cut of No. 636.   Patented.

Full Size Cut of Nos. 501, 505, 506 and 507.   Patented.

Full Size Cut of Nos. 455 and 456.

Full Size Cut of No. 475.

Full Size Cut of No. 61.

Full Size Cut of Nos. 150, 150½, 151, 152, 153 and 155.

Full Size Cut of Nos. 71, 75, 76 and 77.   Patented.

No. 386, Berlin Bronzed,   per gross, $13 75
387, Bronze Metal   "   100 00

Full Size Cut of Nos. 50, 50½, 51, 52, 53 and 55.

No. 50,   Japanned,   51, Copper Bronzed
No. 50½, Coppered,   52, Gold Bronzed,
No. 55,   Tuscan Bronzed,   53, French " "

Full Size Cut of Nos. 61 and 66.

No. 61, Copper Bronzed,   per gross, $9 00
No. 66, Berlin Bronzed,   "   10 00

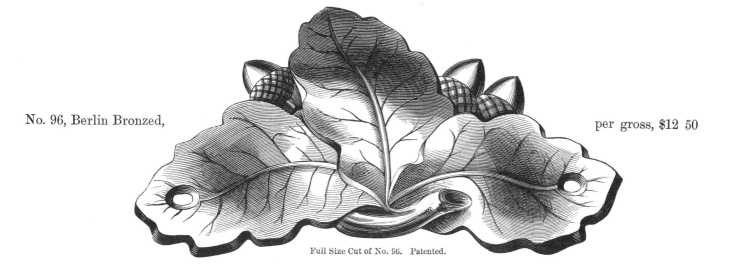

Full Size Cut of No. 96. Patented.

No. 96, Berlin Bronzed,   per gross, $12 50

RUSSELL & ERWIN MANUFACTURING CO.

BRONZED IRON CHEST HANDLES.

No. 50, per dozen pairs, $4 50
No. 60,   "   "   5 87
No. 70,   "   "   8 00

Packed complete with Screws.

FANCY PATTERN DRAWER PULLS.

Per dozen.

No. 110, Dark Antique Plate, Bronze Metal Handle, . . . $6 64

No. 111, Light Antique Plate, Bronze Metal Handle, . . . 6 64

Packed complete with Screws.

HAMMACHER SCHLEMMER & CO.
SINCE 1848 — NEW YORK

Furniture Pulls
Full Size Cuts

"Colonial"

No. 864 Wrought brass, boring 3 inches
Polished, dozen .... $ .72
Oxidized, dozen 90

No. 775 Wrought brass, with plated wire bail, boring 3 inches.
Polished, dozen.. $ .90
Oxidized, dozen.. 1.08

No. 1952 Cast brass, polished, plated wire bail, boring 3 inches, dozen... $1.20

Cast Brass, Polished

No. 5180 Cast brass, polished, boring 1¾ inches, dozen................................. $1.80

No. 4417 Size boring 2½ inches, dozen.................................... $2.40
No. 4418 Size boring 3 inches, dozen...................................... 2.40

Sargent & Co.'s Thermometers.

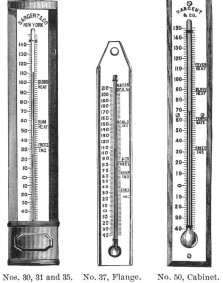

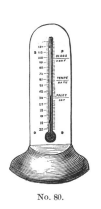
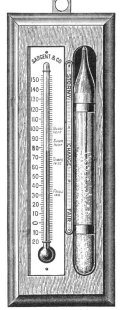

Nos. 30, 31 and 35.   No. 37, Flange.   No. 50, Cabinet.   No. 60, Window.   No. 70.   No. 80.   No. 100.

Cabinet.

COTTAGE.   Fancy Bronzed.   DISTANCE.

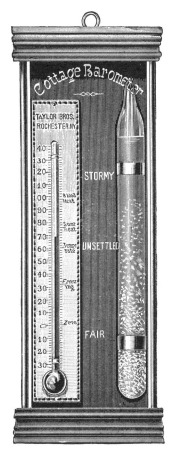
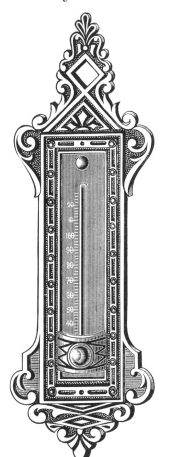
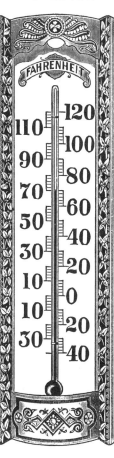

Dairy

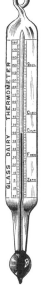

NO. 14.

No. 58, Cottage Barometer, Thermometer and Storm Glass, Mounted on a Black Background, Cherry Mouldings, Ornament either end, United at the sides by Bright Wire Rods,   -   -   per dozen, $5 00

The Universal Favor with which this Instrument has met, since placed on the Market at a Popular Price, proves conclusively the interest taken by all classes of people in knowing something of the daily condition of the atmosphere

LAMPS.
Magic Hand Lamp

C467, 98c Doz. C468, $1.20 Doz. C469, $2.20 Doz.

C467—Full nickel on brass, nickel burner, complete with chimney and wick. 1 doz. in box..........................Doz. **98c**

C468 — *Brilliant lacquered brass.* 7 in. high, body 4 in. Complete with burner, wick and large chimney. ⅙ doz. in box. ....Doz. **$1.20**

C469, "Bridgeport"—Full nickel plated, fount has screw cap, adjustable reflector, burner chimney, finger holder and hanger. Full ht. 8 in. Burns all night without odor. 1 doz. in pkg.
Doz. **$2.20**

C448, 92c Doz.   C450, 96c Doz.   C451, 96c Doz.

C448, "Twinkle"—Fancy shape fount with beaded globe shape chimney, ht. 6⅛ in., complete with brass nutmeg burner and wick, asstd. crystal, blue and green. 1 doz. in box.............................Doz. **92c**

C451, "Fairy" — Opal fancy melon shape body, footed, decorated in rich solid tints such as pink, green, canary, etc. Complete with nutmeg burner, wick and decorated opal chimney, full ht. 7 in. 1 doz. in pkg.....................Doz. **96c**

(4 inch Dish.)

Magic Hand Lamps, Brass,	each, $ .50
Magic Hand Lamps, Nickel Plated,	" .60

Lanterns.

Gipsy, With Bale. Great Western. SQUARE LIFT

(5 inches high.)

Gipsy Lanterns, Brass, with Nickel Top, Coal Oil,	each, $ .75
Great Western Lanterns, Brass, with Chain, Coal Oil,	" .85

LANTERNS.

No. 1.   No. 6, Closed.   No. 4.

No. 1—Lanterns, with Common Oil Burners and Candle Sockets, Lard Oil or Candle,	each,	$ .45
No. 6—Lanterns, with Common Burners and Candle Sockets, Lard Oil or Candle,	"	.60
No. 4—Railroad Lanterns, Lard Oil,	"	.90

Nail City.

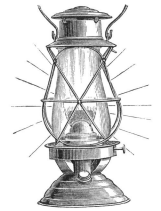

Senior.   Gem.

Nail City Senior Lanterns, with Double Globe, Coal Oil,	each,	$1.15
Nail City Gem Lanterns, with Single Globe, Coal Oil,	"	1.00

Tubular.   Buckeye Senior, Closed.   Police Lanterns.

No. 0—Tubular Lanterns, with Guards, Coal Oil,	each,	$1.25
Buckeye Senior Lanterns, with Double Globe, Coal Oil,	"	1.25
Police Bull's Eye Lanterns, 2¾ inches,	"	.90
Police Bull's Eye Lanterns, 3 inches,	"	1.00

LANTERNS.

DASH.

NO. 0.

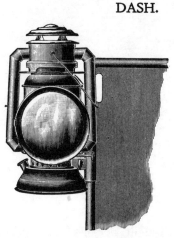

NO. 18 AS ATTACHED TO DASH.

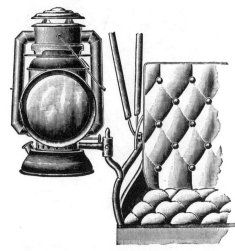

NO. 18 AS ATTACHED TO SEAT ROD OR BRACKET.

No. 0, tin dash lamp, hood reflector, side spring, No. 1, high cone burner, ⅝ inch wick, No. 0 globe, weight per dozen, 80 pounds, per dozen... $22.00

No. 18, cold blast tubular driving lamp, black enameled tin, with dash clamp, 4 inch silvered reflector, 4 inch double convex lens, ½ inch kerosene burner, weight per dozen, 100 pounds, per dozen............................. 42.00

SEARCHLIGHT.

No. 20, tin tubular, cold blast, searchlight, No. 2 burner, 1 inch wick, No. 0 tubular globe, reflector 12 inches in diameter, 7 inches deep, japanned, bulls eye globe, weight per dozen, 100 pounds, per dozen..................... $30.00

KITCHEN.

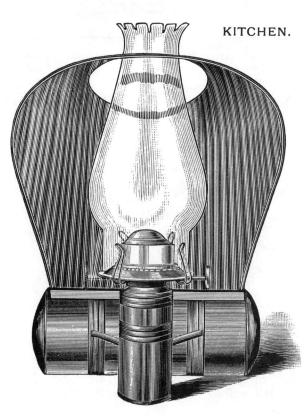

Star Kitchen Lamp, complete with Chimneys, each, $8 50

Wm. Frankfurth Hardware Co.

LANTERNS.

BUCKEYE, JR.   RAILROAD.   RAILROAD.

No. 3.   No. 43.   No. 43½.

No. 3,	Buckeye, Jr., with Double Globes,	per dozen, $10 00
43,	Railroad, with Single Guard for Oil,	" 9 00
43½,	" " Double " "	" 10 00

DASH BOARD TUBULAR.   DASH BOARD QUEEN.   POLICE.

No. 13.   No. 1.

No. 13,	Tubular Side and Dash Lamp, Blued Japanned,	per dozen, $15 00
1,	Queen, Dashboard Lantern, Black "	" 15 00
1,	2¾-inch Police Lantern, " "	" 6 00
2, 3	" " " " "	" 6 50

No. 8440 The Hornet Brass Hand Lamp, complete with burner, chimney and wick, each, 15 cents; per dozen $1 60

C465, 89c Doz.   C466, 89c Doz.

C465, "Southern" Brass Hand Lamp.— Large size, lacquered brass front, brass fluid rachet burner and round wick, requires no chimney. 1 doz. in box with wicks...Doz. **89c**

Iron Hand Lamps.

. . . . per dozen, $5 00

Lamp Fillers.

Pint.   Quart.
$ .20   .25

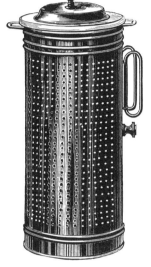

Jacket Lamp with Tubes and Feeder.

Japanned Nursery Lamps and Tea Kettles, Complete, ... each, $2.00

Little Will.   Queen.

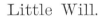

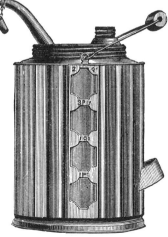
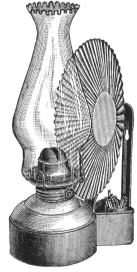

Little Will, 5 Gallon Coal Oil Cans, ...
Queen Oil Cans, . . . { Quarts, . . .
Non-Explosive Oil Cans, . . { Each, . . .

No. 8430 Eureka Side Lamps, complete with No. 2 Banner burner, chimney and wick. each $0 29

ADAMS, QUINCY, FRANKLIN, FIFTH AVE., CHICAGO.

No. 719.
Complete as shown in cut.
7½ inches high to top of Burner.
Central Draft Burner.
In Nickel only, $1 50

No. 2. The Pittsburgh.
Height 11½ in. to top of Burner.
Plain, Brass .. $4 00
Plain, Nickel. 4 50

No. 711. Little Royal.
Complete as shown in cut.
Fount with Bracket.
Central Draft Burner.
Fount same size as No. 719.
In Nickel only, $1 50

No. 2. The Pittsburgh.
Height 11½ inches to top of Burner
Embossed, Brass........... $4 50
Embossed, Nickel......... 5 00

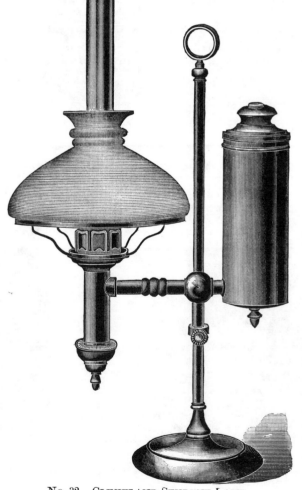

No. 32. Cleveland Student Lamp.
Complete as shown in cut.
The Best Student Lamp in the Market.
In Nickel only, $6 00

STUDENTS' LAMPS.

Library. Study.

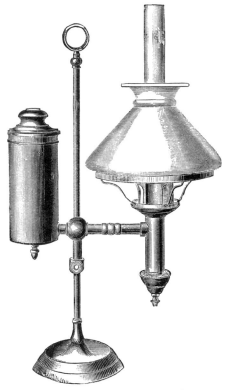
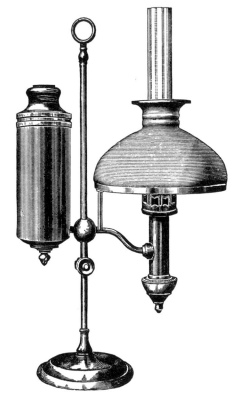

No. 1—Library Student Lamps, Nickel Plated, Complete, with White Shades and Chimneys,   each, $6.50
No. 2—Study Student Lamps, Nickel Plated,   " 5.50
No. 3—Study Student Lamps, Extra Large Reservoir,   " 6.00

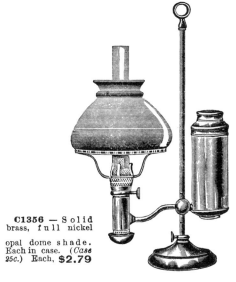
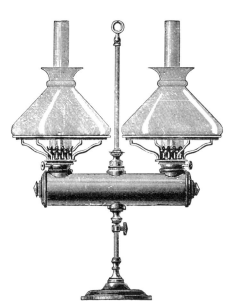

C1356 — Solid brass, full nickel opal dome shade. Each in case. (Case 25c.) Each, **$2.79**

No. 4. No. 5.

No. 4—Manhattan Student Lamps, Complete, with White Shades and Chimneys,   each, $ 5.50
No. 5—Double Library Student Lamps, Complete, with White Shades and Chimneys,   " 12.00
Student Lamp Wicks,   per doz. .40

PARLOR TABLE LAMPS.

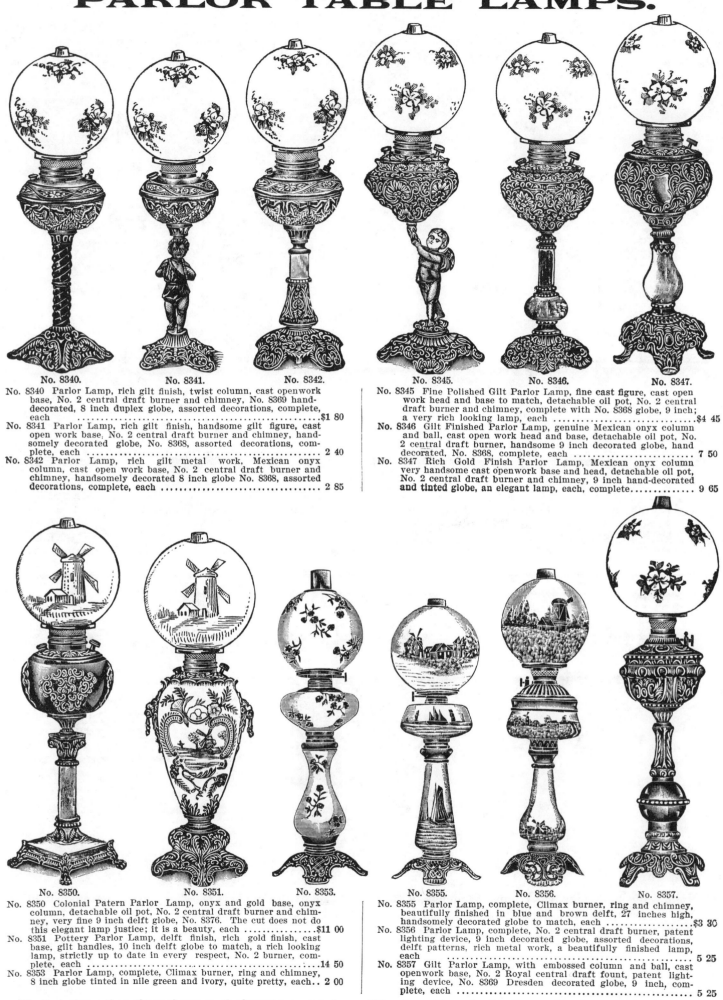

No. 8340.   No. 8341.   No. 8342.   No. 8345.   No. 8346.   No. 8347.

No. 8340 Parlor Lamp, rich gilt finish, twist column, cast openwork base, No. 2 central draft burner and chimney, No. 8369 hand-decorated, 8 inch duplex globe, assorted decorations, complete, each $1 80

No. 8341 Parlor Lamp, rich gilt finish, handsome gilt figure, cast open work base, No. 2 central draft burner and chimney, handsomely decorated globe, No. 8368, assorted decorations, complete, each 2 40

No. 8342 Parlor Lamp, rich gilt metal work, Mexican onyx column, cast open work base, No. 2 central draft burner and chimney, handsomely decorated 8 inch globe No. 8368, assorted decorations, complete, each 2 85

No. 8345 Fine Polished Gilt Parlor Lamp, fine cast figure, cast open work head and base to match, detachable oil pot, No. 2 central draft burner and chimney, complete with No. 8368 globe, 9 inch; a very rich looking lamp, each $4 45

No. 8346 Gilt Finished Parlor Lamp, genuine Mexican onyx column and ball, cast open work head and base, detachable oil pot, No. 2 central draft burner, handsome 9 inch decorated globe, hand decorated, No. 8368, complete, each 7 50

No. 8347 Rich Gold Finish Parlor Lamp, Mexican onyx column very handsome cast openwork base and head, detachable oil pot, No. 2 central draft burner and chimney, 9 inch hand-decorated and tinted globe, an elegant lamp, each, complete.............. 9 65

No. 8350.   No. 8351.   No. 8353.   No. 8355.   No. 8356.   No. 8357.

No. 8350 Colonial Patern Parlor Lamp, onyx and gold base, onyx column, detachable oil pot, No. 2 central draft burner and chimney, very fine 9 inch delft globe, No. 8376. The cut does not do this elegant lamp justice; it is a beauty, each $11 00

No. 8351 Pottery Parlor Lamp, delft finish, rich gold finish, cast base, gilt handles, 10 inch delft globe to match, a rich looking lamp, strictly up to date in every respect, No. 2 burner, complete, each 14 50

No. 8353 Parlor Lamp, complete, Climax burner, ring and chimney, 8 inch globe tinted in nile green and ivory, quite pretty, each.. 2 00

No. 8355 Parlor Lamp, complete, Climax burner, ring and chimney, beautifully finished in blue and brown delft, 27 inches high, handsomely decorated globe to match, each $3 30

No. 8356 Parlor Lamp, complete, No. 2 central draft burner, patent lighting device, 9 inch decorated globe, assorted decorations, delft patterns, rich metal work, a beautifully finished lamp, each 5 25

No. 8357 Gilt Parlor Lamp, with embossed column and ball, cast openwork base, No. 2 Royal central draft fount, patent lighting device, No. 8369 Dresden decorated globe, 9 inch, complete, each 5 25

If other globes than those shown on the lamps are wanted we will be pleased to make the change charging only the difference in cost of globes.

C1498 Asst—2 lamps, both popular low shape, one with 10 in. globe, both body and globe in rich green blended ground with large pink hydrangeas and green leaves, black Florentine design metal trimmings, spun crown, the other with large 11 in. globe, rich pink blended ground with large wild flower decoration, heavy gilt finished metal base and crown, both complete with removable founts, No. 2 Royal center draft burners, globe rings, chimneys and wicks, full ht. 22½ in. 2 in bbl. (Bbl. 35c)...............Each, **$3.30**

C1500, $3.65 Each.

C1502—Comprising 2 styles, both large vase shapes with 10½ in. globes, both bodies and globes in "Kopps" genuine cardinal red, one in allover embossed bullseye effect and the other in the new fancy melon design, both have heavy metal bases, ornamental crowns, best removable brass founts, No. 2 Success center draft burners, globe rings, chimneys and wicks, all metal trimmings in the new rich gun metal finish, full ht. 25¼ in. Total 2 lamps in bbl. (Bbl. 35c.)
Each, **$3.25**

Three Styles C1493, $2.75 Each.

483 Asst—4 large lamps with fancy shape bodies

NOTICE! All our decorated lamps are complete, **INCLUDING CHIMNEYS.** All our founts are of solid brass, not the cheap tin affairs used by some makers.

C1504, $4.20 Each.

C1481—Comprising 6 lamps—4 with large 9 in. globes, 2 with 8½ in. dome shades, all with large fancy shape bodies in rich solid tints, very elaborate and heavily enameled floral decorations, gilt finished metal bases, No. 3 Climax burners, globe rings and chimneys, average ht. 19¾ in. 2 pink and ivory, 2 blue and 2 green, total 6 lamps in bbl. (Tierce 45c). Each, **95c**

C1499, $3.50 Each.

NICKEL TABLE OR SEWING LAMP.

Ordinarily you would pay 65c for a lamp as good as this.

C460—Full nickel plated No. 2 nickel plated burner, 10 in. high without chimney. Outside filling device, extension wick raiser, broad safety base. Each in pasteboard box, *no pkg. charge.* Each, **.44c**

"FANCY PATTERN" HAND LAMP ASST.

Best quality, cream patterns, lowest price.

C1326—Three staple sellers in bright, attractive patterns,

"DEFIANCE" LAMP ASST.

High footed handled lamp assortment. Up to date patterns. Heavy, extra well made and finished. *Always popular sellers.*

C1328—Comprising 5 styles all in pure crystal and in best new patterns, height 5½ inches. All with large convenient handles and all fitted with No. 1 brass collars. 1 doz. each of the five styles. Total of 5 doz. in bbl. (*Bbl. 35c.*) Per dozen, **82c**

OUR "PIONEER" GLASS HAND LAMP ASST.

First class quality, bright patterns, large sizes.

C1325—Asst. comprises 2¼ doz. each of 4 styles, 2 being fancy and 2 almost plain, all in pure crystal, well made and finished. large stuck handles, best No. 1 brass collars. Total 9 doz. in bbl. (*Bbl. 35c.*) Per dozen, **61c**

OUR "SEWING LAMP" ASST.

C1347—Comprising 4 styles, all with No. 2 brass collars and of extra large low shape, with large founts. All of pure crystal brilliantly finished and brightly up to date patterns. ½ doz. of each style. Total 2 doz. in bbl. (*Bbl. 35c.*) Per dozen, **$1.80**

OUR "TABLE LAMP" ASST.

Popular in pattern, unsurpassed in quality, unequaled in price.

C1331—These lamps are all of the best crystal glass, bright

Mammoth Lamp.

all places where a large and steady light is required; 20 inch embossed tin shade harp,

STORE OR HALL LAMPS.

"Royal"   "Banner."

"TINTED" COMPLETE LAMP ASST.

A great 50c offering. All clinch collar lamps.

C1339—Comprising ⅙ doz. each of 3 extra large lamps, all with No. 2 collars and with stems decorated in royal blue, bases in ruby. ⅙ doz. each of 3 styles fitted complete with burners, chimneys, and wicks. Total 1 doz. in bbl., wt. 75 lbs. (*Bbl. 35c.*) Per dozen, **$3.95**

"BIG AND SAFE" COMPLETE LAMP ASST.

The best 50c complete lamps ever offered. All with patent clinch collars. No plaster, no cement, no dirt, no working loose.

C1338—½ doz. each of two patterns of high stand lamps, measuring 19½ in. to top of chimneys and ½ doz. of large sewing lamps measuring 18 in. in height. All complete with No. 2 burners, chimneys and wick. Total of 1 doz. in pkg., wt. 70 lbs. (*Bbl. 35c.*) Per dozen complete, **$3.85**

OUR "DECORATED" COMPLETE LAMP ASST.

Priced much below regular value. It took a lot of work to get these into the 25 and 50c classes.

C1343—Well made, crystal glass, fancy pattern, with tinted satin etched bodies and hand painted floral decorations large globe chimneys tinted and decorated to match. Asstd blue, green and yellow blends. Decorations burnt in and will not wear off. Complete with brass collars and "Sun" burners. Asst. comprises:

¼ doz. footed hand, No.1 burner and chimney, full ht. 14¼ in. @ $2 20 $0 55
¼ doz. stand, No. 1 burner and chimney, full ht. 16 in........ " 3 10 78
¼ doz. stand, No. 1 burner and chimney, full ht. 16¾ in...... " 3 75 94
¼ doz. B stand, No. 2 burner and chimney, full ht. 18¾ in...... " 4 20 1 05

Total 1 doz. in bbl. (*Bbl. 35c.*)   Total, **$3.32**

"DIME LEADER" LAMP ASST

C1348—Large sizes on high foot,

C1329—All purest full finished crysta

BRACKET LAMP.

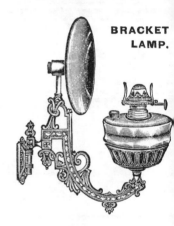

330—All of the best full finished crysta

C1327—All of the best full finished crystal

Library Lamps

ADAMS, QUINCY, FRANKLIN, FIFTH AVE., CHICAGO.

Piano Lamps with Pittsburgh Burner.

Prices given are for Lamps without Shades.

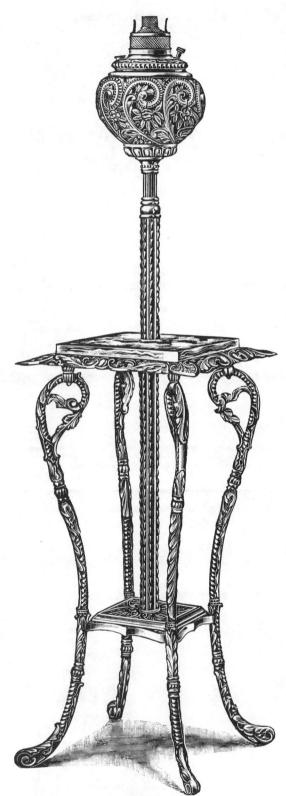

No. 2598.
Fine Cast Metal.
10 inch Mexican Onyx Top
Silver only, $47 00

No. 2597.
Solid Brass Legs. Cast Metal Head.
Brass, $25 00
Silver, 27 00

MARSHALL FIELD & CO.,

Pittsburgh Lamps.

No. 3409.
Height to top of Burner, 10 inches.
Complete as Shown in Cut.
Brass....$7 00
Nickel.. 8 00

No. 3467.
Height to top of Burner, 11½ inches.
Complete as shown in Cut.
Fine Cast Metal.
Silver Finish........$11 50
Gilt Finish 11 00

No. 3442.
Height to top of Burner, 11½ inches.
Complete as shown in Cut.
Brass only, $12 00

Lamp Shades.

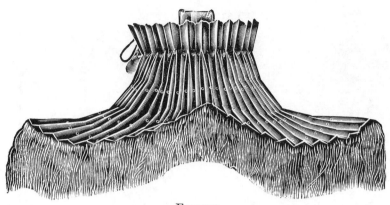

ECLIPSE.
Linen Shade with Silk Fringe.
18 inch for Banquet Lamp, $2 50
24 inch for Piano Lamp... 3 20

No. 1130.
Florentine Silk, Deep Lace.
Colors: Red, Orange, Canary, Pink.
15 inch, $13 00
22 inch, 28 00

No. 520.
Florentine Silk.
15 inch, $17 50
18 inch, 33 00
22 inch, 36 00

No. 1430.
Florentine Silk.
Colors: Red, Orange, Canary, Pink.
15 inch, $10 00
20 inch, 15 00

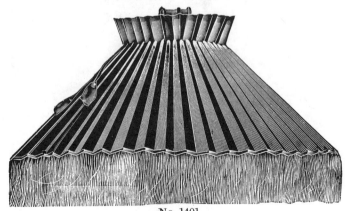

No. 1401.
Linen Shade with Silk Fringe.
10 inch for Table Lamp....$0 90
12 inch for Banquet Lamp. 1 50
14 inch for Piano Lamp.... 1 70

No. 1220.
Florentine Silk with Black or White Lace.
Colors: Red, Orange, Pink, Canary.
15 inch, $ 7 50
20 inch, 13 00

Pittsburgh Lamps.
PRICES EACH.

No. 201.
Height to top of Burner, 22 inches.
Complete as shown in cut.
Gilt, $7 50   Silver, $8 00

No. 204.
With "Pittsburgh" Burner.
Height to top of Burner, 23 inches.
Complete as per Cut.   Gilt, $13 00   Silver, $13 50

No. 3480.
Rich Cast Metal, with Open Work Cast Head.
Height to top of Burner, 16 inches.
Complete as shown in cut.
Gilt, $18 00   Silver, $19 00

Fine Bronze or Silver Mantel Clocks.

Silver or Bronze. Height 19½ in. Width 11¾ in.
8 Day ½ Hour Strike, Visible Escapement, Cathedral Bell,
White or Gilt Dial.................................$52.00 List.

ANT. OAK
Height 23 in. Width 12 in.
8 Day ½ Hour Strike, Cathedral Bell........$20.50 List

Silver or Bronze. Height 19 in. Width 16 in. 8 Day ½ Hour Strike, Visible Escapement, Cathedral Bell, Gilt or White Dial..$61.50

NOVELTY CLOCKS.

1 Day Time, 2-inch Dial. Base 7 Inch 's $5.25
Finished in Silver 20 per cent. extra.

LAWN TENNIS No. 2. Japanese Bronze or Silver.
Height, 7½ inches. One-Day Time, Plain Dial, $4 50;

1 Day Time, 2-inch Dial. Height 4 Inches.. $4.25

One Half Size.
No. 11, Fitted with No. 1 Movement, Fine
Gold Gilt Case, each $7.00

SOUBRETTE. Silver Finish.
Height, 4¼ inches. One-Day
Time, Fancy Dial, $4 00

Height 7½ in.
No. 33, Fitted with No. 1 Movement, Fine
Gold Gilt Case, each $7.00

1 Day Time, 2-inch Dial. Height 5 Inches Base 7¾ Inches............ $6 25
Finished in Silver 20 per cent. extra.

Don Cæsar. Don Juan.

Eight-Day, Half-Hour Strike, Cathedral Gong.
Height, 20½ inches; width, 25½ inches.

Bronze Finish $55 00
Silver Finish 60 50

Commerce

Half-Hour Strike, Cathedral Gong. Height, 20½ in Bronze or Silver Finish, $63 50

Enameled Iron Clocks.

Pompeii.

Neptuno.

Otranto.

Werra.

Nubia.

Eight-Day, Half-Hour Strike, Cathedral Gong. Bronze Trimmed. $22.50

Louis.

Manufactured by the Waterbury Clock Co.

CHESTER. (WALNUT AND OLD OAK.)

8 Day, ½ Hour Strike, Cathedral Bell.. $10 00
5 inch Dial, Roman or Arabic. Height, 16 inches. Base. 11½ in.

SUSSEX. (WALNUT.) Inlaid Wood.

1 Day, Strike $5 25
8 Day, Half-hour Strike, ... 6 75
"   "   Slow " Gong 7 25
Alarm, 60c list extra. Dial 6 in. Height 22¼ in.

HOMER. (WALNUT.)

8 Day, Half-hour Strike, $6 00.
8 Day, Half-hour Strike, Gong, $6 50
Alarm, 60c. Extra, List.
Dial, 6 Inches. Height, 21¾ Inches.

WAYNE. (WALNUT.)

8 Day, Half-hour Strike, $5.75.
8 Day, Half-hour Strike, Gong. $6.25.
Alarm, 60c. Extra, List.
Dial 6 inches. Height 21⅝ inches.

LYONS CALENDAR. (WALNUT.)

Dial, 6 inches. Height, 22¼ inches.
Eight Day, Half-Hour Strike, Calendar ___ $7 25
Cannot be fitted with Alarm.

MERWIN (WALNUT.)

Dial 6 inches. Height 24 inches.
8 Day, Half-hour Strike, $7.00.
8 Day, Half-hour Strike, Gong, $7.50.
Also with Alarm, 60c. Extra, List.

Manufactured by the Seth Thomas Clock Co.

CONCORD. (WALNUT.)

8 Day, Spring, Strike.....................$ 11 00
8 "   "   "   Alarm..............   11 70
Height, 22¼ inches.   6 inch Dial.
Cathedral Bell.

Height 23½ inches.
8 Day, Spring, Time...........................$ 8 00
8 "   "   Strike.......................... 9 60
8 "   "   Calendar, Time................. 9 10
8 "   "   "   Strike 10 70

ATLAS.
(QUARTER STRIKE) MADE IN WALNUT, OAK AND CHERRY.

Height, 22½ inches.
8 Day, strike ¼ hour $20 00
Striking quarter hours on two Cup Bells, and hours on Cathedral Bell.

TAMPA. (WALNUT.)

8 Day, Cathedral Bell.......................$9 50
Height 22 inches.   6 inch Dial.

O. G.

Heights, 25 and 29½ inches.
1 Day, Weight, Strike........................$ 7 50
8 "   "   " 11 40
Alarms, 70 cts. extra

(WALNUT CASE.)

Height, 22 inches. 6 inch Dial. 8 Day, Spring, Half Hour Strike, Cathedral Bell..........$9.20
Alarms 70c., list additional.

½ Size Nickel. 4 inch Dial.

1 Day Alarm $2 75

Nickel. 2 inch Dial.

1 Day Time $2 25

½ Size Nickel. 4 inch Dial.

1 Day Alarm $2.75

½ Size Nickel. 4 inch Dial.

1 Day Alarm $2.75

ONE-HALF SIZE. VISIBLE BALANCE.

1 Day, Lever, Time, Alarm $2.50

ONE-HALF SIZE. VISIBLE BALANCE.

Lever, Time $2.00

1 Day Time, 2-inch Dial. Base 7 Inches. Height 5¾ Inches $6.00
Finished in Silver 20 per cent. extra.

LODGE LEVER.

Gold Gilt Front, Nickel-Plated   3.50

UNION HARDWARE AND METAL COMPANY,

CARVED WOOD CLOCKS.

WHITE STAR LINE, D.

Carved wood, height 24½ inches, dial 6 inches, oak or walnut, eight day strike. Each.. $4.50

BELTON.

FLORIST. Finished in Silver. Dial, 2 inches. Height, 8½ inches.
One-Day Time. List, $4 50

ELDORA.   FENWICK.   ELDON.

Gold plated, height 5½ inches, width 4¾ inches, dial 2 inches.   Each $4.60
One day time, ivorized porcelain dial.

CALENDAR REGULATOR

FIG. 2643

OFFICE REGULATOR

FIG. 2646

CALENDAR CLOCK

FIG. 2644

PRICES, FIG. 2643
Black walnut or ash case, 32 inches high; dial 12 inches.
8 day time and calendar.............$9 90
8 day time, calendar and strike.......11 00
Gong, 30 cents list additional.

PRICES, FIG. 2644
Perpetual calendar.
Rosewood, walnut or oak case, 32 inches high, time dial 12 inches, calendar dial 10 inches.
8 day spring, time.................$18 50
8   "   " strike, cup bell.... 20 00

PRICES, FIG. 2645
Perpetual calendar.
Walnut, cherry, oak and old oak case, 49 inches high; both dials 12 inches.
8 day weight, time.................$44 00

PRICES, FIG. 2646
Rosewood, walnut and oak case, 32 inches high; dial 12 inches.
15 day spring, time................$11 50
8 day spring strike, cathedral bell... 12 70

PRICES, FIG. 2647
Rosewood case, height 21½ inches; gilt frame, 10 inch dial.
8 day time.........................$5 75
8 day time strike 6 85

WITH CALENDAR
8 day time.........................$6 25
8 day strike 7 35
Alarm 50 cents, gong 30 cents, list additional

CALENDAR REGULATOR

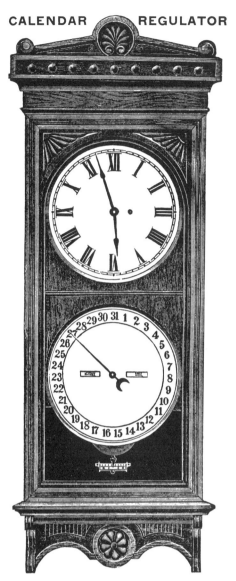

FIG. 2645

OCTAGON DROP

FIG. 2647

"SOLID COMFORT" LOAFER'S LOUNGE.

The Most Comfortable Lounge Made.
Affording a Position of Perfect Comfort and Ease.

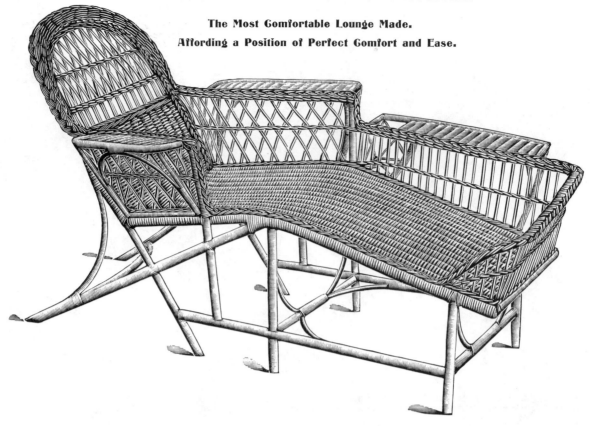

No. 1200. SOLID COMFORT LOUNGE. MADE OF REED AND FINISHED AS DESCRIBED BELOW.

Our "Solid Comfort" Loafer's Lounge is undoubtedly the most comfortable Lounge ever made. The position is a compromise between lying and sitting, and affords absolute rest and comfort, especially when a few soft cushions are used in connection. The Arm Rests add greatly to its comfort, and the Foot Guard affords a brace which also increases the comfort of the occupant. The Side Pocket is a convenient receptacle for book, paper, articles of sewing, etc. For a reading, napping or general resting couch it has no equal. It is suitable alike for Library, Sitting Room or Veranda.

	EACH.		EACH.
No. 1200—Shellaced and Varnished	$22.00	No. 1202—Mohogany Stained and Varnished	$22.00
No. 1201—Oak Stained and Varnished	22.00	No. 1203—XVI Century and Varnished	22.00

FANCY REED SIDEBOARDS.

The desired Finish should always be specified on orders.

Shellaced Finish will always be sent unless otherwise ordered.

No. 2123/F. FANCY CABINET OR SIDEBOARD.

With Polished White Maple Shelves.
Height, 70 in., Width, 33 in.

No. 2124/F. FANCY CABINET OR SIDEBOARD.

With Oval Bevel Mirror and Polished White Maple Shelves.
Height, 70 in., Width, 29 in.

FANCY REED PARLOR STANDS.

With Polished Hardwood Tops.

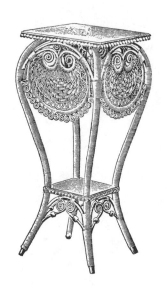  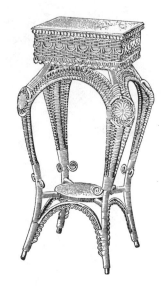 

No. 2108 SQUARE STAND.	No. 1149 ROUND STAND.	No. 2106 SQUARE STAND.	No. 1150 SQUARE TABLE.
Top of Polished White Maple.	Top of Polished Oak.	Top of Polished White Maple.	Top of Polished Oak.
Top, 14 x 14 inches. Height, 32 inches.	Top, 17 in. diameter. Height, 33 in.	Top, 14 x 14 inches. Height, 32 inches.	Top, 17 x 17 inches. Height, 30 inches.
No. 2108 —Shellaced or Stained..$ 7.00	No. 1149 —Shellaced or Stained..$ 8.00	No. 2106 —Shellaced or Stained..$ 9.00	No. 1150 —Shellaced or Stained..$11.00

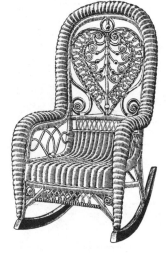

No. 1183 GENTS'.
EACH.
No. 1183 —Shellaced or Stained $13.00
No. 1183 —Enameled and Gold or all Gold............... 17.50

No. 1182 LADIES'.
EACH.
No. 1182 —Shellaced or Stained $11.00
No. 1182 —Enameled and Gold or all Gold............... 15.00

No. 1170 GENTS'.
EACH.
No. 1170 —Shellaced or Stained $14.50
No. 1170 —Enameled and Gold or all Gold............... 19.00

No. 1183 GENTS'
No. 1182 LADIES'

No. 1170 GENTS'

No. 2132 LADIES' ROCKER.	No. 2130 SETTEE.	No. 2131 GENTS' ROCKER.
EACH.	EACH.	EACH.
No. 2132 —Shellaced or Stained ....$ 7.50	No. 2130 —Shellaced or Stained $13.50	No. 2131 —Shellaced or Stained ....$ 9.00
No. 2132 —Enameled and Gold or all Gold..... 11.50	No. 2130 —Enameled and Gold or all Gold..... 19.50	No. 2131 —Enameled and Gold or all Gold..... 13.00

REED FURNITURE.—Separate or in Suits.

Any of these Pieces will be Sold Separately if desired. See below.

No. 851 LADIES' ROCKER.	No. 850 SETTEE.	No. 852 GENTS' ROCKER.
EACH.	EACH.	EACH.
No. 851/F—Shellaced or Stained.................$15.25	No. 850/F—Shellaced or Stained..................$28.00	No. 852/F—Shellaced or Stained..................$17.25
No. 851/E—Enameled and Gold or all Gold....... 19.50	No. 850/E—Enameled and Gold or all Gold....... 33.50	No. 852/E—Enameled and Gold or all Gold....... 21.50

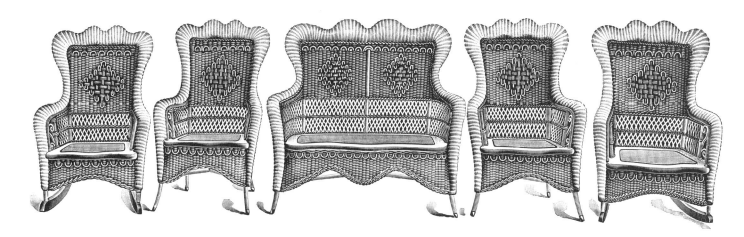

No. 846 LADIES' ROCKER.	No. 845 SETTEE.	No. 847 GENTS' ROCKER.
EACH.	EACH.	EACH.
No. 846/F—Shellaced or Stained.................$10.80	No. 845/F—Shellaced or Stained..................$22.00	No. 847/F—Shellaced or Stained..................$12.80
No. 846/E—Enameled and Gold or all Gold....... 14.00	No. 845/E—Enameled and Gold or all Gold....... 28.25	No. 847/E—Enameled and Gold or all Gold....... 16.60

No. 2126 LADIES' ROCKER.	No. 2125 SETTEE.	No. 2127 GENTS' ROCKER.
EACH.	EACH.	EACH.
No. 2126/F—Shellaced or Stained.................$5.00	No. 2125/F—Shellaced or Stained..................$10.50	No. 2127/F—Shellaced or Stained..................$6.00
No. 2126/E—Enameled and Gold or all Gold....... 8.00	No. 2125/E—Enameled and Gold or all Gold....... 15.00	No. 2127/E—Enameled and Gold or all Gold....... 9.00

FANCY WORK STANDS.

 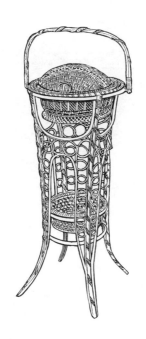 

LADIES' WORK STAND.

Height 29 inches, Oblong Covered Top Basket 14 x 9½ inches; Lower Basket 10 x 7 inches; Stand Extra White Willows trimmed with Fancy Straw Braid; Baskets very fine Willows and White Fancy Straw Braid.

PER DOZEN.
No. 353.................................$46.65

LADIES' WORK STAND.

Height 29 inches, Round Covered Top Basket 13 x 5 inches; Lower Basket 11 x 3 inches; Stand Extra White Willows trimmed with Fancy Straw Braid; Baskets extra fine material; a very fine new design.

PER DOZEN.
No. 354.................................$53.35

LADIES' WORK STAND.

Oblong Covered Top Basket 17 x 12½ inches; Lower Basket 10⅛ x 8 inches; Stand very fine White Willows trimmed with fine White Straw Braid and Wood Balls; Baskets very fine White Straw Braid.

PER DOZEN.
No. 355.................................$60.00

FANCY INFANTS' OR WORK STANDS.

INFANTS' OR WORK STAND.

Height 27 inches, Oblong Open Top Basket 18 x 12 inches; Oblong Open Lower Basket 14⅜ x 12 inches; Stand Extra White Willows trimmed with White Straw Braid; Baskets very fine white material.

PER DOZEN.
No. 356.................$40.00

INFANTS' OR WORK STAND.

Height 28 inches, Oblong Open Top Basket 18½ x 15½ inches; Oblong Open Lower Basket 14½ x 12 inches; Stand extra fine White Willows trimmed with fine White Straw Braid and Wood Balls; Top and Lower Baskets extra fine white material.

PER DOZEN.
No. 357.................$53.35

INFANTS' OR WORK STAND.

Height 28 inches, Square Open Top Basket 18 x 13 inches; Square Open Bottom Basket 13⅜ x 10 inches; Stand is a combination of fine White Willows and Reed, Baskets are made of extra fine White Willows and Straw Braid.

PER DOZEN.
No. 359.................$64.00

INFANTS' OR WORK STAND.

Height 28 inches, Oblong Open Top Basket 16 x 13 inches; Oblong Open Lower Basket 11 x 7½ inches; Stand extra fine Straw and Willow Work, all White; Baskets Upholstered with French Satin, Assorted Colors; Baskets very fine material.

PER DOZEN.
No. 358.................$71.10

CHILDREN'S HIGH CHAIRS.

No. 1187/F. CHILD'S HIGH CHAIR.

EACH.
No. 1187/F—Shellaced or Stained $4.50
No. 1187/E—Enameled and Gold or all Gold 7.50

No. 900 WITH SHELF.

Wood Seat and Frame, filled in with Reed and Willow work and varnished. Seat and Frame in Mahogany finish with Balls and Woven Work in Olive Green and Yellow; Height, 38 inches; Seat, 13 x 14½ inches.

EACH.
No. 900 ....$2.00

No. 910. WITH SHELF.

Wood Seat and Frame, filled in with fine Reed work and varnished. Seat and Frame in Olive Green, with Knobs in Mahogany finish, frame work in fancy rope design; Height, 39 inches; Seat, 13 x 14½ inches.

EACH.
No. 910 ....$2.25

No. 1186/F. CHILD'S HIGH CHAIR.

EACH.
No. 1186/F—Shellaced or Stained.....$5.00
No. 1186/E—Enameled and Gold or all Gold 8.00

NURSERY CHAIRS.

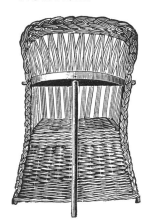

WILLOW NURSERY CHAIR.

PER DOZEN.
No. 80 —Seat, 8 x 11 in.; Back, 13 in. high $12.95
No. 80½—Extra large, Seat 9 x 12 in.; Back 14 in. high... 15.55

FANCY WILLOW NURSERY CHAIR.

PER DOZEN.
No. 81—Seat, 8 x 11 in.; Back, 13 in. high........$15.55

FANCY WILLOW NURSERY CHAIR.

Made of Extra Fine White Willows.

PER DOZEN.
No. 82—Seat, 8 x 11 in.; Back, 13 in. high........$18.90

FANCY NURSERY CHAIR.

Made of Extra Fine White Willows.

PER DOZEN.
No. 83—Extra Large; Seat, 9 x 12 in.; Back, 14 in. high........................$23.55

No. 90.
In Nests of 3. ...$2.00
FANCY DOLL HOOD CRADLE.
Small, 18 inches long; Medium, 21 inches long; Large, 23 inch

FANCY WILLOW AND REED NURSERY CHAIR.

PER DOZEN.
No. 520—Willow Nursery Chair; Seat, 9 x 11 in.; Back, 21 in. high.............$24.00
No. 510—Reed Nursery Chair; Seat, 9 x 11 in.; Back, 21 in. high 16.00

CHILDREN'S HIGH CHAIRS.

Wood Seat. Gloss Finish.	Wood Seat. Gloss Finish.	Cane Seat. Gloss Finish.	Cane Seat. Veneer Back. Polished.
No. 735	No. 745	No. 705	No. 185
Golden Oak.	Golden Oak.	Golden Oak.	Quartered Golden Oak.
Per Dozen $30.00	Per Dozen $30.00	Per Dozen $30.00	Per Dozen $42.00
Shipped with Base Detached.	Shipped with Base Detached.	Shipped Set Up.	Shipped with Base Detached.

CHILDREN'S CHAIRS.

Child's Rocker. Cane Seat. Gloss Finish.	Child's Rocker. Wood Seat. Gloss Finish.	Child's Rocker. Wood Seat. Gloss Finish.	Child's Rocker. Wood Seat. Gloss Finish.
No. 724	No. 164	No. 744	No. 734
Golden Oak..	Golden Oak.	Golden Oak.	Golden Oak.
Per Dozen $24.00	Per Dozen $20.00	Per Dozen $24.00	Per Dozen $24.00

ROCKERS AND CHAIRS.

Mammoth Rocker, Double Cane Seat and Back.

Rattan Seat and Back. Gloss Finish.

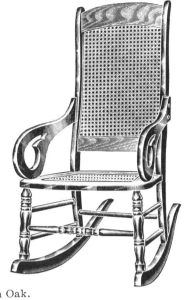

Golden Oak.
Cane Seat and Back. Gloss Finish.

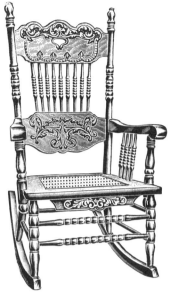

Cane Seat. Gloss Finish.

FOUR PIECE PARLOR SUITS.

Veneer Seat and Back, Marquetry Effect. Polished.

Veneer Seat and Back, Marquetry Effect. Polished.

Veneer Seat and Back, Marquetry Effect. Polished.

Mahogany
Veneer Back, Marquetry Effect, Hand Turned. Polished.

Veneer Back, Marquetry Effect, Hand Turned. Polished.

Veneer Back, Marquetry Effect, Hand Turned. Polished.

Veneer Back, Marquetry Effect, Hand Turned. Polished.

LADIES' ROCKERS.

  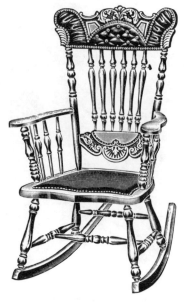 

Veneer Back, Mahogany Finish, Upholstered in Olive or Maroon Fabrikoid Leather. Polished.	Upholstered Seat and Tufted Pad Back in Olive or Maroon Fabrikoid Leather or Green, Red, Blue or Brown Fancy Velours. Gloss Finish.	Upholstered Seat and Back in Olive or Maroon Fabrikoid Leather or Green, Red, Blue or Brown Fancy Velours. Gloss Finish.	Genuine Leather. Gloss Finish. COBBLER SEAT.
No. 2594 Quartered Golden Oak. Per Dozen $114.00	**No. 2544** Golden Oak Finish. Per Dozen $72.00	**No. 2534** Golden Oak Finish. Per Dozen $78.00	**No. 754** Golden Oak Finish. Per Dozen $44.00
No. 2594 Curly Birch, Veneer Back, Mahogany Finish. Per Dozen $114.00			**No. 754** Mahogany Finish. Per Dozen $44.00

ROCKERS.

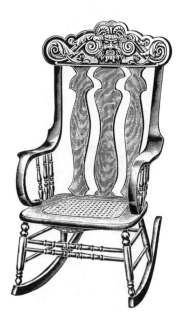 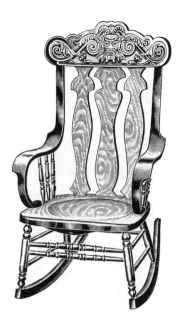  

Cane Seat. Gloss Finish.	Wood Seat. Gloss Finish.	Cane Seat. Gloss Finish.	Wood Seat. Gloss Finish.
No. 464 Golden Antique. Per Dozen $54.00	**No. 464** Golden Antique. Per Dozen $48.00	**No. 434** Golden Oak. Per Dozen $40.00	**No. 434** Golden Oak. Per Dozen $36.00

MORRIS CHAIRS.

| Reversible Cushions in Green, Red, Blue or Brown Fancy Velours. Polished.

No. 2664

Golden Oak Finish.

Per Dozen..........................$120.00 | Reversible Cushions in Green, Red, Blue or Brown Fancy Velours. Polished.

No. 2674

Golden Oak Finish.

Per Dozen..........................$192.00 | Reversible Cushions in Red, Green, Blue or Brown Fancy Velours. Polished.

No. 2654

Golden Oak Finish.

Per Dozen..........................$288.00 |

LADIES' ROCKERS.

 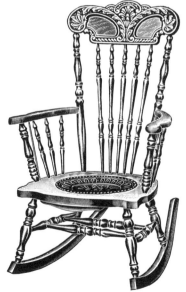  

| Heavy Turnings, Beautifully Embossed. Gloss Finish.

VENEERED SADDLE SEAT.

No. 544

Golden Oak.

Per Dozen................$50.00 | Genuine Leather, High Back. Gloss Finish.

COBBLER SEAT.

No. 264

Golden Oak.
Per Dozen................$40.00

No. 264/CM.
Mahogany Finish.
Per Dozen................$40.00 | Genuine Leather, High Back. Gloss Finish.

COBBLER SEAT.

No. 224

Golden Oak.

Per Dozen................$42.00 | High Back. Gloss Finish.

WOOD SEAT.

No. 264

Golden Oak.

Per Dozen................$32.00 |

DINING ROOM CHAIRS.

   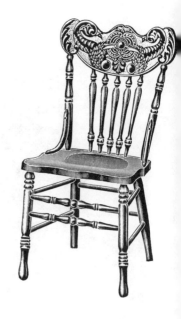

Golden Oak.

| Wood Seat. Gloss Finish. | Cane Seat. Gloss Finish. | Wood or Cane Seat. Gloss Finish. | Wood Seat. Gloss Finish. |

   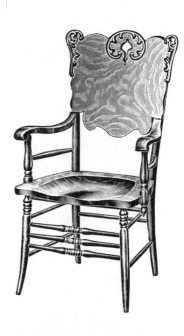

| Upholstered Seat in Olive or Maroon Fabrikoid Leather. | Upholstered Seat in Olive or Maroon Fabrikoid Leather. | Saddle Seat, Veneer Seat | Saddle Seat, Veneer Seat |

| Cane Seat or Upholstered Seat over Cane in Olive. | Cane Seat or Upholstered Seat over Cane in Olive. | Cane Seat or Upholstered Seat over Cane in Olive. | Cane Seat or Upholstered Seat over Cane in Olive. |

FANCY ROCKERS AND ARM CHAIRS.

| Hand Carved, Polished. | Hand Carved, Polished. | Hand Carved, Polished. | Hand Carved, Polished. |

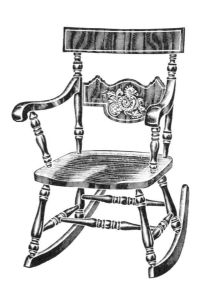

| Veneer Seat and Back, Hand Carved, Polished. | Veneer Seat and Back, Hand Carved, Polished. | Veneer Seat and Back, Marquetry Effect, Polished. | Veneer Seat and Back, Marquetry Effect, Polished. |

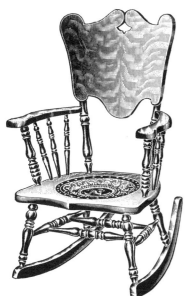

| Veneer Back, Polished. | Hand Carved, Polished. | Polished. | Raised Carving, Polished. |
| COBBLER SEAT. | COBBLER SEAT. | COBBLER SEAT. | COBBLER SEAT. |

OFFICE CHAIRS.

Veneer Seat and Back. Polished.
Revolving and Tilting.

SADDLE SEAT.

No. 567/SG.

Quartered Golden Oak.

Per Dozen $120.00

Wood Seat. Gloss Finish.
Revolving and Tilting.

No. 467/XG.

Antique Finish.

Per Dozen $78.00

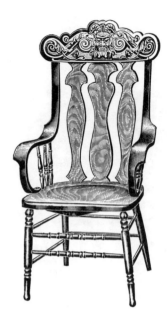

Wood Seat. Gloss Finish.

No. 465/XG.

Golden Antique Finish.

Per Dozen $42.00

OFFICE CHAIRS.

Wood Seat. Gloss Finish.
Revolving and Tilting.

No. 437/XG.

Golden Oak.

Per Dozen $72.00

Shipped with Base Reversed.

Wood Seat. Gloss Finish.

No. 435/XG.

Golden Oak.

Per Dozen $32.00

Shipped with Base Detached.

Wood Seat. Gloss Finish.
Revolving and Tilting.

No. 17/XG.

Golden Oak.

Per Dozen $72.00

Shipped with Base Reversed.

Wood Seat. Gloss Finish.

No. 15/XG.

Golden Oak.

Per Dozen $36.00

Shipped with Base Detached.

OFFICE CHAIRS.

Veneer Seat and Back. Polished. Revolving and Tilting. **No. 647** Quartered Golden Oak. Per Dozen...........$102.00	Upholstered Seat in Olive or Maroon Fabrikoid Leather. Veneer Back. Polished. Revolving and Tilting. **No. 1807** Quartered Golden Oak. Per Dozen...........$120.00	Veneer Seat and Back. Polished. **No. 645** Quartered Golden Oak. Per Dozen...........$66.00

OFFICE CHAIRS.

  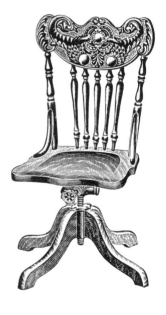 

Veneer Seat and Back. Polished. Revolving and Tilting. **No. 646** Quartered Golden Oak. Per Dozen........$72.00	Wood Seat. Gloss Finish. Revolving and Tilting. **No. 636** Golden Oak. Per Dozen........$60.00	Wood Seat. Gloss Finish. Revolving and Tilting. **No. 626** Golden Oak. Per Dozen........$57.00	Wood Seat. Gloss Finish. Revolving and Tilting. **No. 86** Golden Oak. Per Dozen........$51.00

ANTIQUE SEWING MACHINES

by Gene Bishop

A new area of collectibles — old sewing machines — is now firmly established in the United States. Although there is only a small band of serious collectors, a much larger group of dealers, auctioneers, and private citizens do recognize the value in these fine examples of a bygone mechanical era. This is due largely to a proliferation of recent articles and reports of record-breaking auction prices abroad, especially in Germany, where precision engineering is almost worshiped as a form of fine art. All this publicity is bringing many old machines out of the woodwork here in America.

However, museums, auction houses, and advanced collectors are interested only in sewing machines which were manufactured before 1875. This includes both treadle and handcrank, or portable, types. Those made after 1875 are simply too common. A highly desirable, very rare machine is defined as an early make that was produced in extremely small numbers, or a low production model of a more commonly found make. For example, the Landfear sewing machine was manufactured in 1857, and only one example is known today. A machine such as this is considered extremely rare, and could command a price upwards of $3,000. Another machine in this class is the Singer Model No. 1, which was manufactured from 1851 to 1858. Although more of these machines are in existence, being the earliest Singer makes it a much sought after example. Another of value is the 1857 Grover & Baker portable. Again, a fair number of these machines is known, but it is considered rare because it is the very first mass-produced portable machine. Wilcox and Gibbs and the American Sewing Machine Company manufactured literally thousands of treadle machines in the 1870's, and the standard iron-base treadle model is abundant today. However, the early all-wood full-length cabinet model of these makes is not. A mint condition version in solid walnut or mahogany is very rare.

A good example of an early machine that is somewhat desirable, but not rare, is the Wheeler & Wilson, manufactured from 1856 to 1876. All models of these machines were produced in great numbers, and are available today in sufficient quantities. They are fine examples of crude, early machines, but are quite common. The following is a list of U.S. machines that fall into the rare category (some of which I may wish to purchase):

The list is not complete, but it is an attempt to publish a basic guide to rarity. This guide will do until we have gathered enough information to publish a comprehensive book-length price guide. The writer of this article requests anyone having any information (printed matter, photographs, or actual examples of the machines listed above) to please write to Gene Bishop at P.O. Box 126, Greenwood, SC 29648.

Aetna, Akins & Felthousen, American Gem, American Magnetic, Atlantic, Atwater, Avery, Bartholf, Bartman & Fanton, Beckwith, Blees, Blodgett & Lerow, Boudoir, Bradford & Barber, Bradshaw Shuttle, Brattleboro, Bruen, Buell, Burnet & Broderick, Chicopee, Clark, Clark Revolving Looper, Clinton, Dorcas, Dulaney, Durgin, Elliptic, Empire, Eureka Shuttle, Excelsior, Fairy, Finkle, M., First & Frost, Fosket & Savage, Gardner, Greenman & True, Grover & Baker, Hancock, Home (w/treadle), N. Hunt, Hunt & Webster, Johnson, Ladd & Webster, Ladies Companion (see Pratt), Landfear, Langdon, Lathrop, Leavitt, Lester, Little Giant, Little Monitor, Manhattan, Meyers, Moore, Morey & Johnson, Ne Plus Ultra, Nettleton & Raymond, New York, Old Dominion, Pardox, Parham, Parker, Pearl, Pratt, Quaker City, Robertson (Brass Dolphin & Cherub), Robinson, Secor, Sewing Shears (Hendrocks Patent), Shaw & Clark, Singer (only the first three models), Taggart & Farr, Watson, Waterbury, West & Willson, A.B. Wilson, Whitney & Lyons, Wickersham, Williams & Orvis, and Windsor.

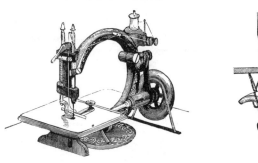

WILLCOX & GIBBS'

On page 322 you will find "A Brief History of the Sewing Machine," but no mention of Isaac M. Singer is made in this 1876 commemorative advertising piece by The Howe Manufacturing Company. Singer was the third American to "invent" a practical sewing machine (patented in 1851). Walter Hunt had developed a straight-seam sewer in 1838, but didn't patent his model because he "did not want to put thousands of seamstresses out of work."

Singer was late on the scene, and was more an "improver" than an inventor. He added a crude wooden foot treadle to Howe's 1845 patent handcrank, and used an up-and-down needle instead of the side-to-side type, thus enabling the easy sewing of curved seams. Singer's supreme talent (in addition to being an accomplished machinist) was that of aggressive promotion. He was one of America's first mass merchandisers. By 1863, Singer was selling 21,000 machines a year. He put on constant demonstrations, and created an innovative installment payment plan. Singer even rented his machines out to reluctant seamstresses.

Although (after many years of patent litigation) the courts made him pay Mr. Howe a royalty of five dollars for each machine he sold, Mr. Singer was able to retire to Europe with a vast fortune in 1863. His estate, at his death, was valued at 13 million dollars. Elias Howe made millions of dollars, too, but spent most of the money on lawyers, and a large personal cash contribution to his Civil War regiment's payroll. He did not (as has been reported) die in poverty, but his young wife did, just before he struck it rich.

The following are additional pre-1875 machines of interest to collectors.

MACHINE	*PAGE	HANDCRANK	TREADLE	VALUE
Singer Turtleback	p. 34		✓	$1500
Grover & Baker	p. 36		✓	$1000
Grover & Baker	p. 37	✓		$ 600
Dolphin Machine	p. 47	✓		$2000
Cherub Machine	p. 48	✓		$2000
Foliage Machine	p. 49	✓		$1000
Horse Machine	p. 49	✓		$2000
Hendricks Sewing Shears	p. 49	✓		$1000
Hook Sewing Machine	p. 50	✓		$ 500
Madame Demorest	p. 53	✓		$ 500
Bartlett	p. 81	✓		$ 300
Beckwith	p. 82	✓		$ 500
McLean & Hooper	p. 83		✓	$ 400
Little Monitor	p. 84		✓	$ 600
Florence	p. 86		✓	$ 200
Empire	p. 88	✓		$ 350
Monitor	p. 89	✓		$ 350
Grant Brothers	p. 90	✓		$ 200
Hancock	p. 92	✓		$ 400
Howe	p. 95			$ 150
Ladd & Webster	p. 97		✓	$ 700
Improved Common Sense	p. 98	✓		$ 100
Johnson	p. 98		✓	$ 500
Lady	p. 99		✓	$2000
Landfear	p. 100		✓	$3000
Lathrop	p. 101		✓	$1500
Leavitt	p. 103		✓	$ 650
New England (This is the most common early portable. Thousands survived.)	p. 107	✓		$ 100
Pratt	p. 108		✓	$2500
Quaker City	p. 109		✓	$1000
Shaw & Clark (top illustration)	p. 112	✓		$ 700
Shaw & Clark (bottom illustration)	p. 112	✓		$ 400
Singer Letter A	p. 114		✓	$ 500
Taggart & Farr	p. 117		✓	$ 750
West & Willson	p. 119		✓	$ 600
Wheeler & Wilson	p. 120		✓	$ 500
Wilcox & Gibbs	p. 123		✓	$ 100
Greenman & True	p. 131		✓	$ 750
Wheeler & Wilson	p. 131		✓	$ 100
Elliptic	p. 131		✓	$ 300
Secor	p. 132		✓	$ 500
Home Shuttle	p. 132		✓	$ 150
Weed	p. 133		✓	$ 100
Singer New Family	p. 134		✓	$ 50
Florence (figure 160)	p. 135		✓	$ 250
Florence Cabinet Model (figure 159)	p. 135		✓	$ 500
Wilson Shuttle	p. 137		✓	$ 100

NOTE:
These are collector values. An antique dealer might ask more money for a nicely finished treadle a collector would deem common.
* *Page numbers refer to illustrations in* **The Sewing Machine — Its Invention and Development** *by Grace Rodgers Cooper (1980 edition), published by The Smithsonian Press, Washington, DC.*

A BRIEF HISTORY
OF THE
SEWING MACHINE,

ELIAS HOWE, JR.
INVENTOR OF THE SEWING MACHINE.

While Howe was attempting to introduce his machine to notice; the attention of inventors began to be turned to the subject of sewing machines, and patents for improvements, modifications, or new arrangements of the parts began to flow in a steady stream from the Patent Office. Between the year when Howe's patent was issued to the year 1871 nearly one thousand different patents relating to sewing machines have been issued.

The above is a correct representation of the first Sewing Machine. It was constructed by Elias Howe, Jr., and in April, 1845, sewed the first seam ever made by machinery.

The earliest patent which appears to have been granted for a machine to improve or facilitate the process of sewing, was granted in England, on the 24th of July, 1755, to Charles F. Weisenthal, for an improved method of embroidering. Under this patent he claimed a needle, pointed at both ends, and having the eye in the middle, so that it could be passed both ways through the cloth without being turned round. The next patent was granted to Robert Alsop, in 1770, for the use of two or more shuttles in embroidery, their purpose being to secure the stitches. In 1804 John Duncan took out a patent for an improved process by the use of barbed or hooked needles, by which the loops were made and secured somewhat as the stitch is made in the single-thread sewing machine. In 1807 James Winter patented in England an appliance for sewing leather gloves, the importance of which here arises only from the fact that the material was held in position by metallic jaws, thus leaving the operator's hands free. On July 17, 1830, a French patent was granted to M. Thimonier for a machine to do crochet work, which could also be applied to sewing. In this machine a hooked needle was used. In 1848 this machine was improved by M. Maguin, a partner of the inventor, and in 1851 was exhibited in the great London World's Fair of that date. None of these machines, however, were intended really for the purpose which the sewing machine performs, and are mentioned here simply because each of them in turn was a partial step in the use of some mechanical process, which was afterwards introduced in the sewing machine.

In the Patent Office at Washington is the model of a "machine to sew a straight seam," which was patented February 21, 1842, by James Greenough, of Washington. This machine made what is known as the "shoemaker's stitch." The needle was made with the eye in the centre, and pointed at both ends, being pushed through and then drawn back by means of pinchers. In 1843 other patents were granted to G. R. Corliss and B. W. Bean. Bean's machine worked by crimping the material, by running it through corrugated rollers, and then sewed by thrusting a needle through the folds, thus, in fact, basting it. Another machine was patented in 1844, by Rogers. The next year, 1846, Elias Howe, Jr., patented his, on September 10. This was the first practicable machine for sewing.

Though not patented until this year, Mr. Howe had invented the machine some years before, and working without the knowledge of what had been done before by others, he had used some devices which others had used, but had so combined them in novel shapes or arrangements that the machine, as a whole, was entirely his own invention. His patent claims, substantially, the use of a needle with the eye in the point, and a shuttle for the purpose of uniting two edges in a seam, or their equivalent, making the stitch by interlocking two threads. He improved his machine as originally invented, but failed in exciting sufficient attention to it, either in the United States or in England, to raise the capital necessary for its successful introduction into popular use. His attempts to do this exhausted his means, and reduced him to great poverty.

Though he afterwards received very large amounts of money from the subsequent inventors, who manufactured their machines under a royalty to him for the use of the appliances governed by his patent, yet the heavy expenses of the lawsuits he was forced to undertake to enforce his claims absorbed so much of the money he received that he died in comparative poverty.

1876

The above cut represents the Folding Cover as it appears when open.

BORDERED TABLE WITH BOX TOP AND SIDE DRAWERS.

BOX TOP AND TABLE.

CABINET CASE.

NEW-YORK, NOVEMBER 1, 1851.

SINGER'S SEWING MACHINE.—Fig. 1.

The accompanying engravings represent a perspective view, figure 1. and a side elevation, figure 2, of Isaac M. Singer's Sewing Machine, which was patented on the 12th of last August. A is the frame, made of cast-iron. and B is a cast-iron standard to support part of the working machinery. C is a large driving wheel, worked by the handle, D. E is a small second wheel, driven by C, and works the shaft that vibrates the needle; E' is another wheel to work the shuttle shaft, a, hung in the bearing straps, b b, fig. 2. F is a round plate on the revolving shaft of E; it has a small roller stud on its inner face fitting into a plate, G, slotted of a heart-shape, to answer the purpose of a cam. This plate, G, is secured to the vibrating arm. H. to which the needle, I, is fastened. The needle performs three strokes up and down during one revolution of the large wheel, C. The thread, J, of the needle is supplied by a bobbin. K. and goes through an eye in the needle, near its point. The cloth is laid flat on the table on the top of a small rough-faced roller S, with the edge to be sewn under the needle. The cloth is held down by a pad, R, acted upon by a coiled spring. P:

THE SINGER
MANUFACTURING COMPANY,
AT THE WORLD'S FAIR,
(CONSTITUTED BY THE HOMES OF THE PEOPLE,)

RECEIVED THE GREAT AWARD
OF THE
HIGHEST SALES!
AND HAVE LEFT ALL RIVALS FAR BEHIND THEM, FOR THEY
SOLD IN 1870
One Hundred and Twenty-seven Thousand
Eight Hundred and Thirty-three Machines!

BLEES
NOISELESS,
LOCK-STITCH
Sewing Machine
Challenges the world in perfection of work, strength and beauty of stitch, durability of construction, and rapidity of motion. Call and examine. Send for Circular. Agents wanted.
MANUFACTURED BY
BLEES
Sewing Machine Co.,
623 Broadway, N. Y.

THE FIRST SINGER

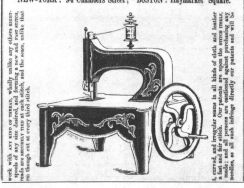

SEWING MACHINES.
Patented Feb. 11, 1851; June 22, 1852; Feb. 22, 1853.
GROVER, BAKER & CO.
PATENTEES AND PROPRIETORS,
NEW-YORK: 94 Chambers Street; BOSTON: Haymarket Square.

work with any kind of thread, wholly unlike any others regarding spools of any size desired, and forming a new and fast seam, reads are securely tied at each stitch, and the seam, unlike that m though cut at every third stitch.

curved and irregular seams in all kinds of cloth and leather a fast and fair stitch. Our patents are upon the several means made; and all persons are cautioned against purchasing sewing needles, as all such infringe directly our patents and will be

THE STANDARD MACHINE OF THE WORLD!

WHEELER & WILSON!

WHEELER & WILSON'S are the Sewing Machines to Buy!

They are the Most Simple Machines. They are the Most Durable Machines. They do the Greatest Variety of Work. They are the Highest Premium Machines. They are Noiseless.

Go and see them operate and you will be satisfied.

BUY NO OTHER.

BOLGIANO'S PERFECTION HOT AIR GAS IRON.

THE BOLGIANO MANUFACTURING CO.
OF BALTIMORE CITY,
Peabody Fire Insurance Building, Room 25,
415 WATER STREET.

JOHN BOLGIANO, - - President.
CHAS. J. TAYLOR, - Secretary.
B. G. WHITE, - - Superintendent.

BOLGIANO'S NO. 1 WATER MOTOR.

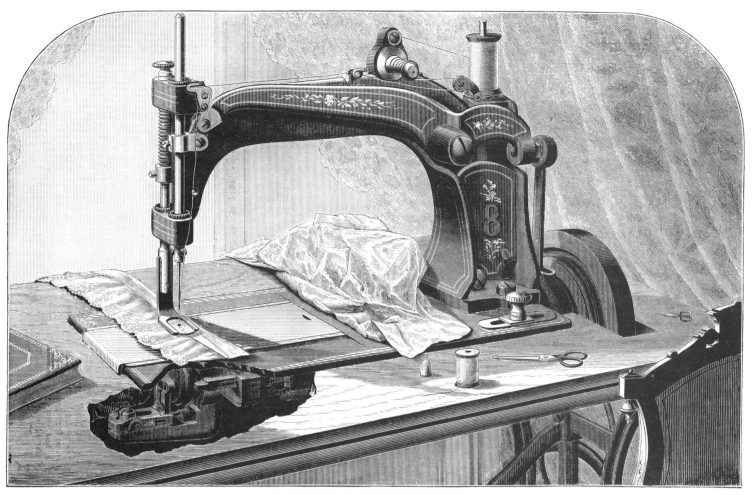

THE WHEELER AND WILSON SEWING MACHINE.

THE WILSON SEWING MACHINE.

A GREAT BOON.
A Good Cheap Sewing-Machine at Last.

We have been offering as a Premium for some months past the

Beckwith Sewing Machine,

which has been fully described in previous numbers of this paper. We have already given and sold some hundreds of these machines, and testimonials of satisfaction are coming from every quarter.

We now offer the **Beckwith Sewing-Machine, Improved,** price $12. A new and very simple braiding foot has been made, by which a child can sew on braid without the least trouble, following any desired pattern with ease; also a new arm, spiral spring and lever for raising the presser foot, all of which are now set in a position that leaves the needle free to be threaded. The joint is much enlarged, and the machine is otherwise greatly strengthened and improved. The use of the braider foot alone will be valued more than the cost of the machine.

THE
Heberling Shirring and Gathering Machine.
(Improved March 1, 1880.)

The only Machine in the world that makes the Hand-Running Stitch. Simple, Practical and Perfect in operation. Doing a large range of material; such as Silks, Satins, Velvets, Alpacas, Cashmeres, White Goods, &c., with one or two rows of stitching at the same time.

Work more regular and beautiful than can be done by hand. Shirrs or gathers adjusted full or scant by drawing thread same as hand work. Will do the work of ten or fifteen hands. For Machines or information in regard to the business, address

HEBERLING B. S. S. MACHINE CO.,
Mt. Pleasant, *Jeff.* Co., Ohio.

J. B. LONG, 42 and 44 Madison St., Chicago, Ill., Gen'l Western Agent. I. A. KINGSBURY, 44 E. 14th St., New York, Gen'l Agt. for State of N. Y.

THE Most Popular OF ALL SEWING MACHINES IS THE Light Running NEW HOME.

Rapidly taking the place of all other Machines wherever introduced.

200,000 SOLD YEARLY

Has more Points of Excellence than all other Machines combined.

LIBERAL INDUCEMENTS TO DEALERS.

NEW HOME SEWING-MACHINE CO.,
30 UNION SQUARE, NEW YORK.

MRS. HENRY WARD BEECHER TO HER READERS.

Of course every farmer must have a reaper and mower, a horse-rake and a drill, and all the other labor-saving inventions, just as rapidly as he can pay for them, even though they are stowed away as dead property the greater part of the year. Such things are necessary, and the wife, perhaps, gets along with a new dress less, that the husband may have a labor-saving machine. But how about labor-saving machines and inventions indoors? A woman who has a sewing machine, a machine to do the churning, and one servant as "help" is usually the most favored housekeeper in the neighborhood. You may look a whole township through, and in nine cases out of ten, the labor-saving affairs are brought to the help of man first. It is no wonder that good mothers are so scarce, stepmothers so common, and invalid women not a rarity. "A woman's work is never done," is a true proverb, with by far too many housewives.

Sewing Machines, &c.

THE LARGEST, THE SIMPLEST. THE DAVIS VERTICAL FEED SHUTTLE SEWING MACHINE. THE CHEAPEST THE BEST

Possesses all the desirable qualities of the Standard Machines in the market. In its *Capacity* — being the LARGEST Family Machine made. In its *Simplicity* — being composed of but THIRTEEN WORKING PARTS. In its *Adaptability to a wide range of work.* In its *Ease of Operation* — running light and quiet, and being easily comprehended. In its *Superior Construction and Beauty of Style and Finish.*

BUT ITS PRINCIPAL FEATURE IS ITS

VERTICAL FEED!!

which is the most practical and desirable device for the purpose possessed by any Machine, giving THE DAVIS the preference, and which the Manufacturers claim makes it

SUPERIOR TO ALL OTHER MACHINES.

THE DAVIS has been before the public nearly Ten Years, and unlike other Machines, has not been puffed into notoriety, but in a quiet way has earned a great reputation on account of its *many desirable qualities*.

☞ Agents are desired in every County in the United States and Canadas, not already occupied, to whom the most liberal terms known to the trade will be given, by addressing the Manufacturers,

THE DAVIS SEWING MACHINE COMPANY,
of Watertown, N. Y.

THE BECKWITH
PORTABLE
Family Sewing Machine,
Price $20.

On 30 days Trial. $20 refunded in 30 days on return of Machine, if desired.

With Strength, Capacity and Speed equal to any, regardless of cost. With Semi-Guiding Feed and Automatic Stitch Fastener. All other Machines require the movement of from 25 to 30 pieces to every stitch—this requires but Two! Hence it is a symbol of symplicity and strength; WITH NO TOILSOME TREAD OF THE TREADLE. For full particulars send for Circular—then buy no other until you see the Machine, for "seeing is believing." Agents wanted *in every town in the country*. If $5 are sent with the order the balance can be C. O. D. Agents must pay full price for single Machines, per centage on first to be deducted when six Machines are paid for. Terms to agents, cash with order, or C. O. D.

BECKWITH S. M. Co., 862 Broadway, N.Y.
Near 17th St. SEND FOR CIRCULAR.

THE FLORENCE

Will sew everything needed in a family, from the heaviest to the lightest fabric.

IT DOES MORE WORK,
MORE KINDS OF WORK,
AND BETTER WORK,

Than any other machine.

If there is a Florence Sewing Machine within one thousand miles of San Francisco not working well and giving entire satisfaction, if I am informed of it, it will be attended to without expense of any kind to the owner.

SAMUEL HILL, Agent,
19 New Montgomery Street,
Grand Hotel Building, San Francisco.

Send for Circulars and samples of the work. Active Agents wanted in every place.

THE PERFECTION AUTOMATIC CHAIN SINGLE STITCH HAND MACHINE.

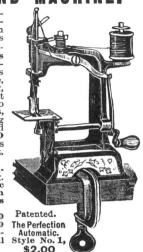

THIS MACHINE IS NOT A TOY OR EXPERIMENT, but a well built machine, a perfect and practical sewer, thousands of which are in use throughout the country. It does the work of the regular chain stitch machines which are usually very high priced; it is built on the same principles and is capable of a great variety of work with any kind of thread, silk or twist, and on all kinds of cloth.

WHERE A FULL SIZED FAMILY MACHINE cannot be afforded or is not needed, this machine will be found a very good substitute, as it answers every purpose of a hand machine. To families possessing machines it is a household necessity, as it can be carried about and is always ready for use on porch, table, arm of chair or window sill, and for fancy sewing such as embroidering, outlining, braiding, etc., in which the chain stitch is so desirable. It is an excellent educator to children in the art of sewing, and is especially needed by unmarried men and women who have not easy access to family sewing machines. **IT MAKES A BEAUTIFUL CHAIN STITCH** and sews muslins, linens, flannels and lawns, in fact, any goods, in a surprising manner. It is well made, strong and durable, and most simple in construction. No getting out of order. Highly improved with automatic tension, stitch and feed regulator, self setting needle, etc. **THERE IS NO COMPARISON** between the Perfection and other hand chain stitch sewing machines; it is absolutely the most perfect, most improved and satisfactory machine of its kind on the market. Will replace free any parts that may prove defective.

FURNISHED COMPLETE READY FOR USE with needle and thread, packed in wooden box with full instructions for operating.

Extra Needles. 10 in package for 25c

STYLE No. 2 IS OUR LATEST MODEL and is fitted with an improvement which makes its speed of operation four times greater than that of style No. 1, and almost equal to that of a stand machine.

No. 26T80 Perfection Hand Machine, style No. 1............................$2.00
No. 26T82 Perfection Hand Machine, style No. 2............................ 3.50

WE GUARANTEE THIS MACHINE to be exactly as represented. If is not found entirely satisfactory it can be returned and we will refund money and charges.

The Perfection Automatic. Style No. 2, $3.50

Patented. The Perfection Automatic. Style No. 1, $2.00

OUR NEW QUEEN $5.95 IMPROVED HAND MACHINE.

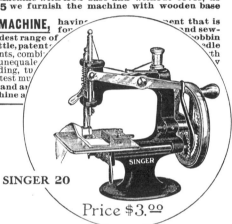

No. 26T90 $5.95 AND $7.85

AT $5.95 we furnish our New Queen Hand Machine with iron base and without cover, as illustrated on the left. For **$7.85** we furnish the machine with wooden base and complete with a fine bent wood cover.

THIS IS A FIRST CLASS, RELIABLE MACHINE, having every improvement that is found on the best and most expensive hand sewing machines, and capable of doing the widest range of work, having automatic bobbin winder, self threading, vibrating steel shuttle, patent needle clamp, tension liberator, all the latest improvements, combined with in construction, speed and light running qualities, unequaled for wide range of working, hemming, felling, binding, tucking, adapted to every variety of sewing, from the lightest muslin.

THE BEARINGS are of the best hardened steel and are so true to the adjustment of this machine and we admit that this machine is equaled by any hand machine on the market, regardless of price or name. One particular point of its superiority lies in its feed, which is the four-motion feed, the same that is used upon our high grade stand machines. This feed is absolutely positive, its movements being regulated by the eccentric lever bar, and does not require the use of coiled springs to obtain the four movements required of the feeding mechanism. The majority of other hand machines use the spring feed, which readily becomes weak and fails to act properly.

OTHER POINTS OF EXCELLENCE lies in its self setting needle, positive stitch regulator, and a device by which the gearing is readily released, thus enabling the operator to wind the bobbin without operating the working parts of the machine. The hand wheel is arranged with a belt groove, so that the machine can be placed on a sewing-machine stand and operated with the treadle in the ordinary manner. We furnish an instruction book and a full set of accessories free of charge with the machine.

No. 26T90 New Queen Hand Machine, with iron base, no cover. Our special price..........$5.95
Full set of attachments, extra... .70
No. 26T95 New Queen Hand Machine, with wood base and cover. Our special price....... 7.85
Full set of attachments, extra... .70

Free on board cars at our factory in Central Illinois. See our liberal terms of shipment on page 657.

SINGER 20

Price $3.00

$10 SPENSER AUTOMATIC SEWING MACHINE $10

The new **Spenser Automatic Hand Sewing Machine** is easily operated and does rapid work, making three stitches at each revolution of the large wheel called the Speeder. **Cloth-plate large**, and sufficient space under the arm for making a skirt or other garment. Full printed instructions with each machine, so that even a child can operate it.

The **Spenser** is attached to a table, arm of chair or other convenient place by a clamp furnished with machine. It does perfect work on thin or thick goods, makes short or long stitch, and uses any number cotton from No. 8 to 300 without change of tension, as the **tension is automatic** and cannot be changed. Extra needles furnished free with machine. The **Spenser** is very strong and durable and simple in construction. Guarantee with each machine. Order now, or descriptive booklet will be mailed on request. Remit by personal check, cashier's check, money order, or registered letter.

If on two weeks' trial you are not fully satisfied, return machine at our expense and we will refund full price paid. Express is prepaid by us.

SPENSER SEWING MACHINE CO.
208 Tremont Street - Boston, Mass.

THE GREATEST HELP EVER OFFERED TO DRESSMAKERS.

FREE ON TRIAL.
BEAUTIFUL WORK.
Write for Particulars.

H. A. HANNUM & CO.,
Cazenovia, N. Y., U. S. A.

AUTOMATIC HAND SEWING MACHINE.

Practical, simple in construction does work nicely. Wt. 2½ lbs. Can be attached to chair, table or other furniture. Guaranteed to give entire satisfaction.

H125. —"Midget" Automatic Tension. Ht. 7½ in., length 4, width 2½. Use Wilcox & Gibbs needles, makes chain stitch as perfectly as a $75 machine. Has feed and stitch regulator, does all kinds of plain sewing, spool of thread on each machine. Frame finished in black japan and decorated with wreath design, sewing plate nickel plated. Each carefully adjusted at factory. Each in sliding cover box. Each, **$1.25**

H100 — "Gem." Automatic tension, speed attachment, stitch and feed regulator, self-setting needle. Machine japanned, bright parts polished and nickel plated, clamp for fastening. A child can operate it, as needle is entirely protected. Each adjusted and tested before leaving factory. Complete with needle, spool of thread and instructions. Does great variety of work with any kind of thread or twist and on all kinds of cloth, also makes chain stitch. In sliding cover wood box. In lots of 5 or more............Each, **$1.35**

Note.—Wilcox & Gibbs needles fit this machine *See Notion Dept.*

F7911 — 6 in. wide, floral decorated, nickel finish wheel.
1 in box..... Each **$1.00**

F7912 — 7 in. wide, floral decorated, polished nickel wheel.
1 in box.... Each **$2.00**

F7913—11 in. wide, gold decorated, polished nickel wheel, large steel tension, turning presser foot for easy threading, double drive gearing. Complete with small tools.

1897 1897

J. H. SUTCLIFFE & CO.'S
NEW IMPROVED
"SUNRAY" SEWING MACHINE

• • • •

New Model High Arm,
Self-Threading Shuttle,
Automatic Bobbin Winder.

• • • •

New and Important Improvements for this Season.

<u>We Warrant each Machine for ten years.</u>

There is no better machine made at any price. We allow you liberal time for trial, and if not satisfactory we refund your money.

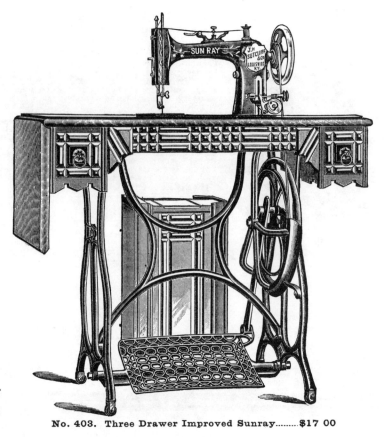

No. 403. Three Drawer Improved Sunray.........$17 00

22,160 IN USE BY 22,160 WELL PLEASED PURCHASERS.

The "Sunray" machine with the recently added improvements stands forward to-day as the peer of any sewing machine made, no matter what name or make. We guarantee it fully and are selling you a machine the finish and construction of which is unsurpassed. Read our very liberal offer; all we ask is a fair trial. Our sales so far this year are double for the same length of time last year. We hope with the recently added improvements to make a still greater increase in our output.

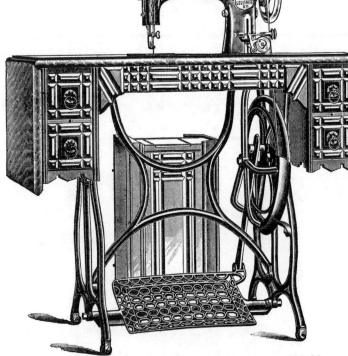

No. 405. Five Drawer Improved Sunray.........$18 00

ALL IMPROVEMENTS TO DATE.
DETAILS OF CONSTRUCTION

THE HEAD—Is five and one-half inches high and eight and one-half inches long (inside measurement). This is as high as any family machine made and must not be confounded with the medium or so-called high arm machines. The bed plate sets into or level with the table. Finely japanned and decorated.

THE NEEDLE—Is straight, self-setting, flat on one side and can not be set wrong.

THE SHUTTLE—Is open on the end, cylinder shape and absolutely self-threading.

THE BEARINGS—Are all steel, well fitted and adjustable.

SELF-THREADING—The machine is self-threading and no hole to put the thread through, except the eye of the needle.

THE FEED—Is positive in action and has no spring to get out of order.

AUTOMATIC BOBBIN WINDER—Will wind the bobbin as smooth as on a spool of thread. A wonderful device. Improved over the old model.

THE STITCH—Is double lock stitch. The same on both sides, and will not ravel. Can be lengthened or shortened from eight to thirty stitches to the inch.

THE TENSION—Is a flat spring tension and is adjustable to all sizes of thread. Easy and rapid adjustment.

HAND WHEEL—Is nickel-plated and has loose wheel attached to operate bobbin winder without running machine.

THE MOVEMENT—Is the celebrated eccentric movement, positive in action and fewer working parts than any machine on the market.

THE STAND—Has large balance wheel, hangs on two adjustable centers. The treadle is large and hangs on cone centers and all lost motion can be taken up. Casters under each end and easily moved. A nice dress guard over wheel to keep dress from getting on wheel.

ATTACHMENTS—Are the famous Johnson's set, in a tin box, and consist of tucker, ruffler, four hemmers, binder, quilter, foot hemmer, and feller, under braider and sheerer, oil can and oil, two screwdrivers, six bobbins, paper of needles, thumb screw, gauge and book of directions.

HOW WE SELL THEM.

No. 403 J. H. S. & Co. Improved "SUNRAY," Gothic wood work, 2 side drawers and one long center drawer, drop leaf and cover, complete $17 00
No. 405 J. H. S. & Co. Improved "SUNRAY," Gothic wood work, four side drawers and one long center drawer, drop leaf and cover, complete 18 00
No. 407 J. H. S. Co. Improved "SUNRAY," Gothic wood work, six side drawers and one long center drawer, drop leaf and cover, complete 19 00
Each machine is accompanied by a full set of attachments. Wood work is either oak or walnut. Price includes packing.

SPECIAL OFFER. READ THIS!

We have confidence in this machine and feel that if you see it you will like it. We are therefore willing to make you this special offer, viz.: Send us your order accompanied by full amount (for either machine) and we will PREPAY THE FREIGHT to any point east of the Rocky Mountains; you may run it 15 days and if it is not satisfactory you may return it and we will refund your money. Seeing is believing. Let us send you a "Sunray."

OUR HIGH GRADE, BALL BEARING, DROP LEAF AND COVER... HOWARD MACHINE

THE GREATEST VALUE EVER OFFERED IN A SEWING MACHINE. A BALL BEARING MACHINE WITH HIGHEST GRADE OF WOODWORK.

ALMOST EVERY TOWN has several dealers who sell sewing machines. There are also dozens of houses that are advertising sewing machines through the newspapers, and by the medium of catalogues and circulars, and to assist you in deciding as to where you should send your order to get the best machines that were ever put out for the money, we would suggest that you make a little inquiry in your own neighborhood among those who are now using our sewing machines. They have had the chance to test them and compare them with other makes and their opinion would doubtless satisfy you, and it will certainly satisfy us.

OUR $1.00 OFFER. To those who do not wish to avail themselves of the extra saving and convenience of sending cash in full with order, we will, on receipt of $1.00, send any sewing machine to any address by freight C. O. D., subject to examination, balance payable after received. For our liberal $1.00 C. O. D. offer see page 657.

THE HOWARD MACHINE cannot be sold by any other dealer, as we control the entire output at the factory and our prices are based on the actual cost of material and labor with our usual one small percentage of profit added.

DESCRIPTION OF THE HEAD.

THE HEAD, as shown in the illustration, has one of the highest arms made, of beautiful design, beautifully enameled and polished and richly decorated with handsome ornamentations. Has highly polished, nickel plated face plate; eccentric action; positive four motion feed; vibrating shuttle; automatic bobbin winder; stitch regulator; in fact, all the latest improvements that are embodied in any high grade machine.

THE STAND

is made of best quality of iron, handsomely japanned, and has a high polished finish. Has the genuine Osgood ball bearing attachment, as shown in the illustration on page 666, built on the same principle as a bicycle hub or pedal, making it one of the most noiseless and lightest running stands possible to produce.

THE WOODWORK.

THE WOODWORK of our Howard Machine is of the best quality quarter sawed oak, beautifully carved, decorated and ornamented. The cover and drawers of latest pattern bent wood veneer. The finish is the same as you will find on such woodwork as is used only on machines that sell at $40.00 to $65.00 and is not to be compared with the cheap cabinet work used by most dealers on machines advertised at $15.00 to $25.00.

ACCESSORIES.

WE FURNISH FREE with this machine a full set of accessories, usually furnished with high grade machines, consisting of one quilter, six bobbins, one cloth guide, one package of needles, one oil can (filled with oil) and two screwdrivers.

ATTACHMENTS.

WE LIST THE HOWARD MACHINE in three styles; three, five and seven drawers, without attachments; which enables you to procure the machine at the lowest possible price in case you do not require the attachments. We furnish them at 75 cents, which is the cost price. When attachments are included with the machines offered by other dealers, they not alone add the cost price of the attachments to the selling price of the machine, but also their profit on the attachments.

IF YOU DESIRE ATTACHMENTS, send 75 cents with your order for the machine and we will furnish a complete set of the latest patent, high grade foot attachments in handsome velvet lined metal box, as follows:
ONE RUFFLER, ONE TUCKER, ONE UNDER BRAIDER, ONE SHIRRING PLATE, ONE S[]
ONE BINDER, AND ONE SET OF HEMMERS OF DIFFERENT WIDTHS UP TO FIVE-EIGHTHS O[]

THIS TELLS JUST WHAT FREIGHT YOU WILL HAVE TO PAY: A sewing machine weighs, crated for shipment, 125 pounds. The railroad companies carry sewing machines at first class freight rate. On pages 7 to 11 you will find the first class freight rate for 100 pounds to a point nearest your town. The freight will be almost, if not exactly the same to your town, so you can tell almost to a penny what the freight will amount to. As a rule the freight on a sewing machine averages about 60 to 75 cents for 500 miles, 80 cents to $1.25 for 1,000 miles.

SUPPLIES
Needles, bobbins, shuttles, at the construction or operation of at any time at lowest prices. Par[] parts, manufactured by the make[] supply of each part constantly on[]

THE SINGER

DO NOT BUY ELSEWHERE BEFORE YOU HAVE SEEN AND TRIED ONE OF OUR

HIGH GRADE, HIGH ARM, BALL BEARING HOWARD MACHINES

and if we cannot convince you that our sewing machines are better than any machines that were ever sold by anyone at anything like the price, we will not expect you to keep the machine you order from us.

20 YEARS' WRITTEN BINDING GUARANTEE is given with every Howard Sewing Machine, by the terms and conditions of which, if any piece or part gives out within twenty years by reason of defect in material or workmanship, we will replace or repair it free of charge.

No. 26T40 Three-drawer and cover, without attachments. Price $14.00
No. 26T41 Five-drawer and cover, without attachments. Price 14.45
No. 26T42 Seven-drawer and cover, without attachments. Price 14.95

These prices are quoted for the machines free on board cars at the factory near Chicago.
A FULL SET OF ATTACHMENTS, AS ILLUSTRATED AND DESCRIBED ON PAGE 666, FOR 75 CENTS.
NOTE TO DEALERS.—Our firm name or trade mark does not appear on our machines, so that dealers may be able to handle them in competition with machines for which they must pay double our prices from other manufacturers. **WE, HOWEVER, ALLOW NO REDUCTION IN PRICES TO ANYONE.**

Wm. Frankfurth Hardware Co.

TAILORS' SHEARS.

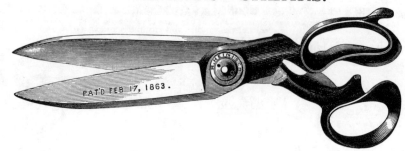

No.	4	5	6	7	8	9	10	11	12
Inches,	12¾	13	13¾	14	14¼	14¾	15¼	15¾	16¼
Per pair,	$5 00	8 00	9 00	10 00	11 00	12 00	13 00	14 00	15 00

Lamp Trimmers.

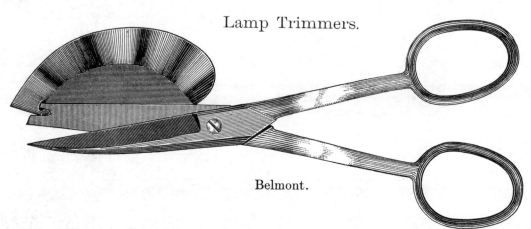

Belmont.

Belmont Lamp Trimmers, Cast Steel, Brass Guard, 5 inches, ... each, $ .50
Barnard's Lamp Trimmers, Solid Cast Steel, Japanned Bows, ... " .50

Heinisch's Pattern, Japanned Handles.

No. 7, Full Length 14 Inches, - - - - - - - each, $10 00
    9,   "   "   15 "   - - - - - - - " 12 00

LAMP TRIMMERS.

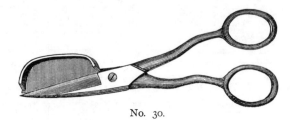

No. 30.

No. 30, Polished Cast Steel, Japanned Handles, - - - - - per dozen $5 00
    40, Barnard's Pattern, Cast Steel, Japanned Handle, - - - - " 6 00
     0, Japanned, - - - - - - - - " 1 25

One-half Dozen in a Box.

SCISSORS.

In Cases.

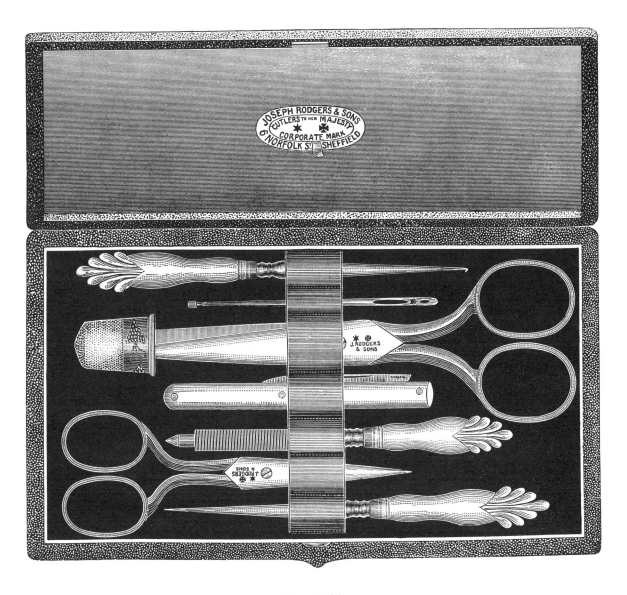

No. 9166.

No. 9166—Carved Pearl Handles, ... per case, $9.00
No. 9931—Plain Pearl Handles, ... " 7.50

A large assortment varying in price from $7.50 to $25.00 each, according to variety, quality and finish.

The Star Scissors and Shears,

Manufactured by
W. Schollhorn & Co.,
New Haven, Conn.

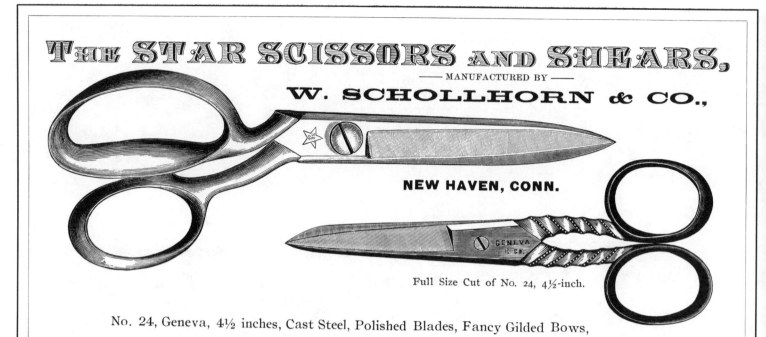

Full Size Cut of No. 24, 4½-inch.

No. 24, Geneva, 4½ inches, Cast Steel, Polished Blades, Fancy Gilded Bows,

---

BUTTON HOLE SCISSORS.

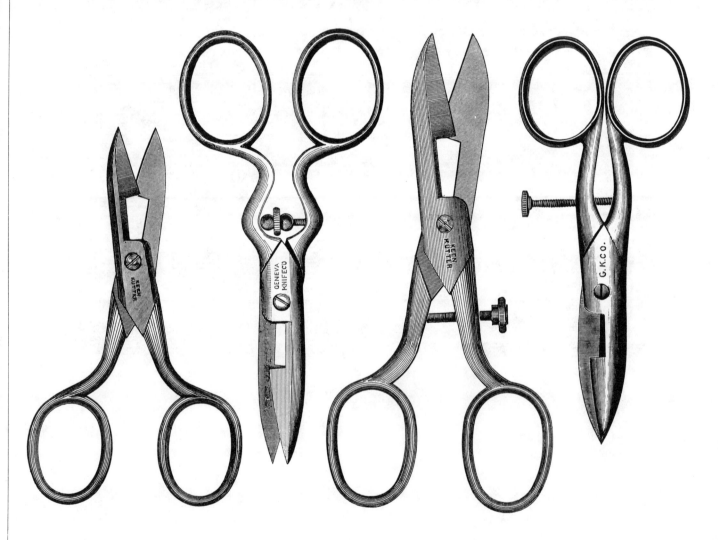

Keen Kutter, Button Hole Scissors, Full Silver Plated, with Set Screw, 4 inches, ... each, $1.00
Belmont, Button Hole Scissors, Best Cast Steel, with Set Screw, 4¼ inches, ... " .75
Belmont, Button Hole Scissors, Best Cast Steel, with Set Screw, 6 inches, ... " .90

SCISSORS.

Highest Grade Solid Crucible Steel Blades, Uniformly Tempered, Carefully Ground and Set. Full Crocus Polished, Plain Screws.

Ladies'
Fitted bows, full crocus polished
No. B1-BBB—
| Length inches.... | 4½ | 5 | 5½ | 6 |
| Per dozen .... | $10.00 | 10.50 | 11.00 | 11.50 |

Ladies'
Shear bow, solid steel, nickel plated.
No. B8—BBB—
| Length, inches .... | 6 | 6½ | 7 |
| Per dozen .... | $8.00 | 8.50 | 9.00 |

Ladies'
Fancy gilt bows, nickel plated blades.
No. B9—BBB—5 inch .... Per dozen $8.00

Ladies'
Fancy silver plated handle, oxidized finish, full crocus polished blades, fitted bows.
No. 2095—
| Length inches.... | 4½ | 5 | 5½ | 6 |
| Per dozen .... | $9.00 | 9.50 | 10.00 | 11.00 |

Ladies'
Fancy gold plated handle, full crocus polished blades, fitted bows.
No. 925—
| Length inches.... | 4½ | 5 | 5½ | 6 |
| Per dozen .... | $7.50 | 8.00 | 8.50 | 9.00 |

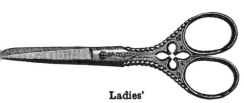

Ladies'
Fancy gold plated bows, polished blade.
No. 900—
| Length inches.... | 4½ | 5 | 5½ | 6 |
| Per dozen .... | $6.50 | 7.00 | 7.50 | 8.00 |

Embroidery or Lace
Fitted bows, full crocus polished.
No. B2—BBB—
| Length, inches .... | 3½ | 4 | 4½ |
| Per dozen .... | $9.50 | 9.70 | 10.00 |

Embroidery or Lace
No. 2095L—3½ in., fancy silver plated handle, oxidized finish, full crocus polished blades, fitted bows. Per dozen $8.00

Embroidery or Lace
No 925L—3½ in., fancy gold plated handle, full crocus polished blades, fitted bows............. Per dozen $7.00

Embroidery or Lace
No. 920—3½ in., fancy handle, full crocus polished.......... Per dozen $6.00

Embroidery or Lace
Fitted bows, polished blades.
No. 915L—Length, inches....... 3½   4
Per dozen $4.00   4.50

Embroidery or Lace
Per dozen
No. 900L—Fancy gold plated bows, polished blades, length 3½ in.......... $6.00

Embroidery or Lace
Cast steel, fancy gilt handle, nickel plated blades, round bows, length 3½ inches.
No. 300L..................... Per dozen $4.00

Stork Pattern
Cast steel, rainbow fancy handle, nickel plated blades, length 4 inches.
No. 301 .... Per dozen $4.00
No. 105—Gold plated fancy handle, 4 inches .... Per dozen 4.00

Pocket
Cast steel, full nickel plated, extra heavy bevel blades. One dozen assorted on a display card, four each, 4, 4½ and 5 in.
No. 20....................... Per dozen $4.50

Embroidery or Lace
Extra quality cast steel, full nickel plated, fitted bows, length 3½ inches.
No. 607L..................... Per dozen $3.50

School Scissors
Hardened cast steel, nickel plated blades, fitted bows, brass nut and bolt.
No. 271—Length, inches.. 4½   5   5½
Per dozen $3.00   3.20   3.50

Button Hole
Cast steel, nickel plated, brass set screw, length 4½ inches.
No. 60 .... Per dozen $4.50

SCISSORS

Full size cut of No. 5153, 3 inches.
Embroidery

Full size cut of No. 5153, 5-inch.
Forged steel, extra quality, crocus polish.

Inches,	3	3½	4	4½	5	5½	6	6½
No. 5153. Per doz.	$3 53	$3 53	$3 75	$4 13	$4 65	$5 40	$6 30	$7 28

Full size cut of No. 7247, 3½ inch.
Gilt bows and shank, polished blades, extra fine quality.

Inches,	3½	5
No. 7247. Per doz.	$8 10	$10 13

Full size cut of No. 4337-O, 6-inch.
Nickel plated, shaped bows, crocus polish blades, extra quality.

Inches,	3½	4	4½	5	5½	6	6½	7
No. 4337-O. Per doz.	$4 58	$5 10	$5 55	$6 08	$6 75	7 43	$8 10	$9 15

Full size cut of No. 2420, 4½-inch.
Nickel plated, fancy bows, extra fine quality.

Inches,	3½	4½	5½
No. 2420. Per doz.	$8 78	$9 90	$11 78

Full size cut of No. 9020, 6-inch.
Queen Scissors, nickel plated, shaped bows, crocus polish blades, extra quality.

Inches,	3½	4	4½	5	5½	6
No. 9020. Per doz.	$7 20	$7 80	$8 40	$9 00	$9 60	$10 20

Full size cut of No. 5153, 4-inch.
Forged steel, extra quality, crocus polish.

	Inches.	4	4½
No. 5153. Pocket Scissors.	Per doz.	$4 28	$4 73

Full size cut of No 7420, 4½-inch.
No. 7420. 4½ inch button hole nickel plated bows, "Queen" bolt and nut patent..........per doz. $10 13
No. 7410. 4½ inch button hole, nickel plated bows, Korn's patent, with screws...........per doz. 7 50
No. 7500. 4½ inch button hole, crocus polish, Korn's patent, with screws................per doz. 6 00

Patent Folding Pocket, Solid Cast Steel,

SCISSORS

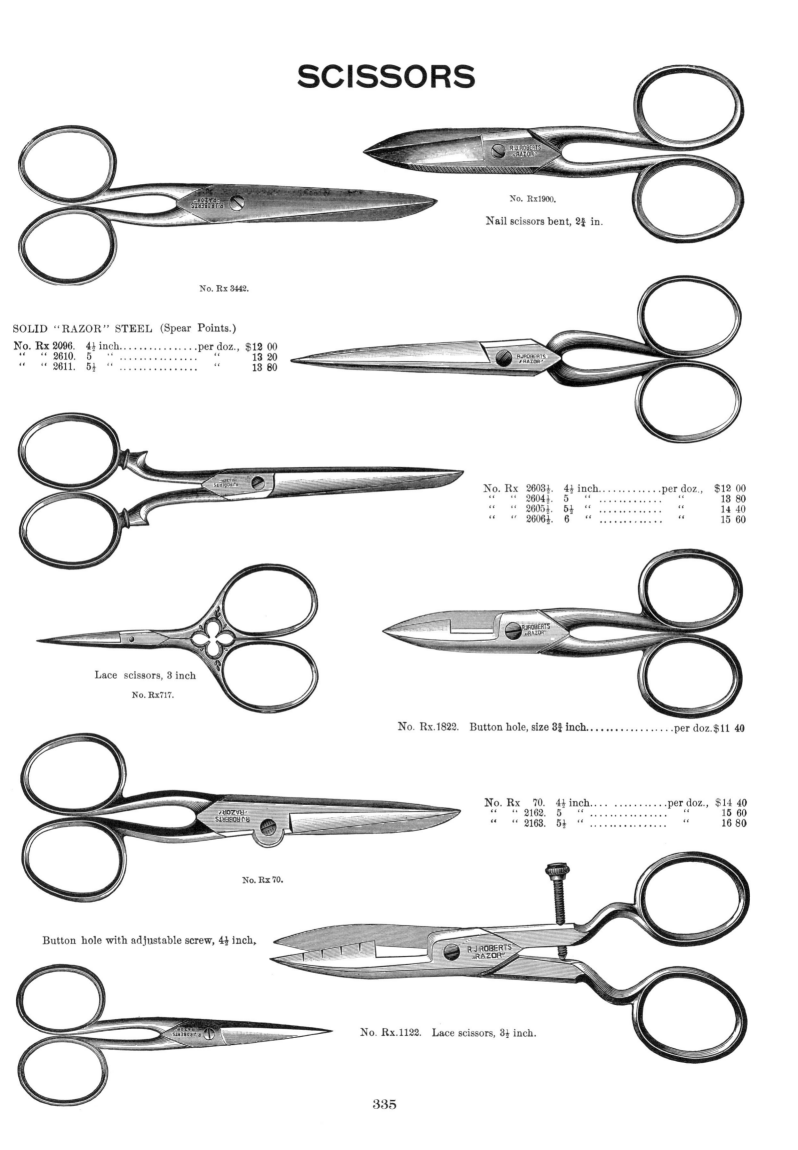

No. Rx1900. Nail scissors bent, 2¾ in.

No. Rx 3442.

SOLID "RAZOR" STEEL (Spear Points.)
No. Rx 2096.	4½ inch	per doz.,	$12 00
" " 2610.	5 "	"	13 20
" " 2611.	5½ "	"	13 80

No. Rx 2603½.	4½ inch	per doz.,	$12 00
" " 2604½.	5 "	"	13 80
" " 2605½.	5½ "	"	14 40
" " 2606½.	6 "	"	15 60

Lace scissors, 3 inch
No. Rx717.

No. Rx.1822. Button hole, size 3¾ inch..........per doz. $11 40

No. Rx 70.	4½ inch	per doz.,	$14 40
" " 2162.	5 "	"	15 60
" " 2163.	5½ "	"	16 80

No. Rx 70.

Button hole with adjustable screw, 4½ inch.

No. Rx.1122. Lace scissors, 3½ inch.

SCISSORS.

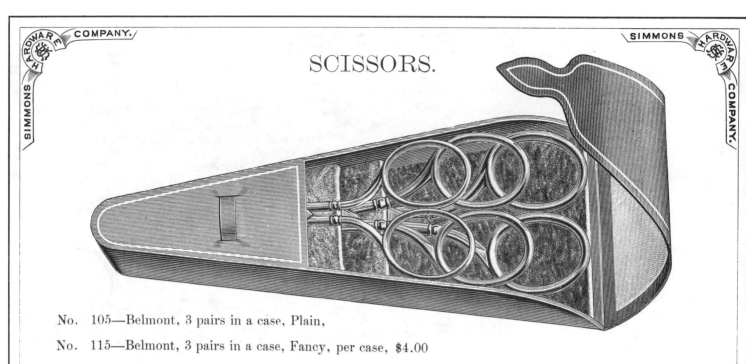

No. 105—Belmont, 3 pairs in a case, Plain,

No. 115—Belmont, 3 pairs in a case, Fancy, per case, $4.00

No. 6429—Jos. Rodgers & Son's, 4 pairs in a case, Flat Bows,

No. 6419—Jos. Rodgers & Son's, 4 pairs in a case, Fancy Bows,

GOLD THIMBLES.

20	21	22	23	24	27
14 kt. Octagon. $8.00 each	14 kt. Engraved, Heavy. $8.50 each.	14 kt. Engraved. $6.50 each.	10 kt. Engraved, Heavy. $7.00 each.	10 kt. Octagon $6.00 each.	10 kt. Plain. $3.70 each.

25	26	28	29	30	31
10 kt. Engraved. $6.50 each.	10 kt. Engraved. $5.00 each.	18 kt. Gold Filled, Oct $30.00 per doz.	18 kt. Gold Filled, Eng $24.00 per doz.	14 kt. Gold Filled, Eng $20.00 per doz.	14 kt. Gold Filled, Plain. $15.00 per doz.

COIN SILVER THIMBLES.

32	37	33	36	34	35
Heavy, Octagon. $9.50 per doz.	Medium Weight, Engraved. $6.50 per doz.	Engraved. $8.00 per doz.	Silver, Gold Band. $20.00 per doz.	Heavy, Plain, Open or Cl $7.00 per doz	Medium Weight.

Solid Gold, Silver and Filled Thimbles.

No. 16513........$1.00
Silver Chased, Round Gold Band.

No. 16514........$.75
Silver Polished, French Gray Rococo Border.

No. 16515........$.35
Silver, Plain and Chased.

No. 16516........$.50
Silver Engraved.

No. 16519........$.20
Sterling Engraved.

No. 16501 10-Kt..$2.00
No. 16502 14-Kt.. 3.00
No. 16503 Silver.. .20
Chased.

No. 16504 10-Kt..$2.50
No. 16505 14-Kt.. 3.25
No. 16506 Silver.. .35
Engraved.

No. 16507 10-Kt..$3.50
No. 16508 14-Kt.. 4.50
No. 16059 Silver.. .50
Raised Border.

No. 16510 10-Kt..$5.00
No. 16511 14-Kt.. 6.00
No. 16512 Silver.. 1.00
Applied Border.

No. 16518........$.75
Faceted Border, Sterling.

No 563.
10 k., $4 00
14 k., 5 50

No. 562.
10 k., $4 50
14 k., 6 00

No. 555.
10 k., $6 00
14 k., 8 00

No. 577
10 k., $6 25
14 k., 8 50

No. 703.
Filled.
14 k., $2 00

No. 323.
Gold Band.
$2 00

No. 301.
Coin Silver.

No. 303.
Coin Silver.

No. 358.
Coin Silver.

No. 305.
Coin Silver.

No. 316.
Coin Silver.

No. 355.
Coin Silver.

8984 1½ oz. .50 each

9005 1½ oz. .50 each

8992 2 oz. .85 each

8985 2 oz. .55 each

8986 1½ oz. .60 each

8987 2 oz. .60 each

8988 2 oz. .65 each

8981 2 oz. .65 each

8980 2 oz. .65 each

8989 2 oz. .65 each

9008 2 oz. .70 each

9007 2 oz. .70 each

8990 1½ oz. .70 each

8991 2 oz. .70 each

Sterling Silver with Gold Bands **10 Kt. Gold**

8993 2½ oz. .95 each

8995 1½ oz. 1.10 each

8997 2 oz. .70 each

8994 2 oz. 1.60 each

8996 2 oz. 1.50 each

8998 24 dwt. 3.75 each

8999 24 dwt. 3.75 each

Sterling Silver Handle Scissors
Made with Good Quality Steel Blades in either the Manicure or Embroidery Style.
Illustrations Exact Size

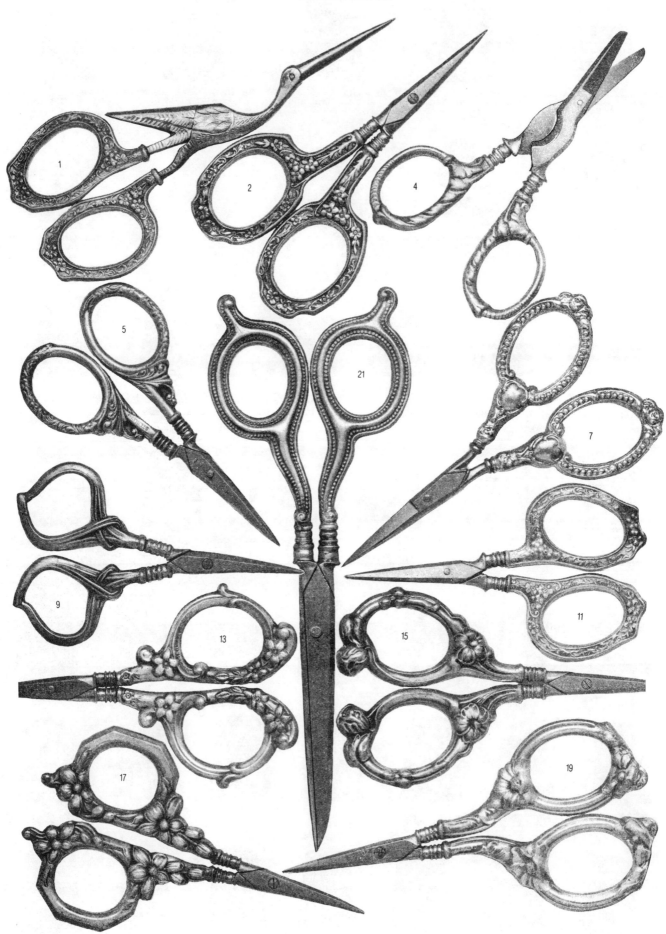

1	Embroidery	2.15 each	09	Manicure	1.00 each
2	Embroidery	1.50 "		Same in Embroidery is 10	1.00 "
	Same in Manicure is 03	1.50 "	11	Embroidery	1.00 "
4	Cigar Cutter	1.90 "		Same in Manicure is 12	1.00 "
5	Embroidery	1.00 "	13	Embroidery	1.50 "
	Same in Manicure is 06	1.00 "		Same in Manicure is 14	1.50 "
7	Embroidery	.80 "	15	Manicure	1.50 "
	Same in Manicure is 08	.95 "		Same in Embroidery is 16	1.50 "
17	Manicure	1.70 each			
	Same in Embroidery is 18	1.70 "			
19	Embroidery	1.50 "			
	Same in Manicure is 20	1.50 "			
21	Work Scissors	2.10 "			
	Same in Embroidery is 22	1.90 "			
	Same in Manicure is 23	1.90 "			

Sterling Silver Sewing Novelties and Sets

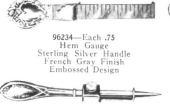
96234—Each .75
Hem Gauge
Sterling Silver Handle
French Gray Finish
Embossed Design

96038—Each 1.00
Stiletto
Sterling Silver Handle
Plain Polished

96230—Each .30
Emery Needle Punch
Sterling Silver Top
Red Silk Tassel

96040—Each .46
Darner
Sterling Silver Handle
French Gray Finish

96035—Each 1.50
Combination Glove
Darner and
Needle Case
Sterling Silver
Polished Plain

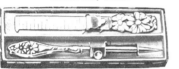
96233—Each .75
Stiletto
Sterling Silver Handle
French Gray Finish

96039—Each 1.00
Stiletto
Sterling Silver Handle
French Gray Finish

96032—Each 1.00
Tape Measure
60 inches
Sterling Silver
French Gray Finish

96033—Each 3.70
Steel Tape Measure
60 inches
Sterling Silver
Plain Polished

96034—Each 3.70
Steel Tape Measure
60 inches
Sterling Silver
French Gray Finish
Engraved

96220—Set 1.50
Sewing Set
Sterling Silver Handles
French Gray Finish
Hem Gauge
and Stiletto

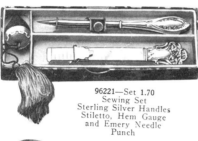
96221—Set 1.70
Sewing Set
Sterling Silver Handles
Stiletto, Hem Gauge
and Emery Needle
Punch

96247—Each 1.00
Shuttle
Sterling Silver
French Gray Finish
Engraved

96246—Each 1.50
Shuttle
Sterling Silver
Plain Polished

96248—Each 2.00
Shuttle
Sterling Silver
Polished
Engraved on One Side

96223—Set 1.20
Sewing Set
Sterling Silver
French Gray Finish
Emery Needle Punch
Stiletto and Thimble

96219—Set 1.50
Sewing Set
Sterling Silver
French Gray Finish
Emery Needle Punch
Scissors and Thimble

96213—Set 1.50
Bodkin Set
3 Pieces
Sterling Silver
French Gray Finish

96216—Set 2.50
Bodkin Set
4 Pieces
Sterling Silver
Plain Polished
Seal Leather Case

96225—Each 2.00
Embroidery Scissors
Sterling Silver
French Gray Finish
Engraved

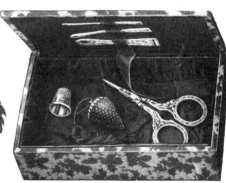
96222—Set 2.20
Sewing Set
Sterling Silver
French Gray Finish
Emery Needle Punch
Thimble Stiletto
and Scissors

96210—Set 3.00
Sewing Set
Sterling Silver
French Gray Finish
Tape Measure, Scissors,
Thimble and Bodkin
Leather Case

96201—Set 3.75
Sewing Set
Sterling Silver
French Gray Finish
Engraved
3 Piece Bodkin Set
Emery Needle Punch
Thimble and Scissors
Sweet Grass Silk Lined Case

96031—Set 4.00
Sewing Set
Sterling Silver
French Gray Finish
Engraved
3 Piece Bodkin Set
Emery Needle Punch
Thimble and Scissors
Tapestry Covered Box

96203—Set 6.50
Sewing Set
Sterling Silver

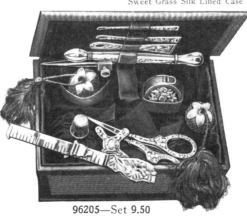
96205—Set 9.50
Sewing Set
Sterling Silver

339

WASHING MACHINES.

WASHING AND WRINGING MACHINES.

FROM time immemorial, among civilized people washing day was a day of terror to men, women, dogs and cats, until the benign

WASHER NO. 1.

offices of mechanical invention had provided a substitute for the muscular system, which it was supposed was the inevitable appliance for the occasion. In "Washer No. 1" and "Wringer No. 1," we have a fair illustration of the old process to which women gave themselves up as inevitable—the washboard being a considerable advance upon the still earlier process of rubbing entirely by hand. I need not enlarge upon the tediousness of this operation.

WRINGER NO. 1.

So it was, that in the course of my travels not long since, I came upon the great factory of the Metropolitan Washing Machine Co., located at the town of Middlefield, State of Connecticut, where it was established in the year 1859, first for the manufacture of the Metropolitan Spring-Dash Washer, as shown in the cut. Soon after this, Mr. DOTY invented a washer so much superior to the spring-dash concern, that the Metropolitan Company adopted it. The old Doty had

ORIGINAL METROPOLITAN.

some faults of construction, which allowed the legs to fall off, and the rusting of the metal parts soon spoiled the whole machine. In the hands of the Metropolitan Company, the Doty was improved in all its construction, as will be seen by the cut. The legs are stout, and receive the box ends into grooves, and are held in place by bolts. The metal parts are all thoroughly galvanized to prevent rust; a balance ball equalizes the tilt of the handles, making the operation particularly easy.

ORIGINAL DOTY WASHER.

To proceed in the order of improvement in Clothes Wringers, Mr. SARGENT invented a concern, of which we give an illustration, still clinging to the old idea of *twisting* the water out of the wet clothes, and twisting the clothes to tatters at the same time. But there was a valuable feature in Sargent's Wringer, in that it had the clamp for holding the wringer to a tub, which original device has become the property of the Metropolitan Washing Machine Co.

The Universal Clothes Wringer, with cogwheels, was thought to be the acme of perfection; but see the bungling way of clamping it to the tub, and as there was

ORIGINAL UNIVERSAL

only a pair of single cogwheels, when a large article of washing was put between the rollers they were pressed apart so as to let the cogs out of touch, when it was no better

SARGENT'S WRINGER.

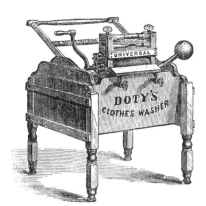

LATEST IMPROVED DOTY.

than a plain friction machine. After several improvements, the cogwheels were made as shown in the cut. Each wheel has a double set of cogs alternating in such manner that when one set of cogs is about to let go, the other set take hold, even if the rolls are three-fourths of an inch apart; and by a shoulder between the two the teeth cannot mash in upon their points.

The latest improved Universal Wringer has movable metal clamps and thumb screws for fastening to any sized tub; a folding shelf or apron, for carrying the clothes over the edge of the tub or machine; compound wooden spring-bars, to equalize the pressure of the rolls; a patent stop, to prevent the rolls from letting the cogs out of gear.

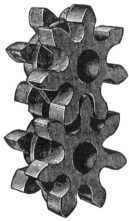

DOUBLE COG-WHEELS.

At the great factory in Middlefield I had an opportunity to see the whole operation, from the wood and metal in the rough to the finely-finished machine, and it will be interesting to the many thousands who use these machines to know something of the place where they are made. At the factory are employed about one hundred workmen in the several departments of manufacture, and the works turn out six hundred machines per day, using five hundred tons of iron and a million feet of timber in a year.

The machinery is of the most improved models, specially adapted for this purpose. Every part of every machine manufactured is a duplicate of itself, so that each member is sure to fit in any machine. The wood work is all brought to an exact size by machinery, and the iron work is cut and turned to a nicety. The rubber rolls are made of the best material and in the most durable manner. The galvanizing of the grey iron is the perfection of the chemic art, and is a secret with the operator. The machinery of the factory is propelled by two twenty-horse power water wheels, and one forty horse power steam engine.

This article was published in 1871.

THE "UNIVERSAL" WRINGER,

So long offered as a PREMIUM by the RURAL NEW-YORKER, the American Agriculturist, and other leading periodicals, still stands unrivaled for **Strength** and **Durability**. Its sale constantly increases, as those who are induced to buy other kinds, which sooner wear out, are sure to get the Universal as their second purchase. It is the only Wringer which has

ROWELL'S PATENT COGS,

with long and strong alternate teeth, which can separate widely, or crowd closely together without binding or losing their power.

These are protected by the **PATENT STOP**, or screw, above the cogs, which prevents them from entirely separating, so the

UNIVERSAL WRINGER NEVER PLAYS OUT OF GEAR,

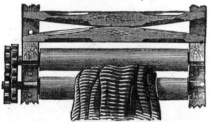

yet the teeth of the Rowell Cogs are **so long**, the rolls can separate far enough to pass the largest article easily.

CAUTION.

The importance of this Patent STOP is not generally understood, and sometimes people buy a wringer *with single or spiral cogs on both ends of the shaft*, expecting to get the same advantage; but as articles can disconnect these single cogs, on either one side or the other continually, one half of the power of the cogs is constantly lost, as in the illustration below, with

COGS SEPARATED AT ONE END.

Or if a large article passes through the wringer, all the cogs are frequently disconnected and rendered *entirely useless*, as in the illustration below, of

A WRINGER WITHOUT THE STOP,

With Cogs Separated at Both Ends, making the machine no better than a *friction wringer*, when the strain is hardest, and the cogs most needed.

Always Try a Wringer before Purchasing, and *buy none which will play out of gear in passing a sheet, blanket, or bedquilt.*

The **UNIVERSAL** is sold as low as any other licensed wringer, and kept by dealers generally.

The celebrated **Doty Washing Machine.** Sold also by the

Metropolitan Washing Machine Co.,
R. C. BROWNING, President,
32 Cortlandt St., New York.

THE ORIGINAL
WRINGER MAN'S MONITOR!
The Only Leading Wooden Frame

"One would suppose that Yankee ingenuity had become well nigh exhausted upon Washing Machines; but we have lately examined a new invention, or rather a modification of an old system, by Mr. James Pullen, of China. His plan is upon the common fluted washboard, but his flutings are made of sheets of copper, zinc, tin— or they may be made of glass, earthen, or similar durable and smooth substances. These sheets or substances are fastened to boards, and used as such boards are commonly used. Now where's the improvement? 1st, In durability — 2d, In its smoothness — and 3d, Ease in performing the operation of rubbing. A common washboard does not last long — it soon gets rough, of course wears out the clothes and makes harder work — they get warped and split, are then patched up with shingle nails, which get rusty, and make a rickety, weak concern — the women scold, and the poor husband has to march off and buy a new one. The metallic rubber will last a long life time, if a little care be taken in drying it after using; for when worn a little on one side it may be turned and worn on the opposite side, and in the contrary direction.

We like the simplicity of the concern. Our washing machines have been too complicated. The inventors seem to have considered it necessary to show their skill in combining the greatest number of mechanical modifications in one piece; and the greater the array of cog wheels, cranks and rollers, the more ingenious has the machine been considered, and the more likely to do its duty without hands.

Disappointment has of course followed. This improvement is simple, it does not promise to indulge idleness, but to lighten and facilitate the labor of washing, and render it a pleasure rather than a dreaded task." From the *New England Farmer*, August 6, 1834.

THE HOUSEWIFE'S FRIEND IS LAMB'S
COMBINED WASHER AND WRINGER.

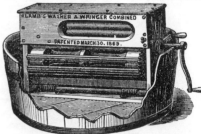

The above cut shows the position of the Washing Machine and Wringer which can be readily adjusted to any common wash tub. Where used, this has given general satisfaction. It can be operated by a child with perfect ease. It needs only to be seen to be appreciated. Retail price only $12. State, County, Town and Territorial Rights for sale. Address, for further particulars, JOHN LAMB, 244 East 27th St., N. Y. City, or Jeffersonville, Sullivan Co., N. Y.

FERGUSON WASHER.

No longer is there any necessity for our housewives to dread that abomination—*wash day*, for by a little exertion, anybody can, in a few hours, secure one of the best machines ever manufactured—the "Ferguson Washer." We have made such arrangements as will enable us to present one of these splendid washers to any person sending us a list of 25 subscribers and $25. The names may be sent in as fast as secured, and when the number is complete, we will at once ship the machine as instructed. Every washer is warranted, and a child ten years old can operate it with ease. Who will have the first one? Special information cheerfully given on application to **Wood's Household Magazine, Newburgh, N. Y.**

Clark's Washing Machine.

LABOR-SAVING implements and machines are a boon to toiling humanity that never fail to arrest the attention of those for whose benefit they are designed. The engraving given herewith is a representation of Clark's Washing Machine, the construction of which may be readily understood by a study of the cut. The corrugated roller operated by the handles shown, presses upon a surface at the bottom of the machine, so constructed as to

be adjusted thereto and held in position by spiral springs, the washing being performed by passing the clothes backward and forward by the rotary motion given to the corrugated cylinder by the crank handles. It certainly has the element of simplicity. The real test of all these inventions is in their practical use. Those interested are referred to the advertisement of the manufacturers.

Wash Boards.

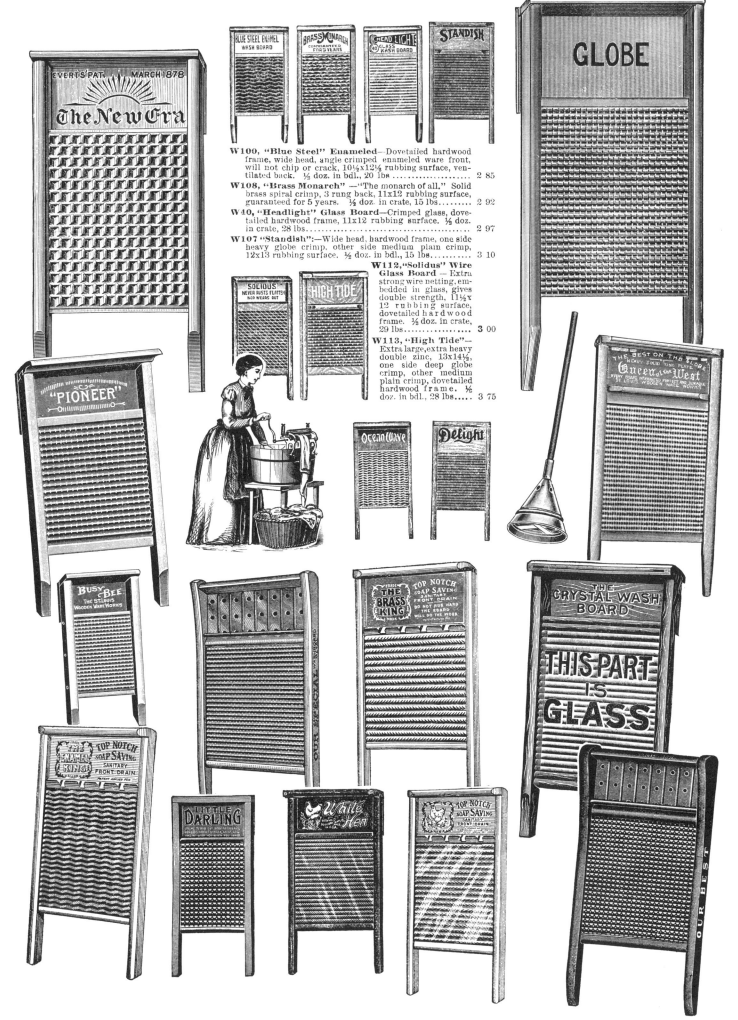

W100, "Blue Steel" Enameled—Dovetailed hardwood frame, wide head, angle crimped enameled ware front, will not chip or crack, 10½x12½ rubbing surface, ventilated back. ½ doz. in bdl., 20 lbs 2 85

W108, "Brass Monarch"—"The monarch of all." Solid brass spiral crimp, 3 rung back, 11x12 rubbing surface, guaranteed for 5 years. ½ doz. in crate, 15 lbs 2 92

W40, "Headlight" Glass Board—Crimped glass, dovetailed hardwood frame, 11x12 rubbing surface. ½ doz. in crate, 28 lbs 2 97

W107 "Standish":—Wide head, hardwood frame, one side heavy globe crimp, other side medium plain crimp, 12x13 rubbing surface. ½ doz. in bdl., 15 lbs 3 10

W112, "Solidus" Wire Glass Board—Extra strong wire netting, embedded in glass, gives double strength, 11½x12 rubbing surface, dovetailed hardwood frame. ½ doz. in crate, 29 lbs 3 00

W113, "High Tide"—Extra large, extra heavy double zinc, 13x14½, one side deep globe crimp, other side medium plain crimp, dovetailed hardwood frame. ½ doz. in bdl., 28 lbs 3 75

BENCH WRINGERS.

GEM.

Domestic Mangle and Wringer,
each, $35 00

No. 3.

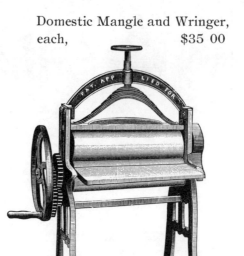

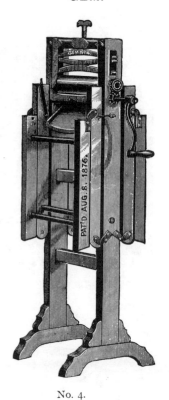

No. 4.

No. 3, Empire Folding Bench Wringer, Rolls 10 x 1¾-inch, - - - per dozen, $108 00
  4, Gem   "   "   "   "   10 x 1¾   "   - - - -   "   84 00

1865. THE PEERLESS CLOTHES WRINGER. 1877.

No. 2, $66·00.

White
Rubber Rolls,
Metal
Journal Boxes,
Patent Crank
Fastening,
Rubber
Fastening Pads,
Maple
Wood Frames.
Simple,
  Durable,
    Efficient.

No. 3, $74·00.

Wrought Iron
Thumbscrews,
Apron or
Clothes Guard,
Hickory
Spring Bar
and
Rubber Springs
Perfect
in Finish.
Send for
  Circular

Warranted Double the Capacity of any Purchase Gear Wringer.

Patent Washing Machines.

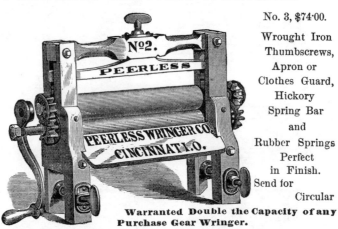

DOTY'S IMPROVED CLOTHES WASHER.

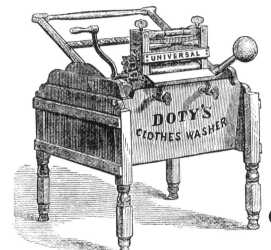

TAKE IT ON TRIAL.

SAVES TIME, LABOR AND CLOTHES.

WITH THE METROPOLITAN BALANCE WEIGHT.

Over Sixty Thousand Families in the United States are now using the DOTY WASHING MACHINE, and as many thousand women thank DOTY for his invention, and will testify that *One* Washing Machine

IS NOT A HUMBUG.

There is *no mistake about it*; this machine will wash clothes well and thoroughly, **much** faster and easier than can be done by hand, and with far *less wear to the garments*.

IT SAVES ITS COST EVERY YEAR BY SAVING CLOTHES.

Any person who will be guided by the printed directions, which are easily understood, can use the washer successfully the first trial, and will be *sure to like it*.

THE PATENT BALANCE WEIGHT

recently invented, does away with the use of springs and makes the work much easier by giving a *heavier blow* with *less effort*. One dollar covers the additional cost above the old style.

WHAT OUR CUSTOMRES SAY ABOUT IT.

NEWARK, N. J., June 10, 1871.

"The Balance Weight improvement in Doty's Machine is a great success. I am selling to many persons who would not buy the other style—I think it will **double the sale,** as it does the *value*, of the washer. Send me five dozen more immediattely, all new style." EZRA AYRES.

ORANGE, N. J., June 8, 1871.

"The improved Doty Washer has been thoroughly tested in my family, and *we like it*. I believe it will increase sales, and make the washer still more satisfactory than before. I can put the balance attachment to the old style, and have been asked to do it in several cases, charging $3.50 each for the job. The sales in Orange are easier and more rapid than ever, though I have sold here over *a thousand Washers*. **Every Washer seems to make Sale for two or three more,** to neighbors. My entire time will now be given to selling the DOTY WASHER and UNIVERSAL WRINGER, which I find is beyond all question the best and most durable. I believe over three-fourths of the 200 Universal Wringers sold by Mr. Keys in Orange **nine years ago, are now in use and in good order.**" MARSHALL L. WARD.

RETAIL PRICES.
(WITH BALANCE ATTACHMENT.)

Family Size....$15.   **Hotel Size....$17.**

☞ **LARGE DISCOUNTS TO THE TRADE.**

METROPOLITAN WASHING MACHINE CO.,
R. C. BROWNING, Pres't.
32 *CORTLANDT ST., NEW YORK.*

Doing a Week's Washing
In 6 Minutes—Read the Proof

THIS woman is using a 1900 Gravity Washer. All she has to do is keep the washer going. A little push starts it one way—a little pull brings it back—the washer does the rest.

The clothes stay still—the water rushes through and around them—and the dirt is taken out.

In six minutes your tubful of clothes is clean.

This machine will wash anything—from lace curtains to carpets, and get them absolutely, spotlessly, specklessly clean.

There isn't anything about a 1900 Gravity Washer to wear out your clothes.

You can wash the finest linen, lawn and lace without breaking a thread.

"Tub rips" and "wash tears" are unknown.

Your clothes last twice as long.

You save time—labor—and money.

You wash quicker—easier—more economically.

Prove all this at my expense and risk.

I let you use a 1900 Gravity Washer a full month FREE.

Send for my New Washer Book.

Read particulars of my offer.

Say you are willing to test a 1900 Gravity Washer. **I will send one to any responsible party, freight prepaid.**

I can ship promptly at any time—so you get your washer at once.

Take it home and use it a month. Do all your washings with it.

And, if you don't find the machine all I claim—if it doesn't save you time and work—if it doesn't wash your clothes cleaner and better—don't keep it.

I agree to accept your decision without any back talk—and I will.

If you want to keep the washer—as you surely will when you see how much time, and work, and money it will save you—you can take plenty of time to pay for it.

Pay so much a week—or so much a month—as suits you best.

Pay for the washer as it saves for you.

I make you this offer because I want you to find out for yourself what a 1900 Gravity Washer will do.

I am willing to trust you, because you can probably get trusted at home. And, if your credit is good in your own town, it is just as good with me.

It takes a big factory—the largest washer factory in the world—to keep up with my orders.

So far as I know, my factory is the only one ever devoted exclusively to making washers.

Over half a million of my washers are in use.

Over half a million pleased women can tell you what my washers will do.

But you don't have to take even their say-so. You can test a 1900 Gravity Washer yourself. Then you will know positively.

Write for my book today. It is FREE.

Your name and address on a post card mailed to me at once, gets you my book by return mail.

You are welcome to the book whether you want to buy a washer now or not.

It is a big illustrated book, printed on heavy enameled paper, and has pictures showing exactly how my Washers work.

You will be pleased with this book. It is the finest even I have ever put out. Write me at once.

Find out just how a 1900 Gravity Washer saves your time and strength—preserves your health—and protects your pocketbook.

Write now—Address—R. F. Bieber, Manager "1900" Washer Co., 371 Henry St., Binghamton. N. Y. Or, if you live in Canada, write to my Canadian Branch, 355 Yonge St., Toronto, Ontario.

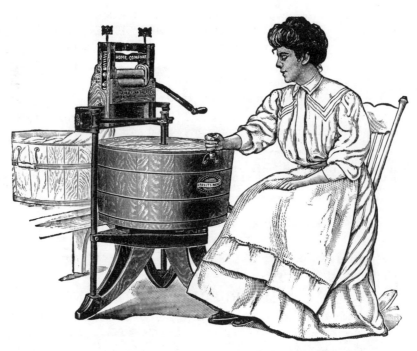

Easy Washing in 6 Minutes

$3.12 MISSISSIPPI CLOTHES WRINGERS $3.12

WARRANTED FOR THREE YEARS.

THE QUICKEST ADJUSTED, THE MOST CONVENIENT AND ONE OF THE STRONGEST AND BEST WRINGERS EVER OFFERED AT ANY PRICE.

OUR MISSISSIPPI WRINGER represents the greatest improvement ever made in the manufacture of clothes wringers. Wringers of this pattern, but not as strongly constructed and without the many improvements of our Mississippi regularly sell for $5.00. Our price of only $3.12 for this great wringer, with all its labor and time saving devices, with its unbreakable malleable iron frame, its pure rubber rolls and its self adjusting steel spring makes it the greatest wringer value on the market, and one which if you need a good wringer, you cannot afford to overlook.

OUR MISSISSIPPI WRINGER can be used with any washing machine or with any tub as the clamps are quickly adjusted to any thickness and once fastened, hold securely. There are no set screws to be tightened every time you put it on or take it off the tub, no screws to adjust the pressure on the rolls; it is all done automatically. As soon as lever is released to take it off the tub, the pressure is immediately removed from the rolls which prevents their flattening, so they always stay smooth and round.

OUR MISSISSIPPI WRINGER IS GUARANTEED TO PLEASE YOU. If it does not, send it back and get your money. Could any offer be fairer? We have sold tens of thousands of these high class wringers under the above guarantee, and they never fail to please.

THE RIGHT KIND TO BUY.

WARRANTED FOR THREE YEARS

THE ADJUSTABLE TOP LEVER FASTENS THE WRINGER TO THE TUB AND AT THE SAME TIME ADJUSTS THE PRESSURE ON THE ROLLS.

BY THROWING THE LEVER to one side, all the strain is taken off the rolls, and the wringer is loosened from the tub. Reverse the lever and the tub is firmly grasped by the clamps, at the same time the pressure on the rolls is automatically adjusted. By turning set screw near top of frame it can be adjusted to the proper pressure on different thickness articles; also regulates the clamps for different thickness tubs. Once adjusted, it requires no further attention, but can be instantly attached or removed.

THE FRAME of our Mississippi Wringer is made of annealed malleable iron, strong enough to stand the hardest service, but not so heavy as to be unhandy. It will not break should you drop it on the floor or otherwise misuse it. All metal parts are heavily galvanized and will not rust, even after years of hard use.

PRESSURE SPRING is of tempered steel and adjusts itself to any unevenness so that you can wring articles that are thick at one end and thin at the other and both ends will be equally dry.

COG WHEELS are double and enclosed so that they cannot catch the fingers.

THE ROLLS. The life of a wringer depends largely on the rolls. These are made of white Para rubber, are soft and elastic, and immovably fastened to the shaft so that they will never twist or become loose. We guarantee the rolls for three years, for should there be any hidden flaw in the rubber it will surely appear in that time, and if it does not you can rest assured they will last at least ten or fifteen years in ordinary family use. Rolls are 11 inches long and 1¾ inches in diameter.

No. 9L2207 Shipping weight, securely boxed, 27 pounds. Price.................$3.12

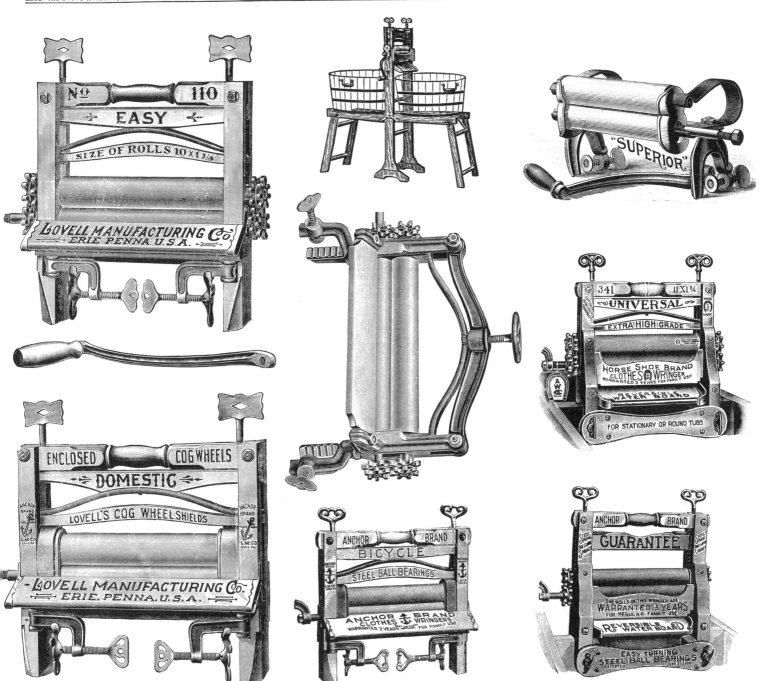

347

Wm. Frankfurth Hardware Co.

COPPER WARE.

Planished Copper.

MEASURES.

Planished Copper, U. S. Standard.

Quarts,		¼	½	1	2	4	Quarts,	⅛	¼	½	1	2	4
Inches,	-						Per Dozen,	$8 00	8 50	12 00	17 00	23 00	30 00

WASH BOILERS.

PLANISHED COPPER.

GALVANIZED, FLAT BOTTOMS, STAMPED TIN COVERS.		TIN, FLAT COPPER BOTTOMS, STAMPED COVERS.

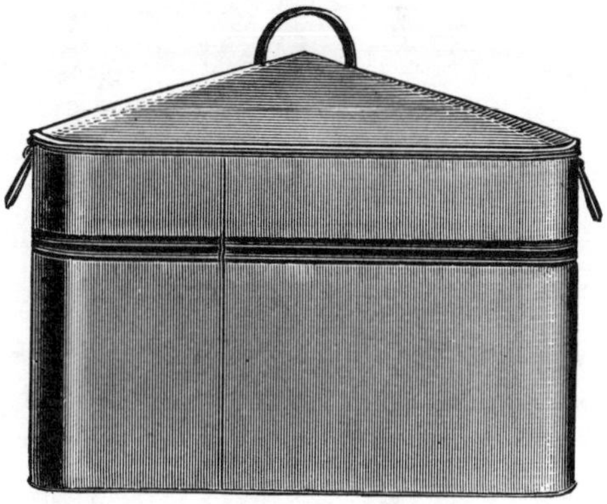

Square. Flat Bottoms.

Nos.	-	-	-	-	-	-	-	-	-	8	9
Inches on Bottom,										12x22	13½x23
Each,										$13 50	15 00

COLD ROLLED COPPER.

TIN, FLAT COPPER BOTTOMS.

Oval, Pit or Flat Bottoms.

Square. Flat Copper Bottoms, Copper Rim.

Nos.	-	-	7	8	9	Nos.	-	-	7	8	9
Inches on Bottom,			10x19¼	11¼x21	12¼x22¼	Inches on Bottom,			10x19¼	11¼x21	12¼x22¼
Each,	-	-	$6 50	7 00	7 75	Each,	-	-	$7 35	7 85	8 60

Tub is made of Cypress, fully corrugated on side and bottom, and in natural finish. The improved, easy-running mechanism is aluminum bronzed, and the post thact onnects dasher to shaft is galvanized.

No. 5, American ....Each $9.00

The tub is (extra large) and is made from Louisiana cypress thoroughly seasoned, fully corrugated on the inside. The fly wheel and hoops are green enameled, the balance of the mechanism is gold bronzed, the tub has three coats of coach varnish.

The fly wheel can be removed easily and a belt pully attached, making it a power machine. The operating lever can be placed in either one of two positions, convenient for a short or tall person. The dolly post is guarded by a galvanized cup, eliminating all possibility of tearing clothes.

No. 22 Miracle...................Each $23.00
One in a Crate
Extra Pulleys for power ....Each $2.50

Bail Movement
Same size and construction as No. 2, only with bail movement which makes it easy to lift rubber out of machine.

Ear Movement
This machine is extra large, having the following inside measurements: length 28½ inches, depth 14 inches, width 22 inches height 25 inches.
Removable Bottom
No. 6 —Red finish........Per dozen $90.00
Twelfth Dozen in a Crate

SUCTION WASHERS

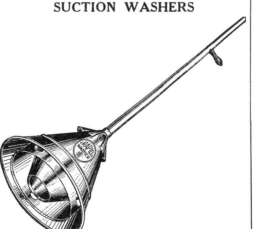

Made of extra heavy tin, handles 40 inches long. Washes heavy pieces, such as rugs, comforters, blankets, etc. Delicate pieces may also be washed without injury. Saves the hands, prevents wear on clothes, and requires but little soap.
Rapid—With handle........Per dozen $15.00

IRONING TABLES
RID-JID SPRINGER OPEN-END FOLDING

The construction brings the rear legs inside the width of the board when folded, but when opened gives them a greater spread than any other board ever designed. This means no tilting or wabbling, and sets solid on any floor, no matter how uneven the floor may be. There is not a screw or nail, but every joint is riveted so that there is no chance of them ever falling out. The open-end space under the top board is suffient to take the longest skirt or a full one-piece suit. Made of clear kiln-dried spruce, free from knots.

No. 1—15x57 inches, weight 14½ lbs.

Per dozen $60.00

Wm. Frankfurth Hardware Co.

MRS. POTT'S SMOOTHING AND POLISHING IRONS.
DOUBLE POINTED.

The above cut represents a Full Set, which consists of Three Irons, One Stand and One Detachable Walnut Handle.

No. 50, Nickel Plated, with one each No. 1, 2 and 3 Iron, - - - - - - per set, $2 50
 55, Extra Polished, " " " No. 1, 2 " 3 " - - - - - - - " 2 00

SQUARE BACK.

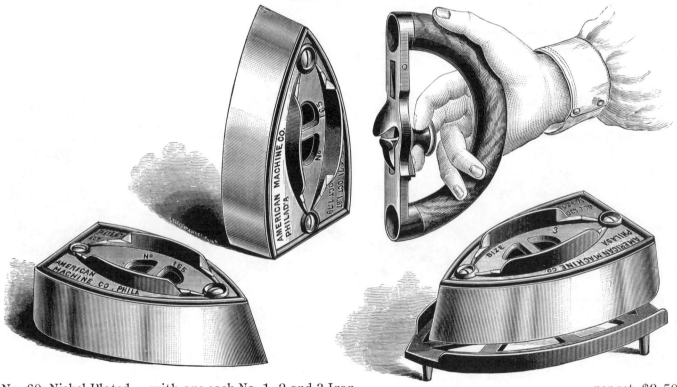

No. 60, Nickel Plated, with one each No. 1, 2 and 3 Iron, - - - - - - per set, $2 50
 65, Extra Polished, " " " No. 1, 2 " 3 " - - - - - - - " 2 00

PATENT SMOOTHING AND POLISHING IRONS.

B. & D.
Per dozen, . . $4 50

Keystone.
Per dozen, . . $5 00

Mrs. Cook's.
Per dozen, . . $6 00

Toilet Iron.

Sad Irons.

GLEASON'S SHIELD

Sad Irons, Extra Finish, ... per pound, $ .06
Sad Irons, Nickel Plated, ... " .10

Geneva Polishing Irons.

Troy Polishing Irons.
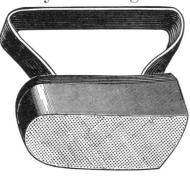

Geneva Laundry Irons.

Geneva Nickel Plated Polishing Irons, ... each, $ .60
Troy Polishing Irons, Perforated Face, ... " .60
Geneva Laundry Irons, Nickel Plated, ... each, No. 1, $ .75 No. 2, .85

Toy Sad Irons.

Sad Iron Holders.

Toy Sad Irons, 2½ pounds, ... each, $ .25
Sad Iron Holders, Finely Woven Cotton, ... " .20

SAD IRONS.

4 to 10 lb. Irons, ... per pound, $

Mahony's Patent Sad Irons.

Rough Face Sad Irons.

Smooth Face Troy Polishing.

Perforated Face Troy Polishing.

Rough Face Troy Polishing.

Smooth Face Sad Irons.					Rough Face Sad Irons.					Troy Polishing Irons.		
Nos.	6	7	8	9	Nos.	5	6	7	8		Smooth,	Perforated,	Rough,
Per pair,	$1 00	1 20	1 40	1 60	Per pair,	$0 80	1 00	1 20	1 40	Per doz.,	$6 00	6 00	6 00

Self-Heating Sad Irons.

Self-Heating Iron.

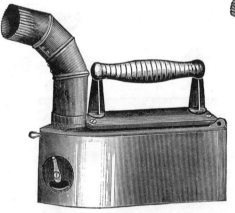
Front View of the Excelsior Tailors' Iron.

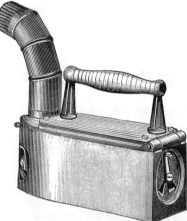
Rear View of the Excelsior Tailors' Iron.

Self-Heating Charcoal Irons.

Nos. 1, 2, 3 and 4, Separate or Assorted, ... per dozen, $10 00

Excelsior Tailors' Charcoal Irons.

Excelsior Tailors' Irons, Weight about 18 pounds each, ... each, $5 00

FLAT IRON HEATERS.

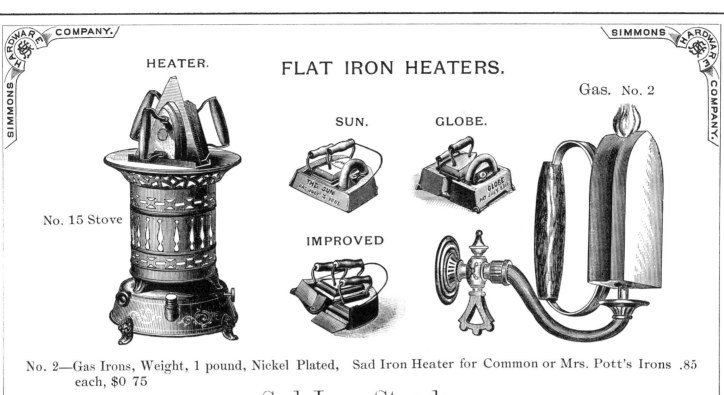

No. 2—Gas Irons, Weight, 1 pound, Nickel Plated, each, $0 75   Sad Iron Heater for Common or Mrs. Pott's Irons .85

Sad Iron Stands.

24841 Nickel plated Toy Sad Iron stand.

No. 3—Grey Iron Sad Iron Stands, ... each, $ .15
No. 2—Grey Iron Sad Iron Stands, ... " .10
Nickel Plated Sad Iron Stands, ... " .25

Coffee Pot Stands.

No. 01½—Nickel Plated Coffee Pot Stands, ... each, $ .20
No. 1½—Japanned Coffee Pot Stands, ... " .10

Sad Iron Stands.

Half Size Cut of Nos. 2, 12 and 22.

No. 2, Plain Finish,	per dozen,	$0 75
No. 22, Coppered,	"	80
No. 12, Japanned, Polished on Top Surface,	"	90

Half Size Cut of Nos. 3, 13 and 23.

No. 3, Plain Finish,	per dozen,	$0 85
No. 23, Coppered,	"	90
No. 13, Japanned, Polished on Top Surface,	"	1 00

Nos. 2 and 3 are loose; other numbers, one dozen in a package.

Coffee Pot Stands.

Half Size Cut of Nos. 1, 2 and 3.

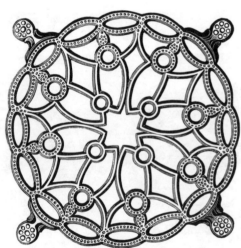

Half Size Cut of Nos. 23 and 25.

No. 1, Japanned,	per dozen,	$1 10	No. 23, Coppered,	per dozen,	$0 75
No. 3, Coppered,	"	1 30	No. 25, Tuscan Bronzed,	"	90
No. 2, Fancy Bronzed and Striped,	"	1 50			

One dozen in a package.

Milwaukee, Wisconsin.

GIRLS' TOY SAD IRONS.

PERFORATED IRON HANDLE.

WALNUT HANDLE.

Handle not Detachable.
No. 105, Plain, Polished, with Stand for Each Iron, - - per dozen, $7 50

Handle Detachable.
No. 90, Nickel Plated, Complete with Handle and Stand for Each Iron, Weight 2 lbs., - per dozen, $7 50

TOY SAD IRON AND STAND.

TOY DUCK AND STAND.

Plain Polished with Coppered Stand.
No. 75, Weight ½ lb. Each, - per dozen sets, $1 50

D, 2¾-inch Long, Fancy Painted, per dozen sets, $1 25
E, 3¼   "   "   "   "   "   1 50

2 1-2 LB. SAD IRON.

Nickel Plated, - - per dozen, $5 50

"GEM" WOOD HANDLE AND STAND.

No. 1, Size 2½ inches high 4 long, 2½ wide.
Polished, Japanned, Bronzed Stand, per doz. sets, $3 00

355

Wm. Frankfurth Hardware Co.

SAD IRONS.

STAR. WITH PERFORATED IRON HANDLE.

DOUBLE POINTED. SQUARE BACK.

 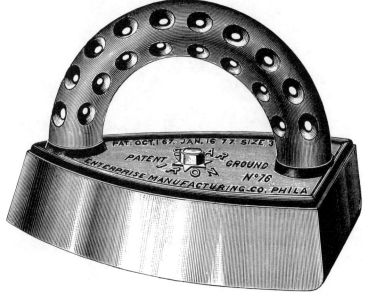

Nos. 70 and 75.   Nos. 71 and 76.

No. 70, Nickel Plated, Assorted Nos.	1, 2 and 3,		per dozen,	$7 25
75, Plain Polished, " "	1, 2 and 3,		"	5 50
71, Nickel Plated, " "	1, 2 and 3,		"	7 25
76, Plain Polished, " "	1, 2 and 3,		"	5 50

POLISHING IRONS.

STAR. MRS. POTTS' PATENT.

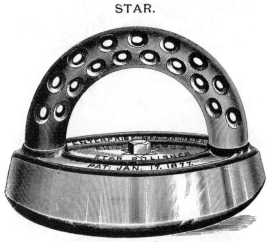 

Nos. 72 and 77.   Nos. 80 and 82.

No. 72, Nickel Plated, Weight, each, 3 pounds,		per dozen,	$7 25
77, Plain Polished " " 3 "		"	5 50
80, Nickel Plated, " " 4½ "		"	15 00
82, " " " " 3 "		"	11 00

Two Dozen in a Case.

Milwaukee, Wisconsin.

SENSIBLE SAD IRONS.

The above Cut represents a Full Set, which Consists of Three Irons, One Stand and One Handle.

No. 2, Nickel Plated, with one each No. 5, 6 and 7 Iron, - - - - per set, $2 50

The Scientific American says: Put a little common white wax in your starch, say two ounces to the pound; then if you use any thin patent starch, be sure you use it warm, otherwise the wax will get cold and gritty, and spot your linen, giving it the appearance of being stained with grease. It is different with collar starch—it can be used quite cold; however, of that anon.

Now, then, about polishing shirts; starch the fronts and wristbands as stiff as you can. Always starch twice, that is, starch and dry, then starch again. Iron your shirt in the usual way, making the linen nice and firm, but without any attempt at a good finish; don't lift the plaits; your shirt is now ready for polishing, but you ought to have a board the same size as a common shirt board, made of hard wood, and covered with only one ply of plain cotton cloth.

this board into the breast of your shirt, then take a polishing iron, which is flat and beveled a little at one end—polish gently with the beveled part, taking care

not to drive the linen up into wave-like blisters: of course this requires a little practice, but if you are careful and persevere, in a short time you will be able to give that enamel-like finish.

No. 4, Nickel Plated Gem Polishing Irons, - - - - - per dozen, $10 00

This Iron is a combined Smoothing and Polishing Iron, and is pronounced by practical laundry men to be the most perfect article of its kind that has ever been placed on the market. The friction point being directly under the hand, makes it easy for the operator.

CHARCOAL AND IRONS.

SELF-HEATING.

FAMILY.   **TAILORS'.**

Nos. 3 and 4.

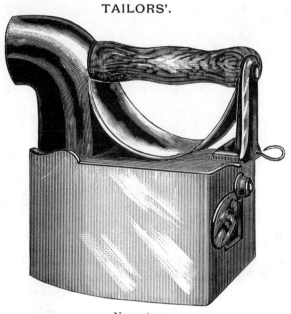
Nos. 1 to 3.

No. 3,	Charcoal,	Family Irons,	Wood Handle,	with Shield,	Weight,	Each,	6½ Pounds,	-	per dozen,	$10 00
4,	"	"	"	"	"	"	"	" 7	"	- -	"	10 00
1,	"	Tailors'	"	"	"	"	"	" 14	"	-	"	22 50
2,	"	"	"	"	"	"	"	" 16	"	- -	"	22 50
3,	"	"	"	"	"	"	"	" 18	"	-	"	22 50

Family Irons, One Dozen in a Box; Tailors' Irons, One-half Dozen in a Box.

DOUBLE FLUE.

LAUNDRY.   **TAILORS'.**

Laundry,	Weight, Each,	8½ Pounds,	- - - - - -	per dozen,	$25 00
Seamstress,	" "	14½ "	- - - - - -	"	35 00
Tailors',	" "	20 "	- - - - - -	"	40 00

These Irons have double Flues at the back which are used to regulate the heat, and a ventilator on top for protecting the hand from the heat.

❦ Milwaukee, Wisconsin. ❦

IMPROVED IDEAL HAIR CURLERS.

Improved Ideal Hair Curler and Frizzer, - - - - - per dozen, $6 00

A new device for curling and frizzing the hair. The hair is wound around the outside shell while cold, avoiding any danger of burning the hands. It will not scorch the hair. The heated iron not being brought in direct contact with the hair, it will not black the hair, face or fingers.

DIRECTIONS: With spring beneath, catch up the hair and wind around shell. When in position insert the heated core, heating shell as desired.

LANGTRY FRIZZING IRONS.

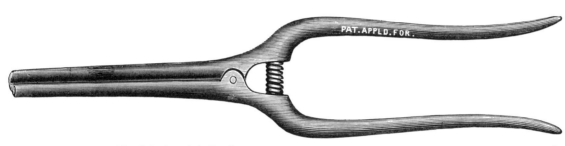

No. 4, Length, 9¼-inch, Half Nickel, with Spring, - - - - - per dozen, $4 00

No. 6, Length, 9½-inch, Full Nickel, Polished Wood Handle, - - - - per dozen, $5 00

PINCHING IRONS.

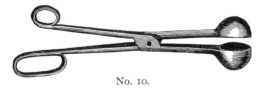

No. 10.

CURLING IRONS.

No. 2, ⅝x10-inch, Solid Steel, Polished, Hardwood Handle, per dozen, $3 00

FLUTING SCISSORS.

Five Prongs, Spring, Wooden Handles.

No. 2, 3-prong Fluting Scissors,

FLUTING MACHINES.

Royal.

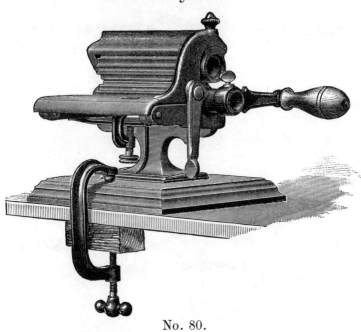

No. 80.

No. 80—6 inch Rolls, 4 Heaters and 1 pair Tongs; 15 and 18 Flutes, ... each, $3.50

American.

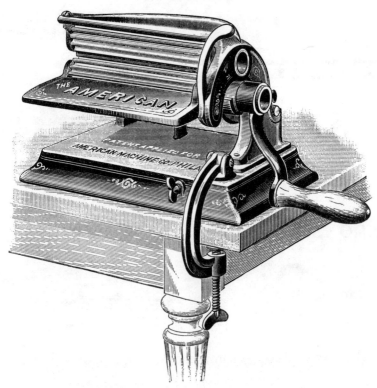

5 inch Rolls, 4 Heaters and 1 pair Tongs; 15, 18 and 22 Flutes, ... each, $3.50
6 inch Rolls, 4 Heaters and 1 pair Tongs; 15, 18 and 22 Flutes, ... " 4.00

HAND FLUTERS.

Shepard's.

No. 85—Hand Fluters, 2½×3¼ inch Rolls, Nickel Plated, Two Heaters with each Fluter, . each, $1.15

Geneva.

Geneva Fluters,	each, $1.00
White Metal Bases, extra,	" .75
Brass Bases, Extra,	" 1.00

Plaiters.

Young's.   Universal.

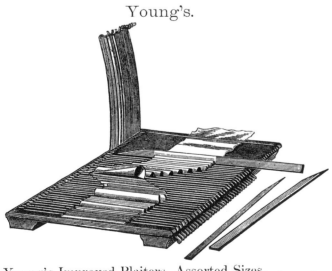

Young's Improved Plaiters, Assorted Sizes,	each, $1.25
Universal Plaiters,	" .60

BIBLIOGRAPHY

Early Reference Material

Appleton's Cyclopaedia of Applied Mechanics, 1896. New York.

The Great Industries of the United States, 1872. Horace Greely. J.B. Burr, Hyde & Co., Chicago.

Hill's Manual of Social and Business Forms, 1887. Chicago. Approved Methods of Speaking and Acting in Life.

The Household, 1871. Brattleboro, VT. Monthly Journal Devoted to the Interests of The American Housewife.

The Metal Worker, 1877, 1892. New York. A Weekly Journal of The Stove, Tin, Plumbing, and House Furnishing Trades.

Moore's Rural New Yorker, 1871. A National Illustrated Weekly Rural Family Newspaper.

Scientific American, 1864–1876. New York. A Weekly Journal of Art, Science, Mechanics, Chemistry, and Manufactures.

Trade Catalogs Represented

Benj. Allen & Co., 1891, Chicago. Silverware and Clocks.

Bridge & Beach Mfg. Co. 1905. St. Louis, MO. Superior Stoves.

Butler Brothers, 1905. General Merchandise Catalog. New York.

Hammacher Schlemmer & Co., 1914. New York. Hardware.

The Howe Exhibition Catalog, 1876. Sewing Machines.

Manning, Maxwell & Moore, 1902. New York. Hardware.

Marples, William & Sons, 1909. Sheffield, England. Tools.

Marshall Field & Co., 1890, 91, 93 and 1915. Chicago. Clocks, Lamps, Sewing Tools, and Silverplated Ware.

Russell & Erwin Mfg. Co., New Britain, CT, 1865 and 1875. Catalogs of American Hardware.

Sargent & Co. New Haven, CT. 1884. Hardware Catalog.

Simmons Hardware Company, 1881 and 1905. St. Louis. House Furnishing Catalogs.

J.H. Sutcliffe & Co., 1896. General Merchandise Catalog.

Union Hardware & Metal Co., 1903. Los Angeles, CA. Tools.

Weed & Company, 1915. Buffalo, NY. Hardware, Houseware.

Wm. Frankfurth Hardware Co., 1886. Milwaukee, WI. Farming Tools and House Furnishing Goods.

Books & Price Guides

American Clocks and Clockmakers, 1989. Robert & Harriett Swedberg. Wallace-Homestead Book Company, Radnor, PA. 19089.

American Copper & Brass, 1908. Henry J. Kauffman. 1979 reprint by Bonanza Books, New York.

American Science and Invention, A Pictorial History, 1954. Mitchell Wilson. Bonanza Books, New York.

Antique & Collectible Thimbles & Accessories, 1990. Averil Mathis. Collector Books, Paducah, KY 42002.

Antique Builder's Hardware, Door Knobs & Accessories, 1982. Maud L. Eastwood. Tillamook, OR 97141.

Antique Irons, 1972. Gerald J. Rihn. Glenshaw, PA 15116.

Antique Wood Stoves, Artistry in Iron, 1975. Will and Jane Curtis. Cobblesmith Pub., Ashville, ME 04607.

The Branding of America, The Forgotten Fathers of America's Best-Known Brand Names, 1987. Ronald Hambleton. Yankee Books, Camden, ME.

The Chronicle of The Early American Industries Association. Reprinted in Two Vols. 1933 to 1973. E.A.I.A., P.O. Box 2128, Empire State Plaza Station, Albany, NY 12220-0128.

Collector's Encyclopedia of Graniteware, 1990. Helen Greguire, Collector Books, Paducah, KY 42002.

Corkscrews for Collectors, 1981. B. Watney and H. Babbidge. Sotheby Publications, Totowa, NJ.

Fine Wicker Furniture, 1990. Tom Scott. Schiffer Publishing, Westchester, PA 19380.

A Guide To American Trade Catalogs 1744–1900. Lawrence B. Romaine, 1960. Reprint by Dover Publications, Mineola, NY 11501.

Handbook of Thimbles & Sewing Implements, 1988. Estelle Zalkin. Warman Publishing Co., Radnor, PA 19089.

Keys, Their History & Collection, 1974. Eric Monk. Shire Publications, Cromwell House, Aylesbury, Bucks, HP179AJ, United Kingdom.

New Light on Old Lamps, 1984. Larry Freeman. Century House, Watkins Glen, NY.

Oil Lamps, The Kerosene Era in North America, 1976. Catherine M. Thuro. Wallace-Homestead Book Co., Radnor, PA 19089.

Oil Lamps II, 1983. Catherine M. Thuro. Thorncliffe House, Inc.

Over The Counter And On The Shelf, Country Storekeeping in America 1620–1920. Laurence A. Johnson, 1961. Charles E. Tuttle Company. Rutland, VT.

The Padlock Collector, 5th edition. Franklin M. Arnall, Claremont, CA 91711.

Railroad Switch Keys & Padlocks, 1990. Don Stewart. Key Collectors International, P.O. Box 9397, Phoenix, AZ 85068.

The Sewing Machine — Its Invention and Development, 1980. Grace Rodgers Cooper. The Smithsonian Press, Washington, DC.

Silverplated Flatware, 1990. Tere Hagan. Collector Books, Paducah, KY 42002.

Smoothing Irons, A History and Collector's Guide, 1977. Brian Jewell. Wallace-Homestead Book Co., Radnor, PA 19089.

Standard Guide to Key Collecting, 3rd Edition, 1990. Don Stewart. Key Collectors International, P.O. Box 9397, Phoenix, AZ 85068.

Thimble Collector's Encyclopedia, 1983. John J. von Hoelle. Dine-American Advertising, publishers. Wilmington, DE 19810.

300 Years of Housekeeping Collectibles, 1991. Linda Campbell Franklin. Books Americana, Florence, AL 35630.

300 Years of Kitchen Collectibles, 3rd Edition, 1991. Linda Campbell Franklin. Books Americana, Florence, AL 35630.

The Vanishing American Outhouse, A History of Country Plumbing, 1989. Ronald S. Barlow. Windmill Publishing Co., El Cajon, CA 92020.

Veteran Scales and Balances, 1978. Brian Jewell. Midas Books, Tunbridge Wells, Kent, England, TN3 0NX.

Periodicals Reviewed

Antiques & Auction News. A weekly advertiser for East Coast shows and auctions. P.O. Box 500, Mount Joy, PA 17552.

Antiques & The Arts Weekly. Reviews auctions, shows, and museum exhibits. Lots of dealer ads. Bee Publishing Co., 5 Church Hill Rd., Newtown, CT 06570.

Antique Trader Weekly. One of the oldest and best. Nationwide buyers and sellers of 19th and 20th century antiques and collectibles fill the pages of this weekly tabloid. P.O. Box 1050, Dubuque, IA 52001.

Antique Week. Shop, show & auction ads, old books, collectibles. One of the best midwestern tabloids. Mayhill Publishing Co., P.O. Box 90, Knightstown, IN 46148.

Country Collectibles Magazine. Articles and photos. 19th and 20th century antiques, collectibles. Harris Publications, 1115 Broadway, New York, NY 10010.

Maine Antique Digest. "The Marketplace for Americana," a monthly newspaper of the antiques trade. Shops, shows, auctions, dealer gossip, and more. P.O. Box 1429, Waldoboro, ME 04572.

Price Guide Section

CAUTION: A word about our prices. We do not buy or sell antiques, we just record prices. Our staff has gathered information from dealers, collectors, shops, shows, auctions, and flea markets all over the country, and we have combed the pages of hundreds of collector publications, including those listed in the bibliography. All of this amounts to a two-year effort to make some sort of sense out of a nationwide market for many items that have no previously recorded price history. Our advice for those who wish to sell antiques and collectibles is to do your own homework and then get several offers on important pieces. Neophyte collectors would be well advised to visit museums and historical society collections, and learn first-hand how to distinguish reproductions and fakes from factory-made items of a century ago.

Page 24
These Brass and Iron "Andirons" of the 1880's are based on designs of 100 years earlier. Factory-made andirons of the Victorian period sell for $150-$400 a pair.

Page 25
More examples of Victorian andirons, value $150-$400 per pair. Older 18th and 19th century andirons can bring from $1,000 to $2,000. Signs of more recent manufacture are the fine concentric rings resulting from the spun brass process, and lightweight stamped-out parts.

Page 26
Brass Candlesticks. Most of the circa 1880's examples on this page can be purchased for $50 to $100 (or less). The No. 40 Snuffer runs about $45-$60. No. 30 Snuffer on a Tray $75-$100. No. 389 Bicycle Lamp $45-$75. Earlier cast brass candlesticks from the 1700's could be worth up to $1,000 a pair, but many styles have been reproduced. (Reproductions start at $5 each).

Page 27
Figural Dragon Candlesticks $40-$60 each.

Page 28
Brass Candlesticks. No. 387 Griffin $65-$95 ea. Nos. 414 and 447 are Dolphin designs which might fetch more.

Page 29
Brass Tea Kettles on stands $75-$150 for this quality. Smaller sizes and reproductions are widely available at $25-$60.

Page 30
Fireplace Tools with gold colored handles can be found separately for $10-$35 each. (Fancy patterns are the highest.)

Page 31
Fire Iron Stands. No. 8 $55-$95. No. 556 $75-$125. No. 134 $55-$95. No. 91 $150-$200. No. 1280 $200-$300. No. 9 $150-$200.

Page 32
Bronzed Fire Iron Stands are fairly common on the East Coast at $25-$50. California prices can be twice as much ($50-$100).

Page 33
Fire Iron Sets came in iron, steel, brass, and bronze plate. Prices vary widely and old brushes are rarely intact. Millions were made, and $35 to $95 is the price range for old and not-so-old alike.

Page 34
Fire Dogs are andirons. These Eastlake designs from the 1870's are worth $150-$300 a pair. They are sought after for use with geometric tile or brickwork.

Page 35
Cottage Fire Iron Sets are 4 to 8 inches shorter than those on preceding pages. These charmers are worth $75-$125.

Page 36
Call Bells were first produced in America by the Gong Bell Manufacturing Co. of East Hampton, Conn. William Barton made his first brass bell in the 1790's, and began casting commercially in 1808 when he moved to Connecticut. Call bells sell for $10 to $40 each in the non-figural designs shown on this page. Add a cupid, an owl, a dove, or a kicking mule, and prices can triple.

Page 37
Call Bells Nos. 6600 & 8800 $30-$60. No. 2400 $20-30. The "silver chime" nickel-plated brass hand bells can be purchased for under $35, and are still being made today.

Page 38
More Hand Bells. None of these are really "electric." Victorian advertising men used this word much as we used the term "atomic" during the 1950's. No. 213 is a $10 bell. The rest vary in value from $12 to $40. All were produced well into the 20th century.

Page 39
"How to conduct yourself on a shopping trip" was part of the many pages of advice offered in *Hill's Manual of Social and Business Forms*, Chicago, 1887. The contrasts between a country store and its uptown counterpart are vividly illustrated.

Page 40
Cow Bells are rarely found with their original clappers, and never with readable labels. That's why we included these. You can cut them out and paste them on your old bell (and triple its value). Rusty $8-$15. Brass or copper brings more — $35 tops.

Page 41
This 1877 advertisement was clipped from a grocery trade publication. The crude cartoon at lower left is worth $15 by itself. Company officials dropped it after only 3 months.

Molasses Pumps in cast iron $45-$75. Plain, undecorated Stoneware Jugs sell for $40-$60.

Page 42 and 43
Coffee Mills became popular after our Civil War. Prior to serving in this conflict, few enlisted men had ever tasted the energizing brew. But by the end of the war most had acquired a habitual taste for caffeine. Pre-ground coffee was available in stores, but most folks felt safer starting with whole beans. Some of the adulterants found in pre-packaged brands were chicory root, waste barley and rye, nuts, shells, and sawdust. Enterprise Coffee Mills often bring higher prices than other less-fancy makes. Two-wheel models are worth twice as much as single-wheel types. Eagle brass finials on lids are top-of-the-line. Original pinstriping is an important contribution to premium prices. No. 0 $60-$85. No. 1 $75-$150. No. 2 $195-$350. No. 3 $300-$400. No. 5-8 $400-$500. Nos. 9-12 $500-$650. Nos. 13-16 $650-$1,500. Floor models $1,500-$3,500. These are very general estimates, and not all numbers are included. We have seen large countertop grinders for up to $3,000 in mint condition, showing no signs of wear. The Lane Bros. Patent Coffee Roaster looks like a wood stove, and might sell for $300 or more.

Page 44
Coffee Mills are now called coffee grinders. Most of the examples on this page date from 1870-1880. The wall-mounted type sell for $35 to $60 each. No. 5 Norton Bros. $50-$100. Diamond Mill $35-$55. Iron Bracket $35-$55.

Page 456
Coffee Mills made of dovetailed wood with iron tops and tin or brass hoppers currently sell for $45-$80 in antique shops. Many have been reproduced. Look for real signs of wear, including old screws and hardware.

Page 46
Coffee Mills from a 1903 catalog. Wall-mounted $30-$50. Lap Grinders $45-$65.

Page 47
Coffee Roasters offered in this 1881 catalog date from as early as 1859. No. 1 Globe $350-$450. No. 1 Oblate $150-$200. No. 2 Hyde's $150-$200. Linden's patent $300-$400. Sperry's patent $100-$150.

Page 48
We call them **String Holders** today, but a hundred years ago they were known as **Twine Boxes** (even though they are round as a ball). Prior to 1870 there were no paper bags used in stores. Purchases were wrapped in paper and bound with string, deftly pulled from a ceiling or counter-mounted fixture. The circa 1870 holders on this page are $125-$200.

Page 49
Twine Boxes from 1875 catalog. (Note details that are not present in reproductions.) Nos. 45 & 55 $250-$350. Nos. 65 & 75 $300-$400. Heart motif makes them worth the price. Nos. 20 & 30 $125-$175. Nos. 25 & 35 $175-$300.

Page 50
Twine Boxes (String Holders) from an 1887 catalog. No. 23 $65-$125. No. 63 $195-$295. No. 76 $75-$150. No. 31 Bee Hive has been reproduced many times, usually smaller than the original. Look for poor casting detail and rust that can be rubbed off. Repros $35-$50. Old examples $75-$150.

Page 51 Match Safes
In 1829 an English chemist discovered that chlorate of potash would ignite by friction. (Earlier matches were lit by dipping them in a vial of sulphuric acid.) Matches were factory-made in the United States beginning in 1836. By 1870 workmen were dipping a million matches an hour into a paste made of glue, phosphorus, chalk, and sul-

phur. These matches were highly flammable and could ignite quite easily by accident. Coat pockets were burned up, and houses were burned down, until somebody invented the match safe. No. 80½ Verde (antique green) $50-$100. No. 10 $50-$75. No. 2 $40-$60.

Page 52

Match Safes, circa 1887. No. 225 $35-$50. No. 1136 $45-$65. No. 2 plate $40-$60.

Page 53

Match Safes, circa 1884. No. 40 $40-$60. No. 50 $40-$60.

Page 54

Match Safes, circa 1887. Square Tin $40-$60. Twin $35-$85. No. 1184 $45-$60. No. 30 $60-$95. Round $50-$95. Daisy $30-$60.

Page 55

Ornamental Match Safes, circa 1884. All figural patterns have wide appeal. These are worth from $55 to $200

Page 56

Hanging Hooks $30-$50. The woodpecker match safe has been widely reproduced, but a highly detailed old one could bring from $150 to $225.

Page 57

Tobacco Cutters, circa 1865-75. Since the early 1700's tobacco had been a country storekeeper's chief stock in trade. Pigtail tobacco was frequently used as a hard currency substitute and sold for three cents a yard in 1801. Tobacco cutters made swift work of chopping off an accurate "chaw." No. 12 Clipper $65-$125. Morse's Patent $65-$95. No. 2 Jester $250-$500. Griswold (not illus.) $95-$125.

Page 58

Oil Stoves were a welcome relief from the constant heat of wood and coal burning stoves in stifling hot summer kitchens. No. 1 and No. 12 Monitor $85-$125. No. 2 with utensils $150-$200. The Daisy ash sifter saved small pieces of coal for reuse $15-$30.

Page 59

Portable Cook Stove with utensils $200-$350. No. 666 closed $10-$20. No. 666 open $15-$30 (with cup). No. 662 pocket $10-$20. No. 675 with boiler $25-$40.

Page 60

Brass Coal Hods in these traditional styles were originally (and still are) imported from England and Europe. Value $85-$165. Plain japanned tin or galvanized hods $5-$15. Fancy funnel or pinstriped tin $20-$60.

Page 61

Antique Stoves. What can we tell you about cast iron cooking stoves that you don't already know?

Well, start with Count Rumford (a.k.a. Benjamin Thompson), born in Woburn, Mass. in 1753. A Tory, he fled to Britain, and then to Bavaria in 1785 after being accused of espionage. Munich had a beggar population of 2,600, and Thompson created Germany's first "poor farm." In order to feed all those folks efficiently, he invented a modern range which enclosed the old-fashioned open hearth with bricks and sheet iron. Count Rumford went on to invent the vacuum thermos and modern fireplace flue.

The evolution of heating and cooking stoves in America (prior to Ben Franklin's 1742 invention of a cast iron fireplace insert) is hard to trace. All box-shaped stoves became known as Franklin stoves. Dutch heating stoves may have been imported here much earlier and copied by New England foundries. But these stoves were not made with cooktops, although they did have ovens for baking pies and bread. In the 1840's we Americans finally caught up with German stove designers by cutting boiling holes in the tops of our parlor stoves and moving them into the kitchen.

When American housewives forsook the open hearth, they did so with a vengeance. In 1877 The National Association of Stove Manufacturers issued a directory of stove, range, and furnace names which listed 4,275 names and trademarks in current use.

The stoves shown on page 61 are circa 1876-1890. American Range No. 88 $900-$1,500. No. 2 Uncle Sam $1,500-$2,000. Richmond, mantel closet $3,000-$5,000. Magic Welcome $2,000-$4,500. Mermaid Cook $300-$500. Laundry Stove $300-$500. Parlor Cook (figural lid) $900-$1,200.

Page 62

Sadie No. 6 $400-$800. Sam and Extra Sam $500-$1,000.

Page 63

New Argand parlor stove with room for a tea kettle and a small oven for scones. The classical female heads in oval relief add value, but full figures or animal motif would boost value even more $1,000-$2,000.

Page 64

This circa 1875 Oven Range would be worth $6,000-$10,000 if indeed any have survived ten decades of scrap-iron drives. The attached polished-copper hot water tank on its original pedestal base would add another thousand or so. It provided warm water to an upstairs bathroom (a rarity in the 1870's).

Page 65

An 1891 version of Ben Franklin's 1742 box stove. Worth from $200-$400. No. 2 Uncle Sam range of 1877 $2,500-$4,000. The portrait of George Washington and the patriotic name are what give this stove extra value.

Page 66

(Top) A Model Kitchen of the 1870's. Complete with running hot water and a built-in pantry. Rural homemakers did not share these luxuries. Water was at a cold pump 25 feet from the back door.

(Bottom) A Giant Size Majestic Coal and Gas Range. Popular from 1890-1920. Today's gourmet cooks admire these nickel-plated steel monsters and will pay $1,000 for one. Smaller stoves are worth more.

Page 67

Hoosier-style Kitchen Cabinets were originated by The Hoosier Manufacturing Co. of Newcastle, Indiana, in 1898. The example pictured here is a top-of-the-line model of 1905. Several Indiana furniture factories made dozens of styles. Well known brands include McDougall, Nappanee, Boone and Hoosier. Some cabinets have built-in flour bins and a slide-out enamelware counter. The most valuable examples have leaded glass windows and a varnished oak or maple finish. Painted Hoosiers are usually worth half as much. Original hardware, labels, flour sifters and glass are important, but reproduction parts are widely available. The McDougall cabinet shown is worth $700-$1,200. The 15-piece German porcelain Spice Set sells for $150-$200.

Page 68

Cast Iron Cook Stove, circa 1890-1905 $800-$1,000 in fully restored condition. (Raised back or warming ovens would increase value).

Page 69

Steel Range with hot water reservoir and high closet $3,000-$4,000 restored.

Page 70

"Flora" **Cast Iron Stove.** As soon as you add a hot-shelf and extras like the tea-shelf (round trivet extension) and some nickel trim, combined with ornate castings and an attractive nameplate, value goes up. Fully restored $2,000-$3,000. The "Laundry" stove would appeal to a sad iron collector for sure $75-$150 (more with a matching set of irons). The "Victor" cannon ball heating stove came in various sizes up to 800 lbs., and this "potbelly" style is quite common $75-$125. "Novel Superior" with overhead warming oven $2,800-$4,000.

Page 71

Heating Stoves. "Clio," "Laurel Todd," and "Pioneer" $150-$300 (more if original urns and nickel trim). The larger "Gem Oak" could be worth up to $1,000 in fully restored condition. A rusty "as found" example might fetch $150-$300.

Page 72

No. 213 "Superior Radiator" is a highly desirable square base baseburner worth from $1,000-$2,500 (stoves are worth whatever you can get for them). The "Sligo" cannonball or potbelly style is very common and worth only $150-$200. Myrtle, Alamo, and Forest are among the very smallest cook stoves $200-$450.

Page 73

"Superior Air Tight" parlor heating stove with lots of decorative nickel trim $400-$1,000. The "Osage" could warm up a wash boiler or a small room $150-$200 "Superior Radiator" $1,000-$2,500.

Page 74

(Top) A Modern Kitchen, circa 1905. (Bottom) An 1890's Hotel Kitchen.

Clear photographs of indoor kitchen scenes of this vintage are very scarce, and worth from $60 to $100 each.

Page 75 and 76

Directions For Cleaning The Kitchen are reprinted directly from an 1881 catalog.

Page 77

Apple Parers were first advertised in the 1860's. This Reading 1878 model is not rare $50-$60 in the East and Midwest. $75-$100 in California. **White Mountain Potato Parer** $35-$45.

Page 78

Herring's Patent Ribbon Slicer for potatoes $75-$100. Saratoga Parer and Slicer (pat. 1870-71) $95-$135., Cherry Stoner, made by famous cutlery firm, Goodell Company, of Antrim, NH $35-$50.

Page 79
  Rotary Potato Parer with lots of great gear wheels $85–$150. **Rotary Peach Parer**, ditto $85–$125. The Family Parer is the same as the one on page 77 $35–$45.

Page 80
  Gold Medal Apple Parer, an $85–$110 value (note heart cutouts). O.K. Potato Parer (peeler) $5–$15. The White Mountain Apple Parer is still being made today, but the gray metal is quite bright compared to 1880's cast iron $35–$50.

Page 81
  Nos. 15 to 35 Enterprise **Lard, Fruit, and Jelly Press**, in rack or screw movement $50–$75. Combination Fruit Press $50–$65.
  (Lower left) Enterprise Save-All Fruit, Lard, and Jelly Press $75–$100. Henis's 1881 patent potato masher or ricer $25–$40.

Page 82
  Egg Beaters Silver's Measuring Glass and Egg Beater $65–$90. Silver's Egg Timer $12–$20. Dover Famly Size Egg Beater, first patented July 14, 1885, was so popular and so universally known that cookbooks referred to all rotary egg beaters as "dover." This size can't be beat as a bargain for $25–45. **Pot Cleaners** have been widely reproduced, but an old one might fetch from $25–$65. **Skewers** like this factory-made example are not hand-forged, so figure $45–$65. (They are still made in Mexico and sold here as antiques.) Lightning Egg Beater $35–$45. Triumph Plate Lifter $25–$45.

Page 83
  Wire Spoon Egg Beater $18–$25. Wire whip style has been made almost identically for the past 150 years. So, unless yours is a huge monster or otherwise very unusual, under $15. Spiral $10–$20. Tumbler insert, Push-Pull style $30–$45. No. 4154 Acme $75–$100. No. 4153 "Easy" letters cast in crank wheel $75–$100. Monroe's Mammoth $100–$135. Peerless $30–$40. No. 3 Dover $50–$70. P.D. & Co. (Paine, Diehl & Company). A classic beauty, much sought after by collectors $85–$120.

Page 84
  Globe Egg Beater $35–$45. Tumbler Size No. 10 $40–$50. Family Size No. 1 $25–$45. Extra Family No. 2 $25–$45. Hotel Size No. 4 $50–$75. Lightning, churn-style $50–$65. Kingston, oval-bladed $40–$55. Glass Egg Beater $45–$60. French Whisk $5–$20. Wire Spoon $18–$25. Glass Cutters $4–$12.

Page 85
  Woodenware of all types is desired by Kitchen Collectors. They are looking for an attractive grain pattern and nice old patina. So if yours is a Plain-Jane, price it accordingly. Steak Maul $30–$60. Rolling Pins $10 for a common solid-handle. Up to $120 for a beautiful birds-eye maple example. Glass with revolving wood handles $40–$70. Pool's Potato Masher $10–$25. Mix's Potato Masher $10–$25. Early Maple or Glass Masher $35–$40. Porcelain Head Potato Masher $35–$50. Common Hardwood (about 12 in.) $10–$35.

Page 86
  Tap Borers (to tap barrels for the goodies inside) $5–$15. **Wood Faucets** $5–$15. (More money for a 16-incher with brass tags.) Cabbage Borers $5–$15. Fruit Augers $85–$125. Everything came to the storekeeper packed tightly in barrels. An auger was the way to get it out. Butter Spade $10–$18.

Page 87
  Vegetable Cutter $20–$40. Nos. 1–2 Slaw Cutters $25–$45. Nos. 3–8 Kraut Cutters $25–$50. Handmade ones with shaped, crested, or "gravestone" boards or heart cutouts, are worth five or ten times as much as these factory-made examples.

Page 88
  Improved Cylinder Churn $35–$65 on the East Coast ($75–$95 Midwest and California). Charm Barrel Churn $40–$95 East Coast ($80–$195 Midwest and California). Swing Churn $40–$95 East Coast ($80–$195 Midwest and California). Wooden Dash Churn $40–$95 Eastern price ($80–$195 Western price). Lightning Churn $40–$100. Milk can $45–$75.

Page 89
  Gem Nutmeg Grater $65–$90. Rajah Patent, box-style $100–$125. Cylindrical Tin handled grater often has both fine and coarse surfaces. Wood handled half-round is a very old style. And the plain flat type was, of course, the cheapest! If yours is a hundred years old, it's worth $15–$45. (Fine Old Brass Graters of English origin are worth considerably more.) No. 712 Houchin's Patent coconut and horseradish grater $75–$125. No. 709 (smaller) Family Size $75–$125. No. 04 Panel-back with lid $7–$12. No. 7 The Edgar Nutmeg Grater $55–$95.

Page 90
  Bread Plates have been widely reproduced (then oiled and aged). A genuine old bread board with a carved motto $65–$100. Knives with B-R-E-A-D carved in handle are also faked and new, or old, they sell for $10–$45. Plain handle bread knives sell for $5–$15 (pg. 160).

Page 91
  Wooden Bowls, machine made. From $30–$75. Many hand carved reproductions have arrived lately from the Third World. They are usually soft woods, stained or aged. **Butter Pats**, machine-made $20–$30 a pair. Biscuit Beaters $75–$125. Wooden Spoons $20–$35 if nice patina. Butter Prints (stamps) machine-carved on machine-turned blanks $30–$50 East Coast, $45–$85 West Coast. Butter Molds, machine-carved on machine-turned blanks $40–$55 East Coast, $45–$85 Midwest and Far West. Older prints, stamps, and moulds that are handcarved sell for $70–$200 in common designs such as wheat sheaf, flowers, acorns, pineapple, thistles, leaves, and stars. Rare patterns sell for up to three times as much. They are eagles, chickens, lollipops, hearts, swans, and cows. All have been reproduced for the past 50 years, so be cautious! **Butter Ladles** $30–$60 if good grain pattern. Wooden Scoops $40–$65 with nice patina.

Page 92
  Hinged Butter Moulds $15–$20. Round (see pg. 91 price listings). Rolling Pins, Opalite white glass $40–$65. Plain handle maple, 12 in. $10–$35. Stained-handle maple with revolving roller $20–$65. Fancy Cake Roller $65–$125.

Page 93
  Salt Boxes, banded woods, $65–$125. Spice Cabinets (Sears sold these for 48 cents each in 1910) $85–$200. German factory-made **Spinning Wheels** $100–$350. Steamer Set (3-pc.) $85–$175. **School Slates** $30–$65 (Repairs to slates are o.k.).

Page 94
  No. 10 Banded Knife Boxes $35–$75. No. 20 Banded Knife Boxes $35–$50. Knife Scouring Boards $45–$85. Skirt Boards $10–35. Pastry Boards $15–$30. Meat Blocks $5–$15. Bosom Boards $10–$30. Pine Stepladder $65–$150. Sugar Kanakins $75–$150. Nested Boxes $75–$150. Snow Shovel $25–$85.

Page 95
  Settee Table (or Chair Table). Factory-made examples from the 1880's sell for $200–$400 in the East & $400–$800 on the West Coast. Original handmade 19th century chair tables sell for thousands and reproductions abound. Folding Kitchen Table $100–$195. Ironing Table $50–$125. Wash Bench $75–$150. Wooden Pails $40–$65.

Page 96
  Can Openers were first patented in 1858, about forty years after an army food supplier, Nicholas Appert, invented a preserving process for Napoleon's troops. Peter Durand, an Englishman, received an 1810 patent for "an iron can coated with tin and its lid soldered on." Tin cans were produced by food packagers in America starting in 1820, and became widely used during the Civil War, where troops used pocket knives and wood chisels to open their canned rations. From 1860 to 1890, dozens of designs vied for consumer acceptance. (Those pictured here are circa 1877–1887.) No. 5 $7–$15. Star $25–$35. No. 20 Can Opener $15–$30. World's Best, Lyman's Patent $15–$30. No. 2 World's Best, wooden handle $30–$45. No. 1 $15–$30.

Page 97
  Sargent Can Openers $20–$50. No. 5 $10–$15.

Page 98
  Streeter's Skewer-Puller $15–$20. Messenger's Adjustable Comet $15–$20. Lyman's 1870 Patent, Revolving $15–$30. Poole's Steel Bar $15–$30.

Page 99
  Lemon Squeezers. Dean's No. 2 japanned/glass $50–$70. Drum, porcelain/iron $50–$70. King's, wood/iron/glass $55–$75. Manny's Crystal, 1885 pat. $35–$45. 1868 Patent, porcelain/iron $45–$65. Wood Muddler, Toddy stick $25–$35.

Page 100
  No. 5 "Eureka" **Lemon Squeezer** $70–$100. No. 4 "Boss" Lemon Squeezer $50–$65. No. 12 Patent (cuts and squeezes) $75–$125.

Page 101
  Mutton Holder $10–$20. Bread Rasp $10–$20. (Not all of these burnt-crust-removers have curved blades or wooden handles.) Paste Jagger (solid brass) $20–$40. Nut Crackers: 4-piece/4 hinge $30–$45. All others $8–$20. (Second from right is also used as a lobster crack.)

Page 102
  Lightning Bar Lemon Squeezer $55–$75.

Modern Cork Puller $65–$90. Champion Cork Squeezer $65–$90. Boss Lemon Squeezer $50–$65. King, with glass insert $65–$90.

Page 103

Lemon Squeezers. Manny's Nos. 1 & 2 glass reamers $35–$50. No. 22 Little Giant $45–$65. Crown, iron with hardwood bowl $50–$65. Nos. 1 & 2 Wood Frame $35–$60. (These were made well into the 1900's, but collectors will pay for nice patina.)

Page 104

Mincing Knives. The (6) models at top of page are mass-produced, but very much resemble earlier hand-forged mincers $15–$45. Smith's Patent, wire-braced type $25–$50. All are circa 1880.

Page 105

Lightning Mincing Knife. $45–$60. No. 0 Polished Maple $15–$25. No. 6 Cast Steel $15–$25. No. 15 Saw Steel $15–$25. No. 5 japanned iron $15–$25. German Rocker $15–$30. (All of these rocker-style mincing knives were used with wooden chopping bowls.)

Page 106

Syllabub Churn $35–$50. A syllabub was a peppy drink of milk and liquor, first mentioned in the late 1700's. (Row 3) Small Cake Cutters $10–$20 ea. (Row 4) Heart Patty Pans $5–$10. Milk Skimmer $8–$15. Apple Corers $10–$20. Star Patty Pans $15–$25. (Row 5) Heart-shaped cutter $10–$25. Diamond $10–$25. Horse $30–$60. Those pictured here were turned out by the thousands in factories. But because of their crude shapes and flat backs, they are often mistaken for earlier handmade cutters, which sell for $100 or more in 6–7 inch sizes, and $500, $600, and $700 in figural 11–12 inch styles such as an 1850's leaping reindeer, a tall Indian girl with topknot, an 11½ inch Indian brave, and an 8 inch fat cat. The world record is $7,400 for a 7½ by 12 inch running slave. Itinerant tinsmiths used a heavier sheet stock and made much deeper cutters (up to 1 inch) than factories. Their crude solder showed plainly. Shiny solder in an almost invisible line indicates machine-made tinware.

Page 107

Fancy Tin *Vegetable Cutters* on cup bases. No. 1 Boat Shape $25–$40. No. 1 Pinwheel $35–$50. No. 3 $35–$50. No. 11 Tube Cutter Set $40–$65. No. 1 Tube Cutter Set $40–$65.

Page 108

Planished (polished) Tinware. Pepper Box $25–$40. Flour Dredge $25–$45. Etnas were not used as much for sharing as for warming hot drinks, with alcohol poured into the saucer base and lit $100–$200. Oval Melon Moulds were in production for 50 years $35–$70. Berlin Moulds $35–$70. Jelly Moulds $35–$75. Rice Moulds $35–$65. Confectioner's, pâte, or French Pie Mould $25–$60. *Reproduction Alert*: Any tinware item can be cheaply reproduced. Factories in Mexico still produce many of the pieces in this section. American manufacturers have made colonial-style tinware artifacts since the 1920's.

Page 109

Stamped Tinware. Turban Cake Mold $25–$35. Octagon $20–$30. Scalloped (no tube) $20–$35. Turk's Head, with tube $25–$35. Scalloped, with tube $25–$40. Milk Strainer $15–$25. Turks Head 6-cup Pan $15–$25. Colanders $30–$45. A fancy pattern hole design would bring twice as much as this Plain Jane.

Page 110

Tin Coffee Pots (Row 1) $25–$65. Etna on stand $30–$60. Polished Tea Pot $25–$65. Two-piece Coffee Maker $35–$75. Straight-sided Coffee Pot $25–$60.

Page 111

More examples of "Stamped Tinware," which was a great improvement over the earlier "pieced" variety, because it contained fewer joints to rust out. (Note that stamped tinware is a much lighter gage of metal and not as desirable to advanced collectors.) Flat Bottom Tea Kettle $20–$35. Gravy Strainer $15–$40. Tall Tea Kettle $25–$40. Wash Bowl with rings $25–$35. Plain style $20–$30. Retinned Wash Bowl with handle $35–$50. Water Dipper $20–$30. Straight sided cup $5–$15. Regular Drinking Cup $5–$10. Patent Cookie Pan $15–$25. Sponge Cake Pan $15–$25. Tea Kettle (bottom right) $20–$40.

Page 112

Copper Tea Kettles (Row 1) polished $35–$75. Nickel-plated $25–$40. (Row 2) Spun Copper $35–$70. Nickel-plated $25–$45. (row 3) Modern shaped "Aladdin" is circa 1910 $25–$65. (Bottom right) Chicago spout $25–$65.

Page 113

Pieced Tinware. Tea Kettle at upper left is identical in shape to early handmade copper ones, which are worth three times its value of $65–$90. (They will have dovetailed joints.) Jelly Funnel $15–$25. Tea Steeper $28–$45. Range Tea Kettle $45–$65. Coffee Funnel $15–$35. Oil Stove Tea Kettle $25–$50. Candle Mold (4 to 12 tubes) $60–$100. The oldest molds of only 2 or 4 tubes are often more valuable to collectors. Decorators go for the newer multi-tube variety, which are still being produced today. Coffee Flask $45–$70. Pudding Mold $45–$75. Egg Poachers $45–$75. (Not at all rare, but very much in demand.)

Page 114

Pieced Tinware Coffee Pots. (Row 1) Bossed (Braced) Handle $35–$50. Solid Lip $45–$65. Hinged Cover $35–$50. (Row 2) Copper Bottom $40–$65. Straight-sided with black handle $35–$50. Copper bottom, straight handle $40–$65. (Row 3) Tea Pots (left) $45–$65 (middle) $50–$75 (right) $45–$65.

Page 115

Brass Kettles. These are lightweight, Hayden's Patent, spun brass; turned out by the thousands starting in 1851 and selling today for $35–$85 each. (Heavier, hand-wrought, brass, or copper kettles are worth much more.) American Broiler, 1868 patent $40–$75. Spun Copper Tea Kettle $45–$85. Epicure Broiler $40–$75. Self-Basting Broiler $35–$50.

Page 116

Deep Stamped Tinware. Flat Handle Cup Dipper $10–$15. Wood Handle Cup Dipper $15–$25. Threaded Handle Soup Ladle $10–$15. Wood Handle Soup Ladle $15–$25. Flat Handle Ladle $5–$10. Wood Handle Ladle $15–$25. Flat Handle Deep Skimmer $7–$15. Wood Handle, Deep Skimmer $15–$30. Flat Handle, Flat Skimmer $7–$15. Wood Handle, Flat Skimmer $15–$25. The flat skimmer was also called an "Egg Slice," but was used only for lifting. Patent Muffin and Corn Cake Pans $15–$25 each.

Page 117

Grocers' Scoop, Open Top $20–$40. Grocers' Scoop, Closed $20–$40. Wood Handled, Retinned $20–$45. Ring Handled, Tin Scoop $15–$25. Solid Steel, Wood Handled $20–$45. Solid Wood Scoop (not shown) $40–$65. Cooks' Wood Handled $15–$25. Pocket Oil Stove $15–$25. Oyster Stand, Complete $55–$100.

Page 118

Cast Iron Hollow Ware. Scotch Bowl $45–$65. Yankee Bowl $50–$75. Tea Kettle $50–$75. Ham Boiler $40–$60. Bake Oven $45–$60. Sad Iron Heater (without irons) $35–$55.

Page 119

Cast Iron Hollow Ware. 3-Legged Pot $30–$50. 3-Legged Kettle $30–$50. Spider (fry pan) $20–$30. Long Pan $20–$30. (This fits into the hole left when 2 burner lids are removed.) Handled Griddle $20–$35. Bailed Griddle $20–$35.

Page 120

Cast Iron Bake Pans. All of these designs fetch from $45 to $65 each. They were first manufactured by Nathaniel Waterman in the mid-19th century (1850's).

Page 121

No. 11 French Roll Pans $45–$65. No. 21 Vienna Roll Pans $45–$65. No. 10 Deep Corn Cake or "Popover" $35–$60. (Still being made today.) No. 12 Oblong Pans $45–$65. No. 20 Vienna Bread Pan is the most unusual of the bunch, and worth from $65 to $100.

Page 122

Waffle Irons. (Top) Raised Base $75–$125. Griswold $75–$125. Short Handles $55–$75. French 8-holer $150–$200. Soapstone Griddles, Oval $50–$65. Round $50–$65 (no grease required). Foot Warmer $35–$45.

Page 123

Early Waffle iron, circa 1860 $100–$150. French Fryers, Large $55–$125. Small with Fry Pan Base and 1879 Patent (not shown) $55–$100. Excelsior Waffle Irons, Round $75–$125. Square $50–$100. Long Handled Wafer Irons $65–$100. (They look older than this 1881 catalog illustration.)

Page 124

Granite Ironware took the country by storm in the late 1870's. Once housewives were convinced the enamel could not poison their food (a rumor circulated by tinware manufacturers), they bought every shape that factories offered. Bright colors and swirling patterns soon followed. The prices we quote here are for grey granite color. Light and dark blue, green, brown, pink, and red are often priced at up to three times the original granite color. Near perfect condition is desired by folks who pay these prices. $12–$20 for common spoons or dippers. $300–$500 for the chafing dish (lower right). $200–$300 for the coffee pots (row four). The slotted broiler at $250 and the 5-piece measure set for $300–

$500 are also pretty scarce. Most other pieces on this page would fall between $45–$100.

Some recent record-shattering auction prices: Dolly's Favorite Toy Stove in grey granite iron (circa 1915) $6,000. Salesman's sample pewter-trimmed lady's cuspidor 2½ x 3 inches $4,700. Caster set with engraved metal trim on grey enamel $3,100. Butter Churn in blue and white swirl pattern $1,400. Child's ABC Plate with raised letters $2,000. Toy kitchen utensil rack with tools $1,300. Child size pitcher and bowl on enameled iron stand $1,450. Handcranked vegetable grater, round throat $170. L & G oval platter $85. Oyster patty $350. Spit cup with hinged lid, iron handle $225. Toy bathtub, oval shape $250. Utensil rack with skimmer and spoons $400. Set of six measures embossed with capacity $600. Washboard $125. Pewter-trimmed spooner $775. Green and white swirled coffee pot $350.

Page 125

Grey Graniteware Prices. (Row 1) Coffee Pot $150–$225. Tea Pot $275–$350. (Row 2) $225–$275. The smallest one in the middle might bring more. (Row 3) The straight angled spouts make these two pots more rare than the others, to the tune of $350–$450 each.

Page 126

Grey Graniteware Syrup Pitcher, with Manning & Bowman's typical Greco-Roman finial and no-spill spout $450–$600. Slop Bowl $275–$325. Cream Pitcher $275–$335. No. 4000 Creamer $300–$350. No. 4000 Slop Bowl $275–$325. Butter Dish $350–$450. Sugar Bowls $300–$365 each. (All pieces have nickel-plated trim.)

Page 127

Grey Graniteware Covered Bucket $75–$95. Slop Jar (for under kitchen sink) $65–$90. Wine Cooler $65–$90. Chamber Pot $65–$90. Water Carrier $250–$275. Pattie Pan $60–$90. Water Bucket $75–$90. Water Bucket with spout $200–$250. Cake Mold $65–$85. Dinner Plate $30–$45. Tumbler $100 (a rare shape, once common, few survived). Cup with Saucer $45–$60. Miner's Cup $30–$50.

Page 128

More Gray Graniteware. (Green and red colors can be worth up to three times as much.) Round Cake Griddle $100–$115. Pitcher and Bowl $125–$150. Spoons $10–$20. Ladles $40–$60. Dippers $40–$70. Fry Pan $60–$70. Jelly Cake Pan $55–$65. Egg Fryer $55–$75. Double Boilers $210–$250. Milk Pan $60–$80. Mountain Cake Pan $40–$55. Dish Pan $50. Pudding Pan $40–$55. Funnel $45–$60. "Belle" Tea Pot $55–$70. Lipped Sauce Pan $80–$110. Wall Soap Dish $20–$35.

Page 129

Graniteware Tea Kettles, pit bottom $150–$200. Heating Stove Size $125–$175. Flat Bottom $125–$150. Cast Iron Granite Pot $125–$175. Ketle $125–$175. Tea Kettle $150–$225. Spider $125–$150. Preserving Kettle $125–$175. Stove Kettle $125–$175. Round Bake Pan $40–$60. Flat Bottom Stove Kettle with Lid $75–$95. Lipped Water Pail $200–$250. Wash Bowl with ring hanger $35–$55. In grey, this Coffee Boiler might be worth $175–$225 in very good condition. In green or brown, mottled with white, twice the money. Red and mottled white may be up to four times as much as the basic gray granite color.

Page 130

"Perfection" Graniteware was made by one of the largest firms, Manning & Bowman, of Meriden, Connecticut. They offered "Handsomely Mottled Colors," on this 1886 catalog page, including pink, maroon and mauve (the prices we quote are for gray). No. 5500 Tea Pot $175–$225. No. 8500 Coffee Pot $175–$225. No. 4100 Tea Pot $150–$200. Coffee, ditto.

Page 131

Perfection Graniteware No. 4700 Coffee Pot $225–$275. No. 2000 Series $225–$275 in gray (up to $400 for exotic colors). Soup Ladles Nos. 113 & 114 $50–$70.

Page 132

French Coffee Pots or "Biggins." $100–$175 (the smaller, the better). "Good Morning" Coffee Makers. No. 6800 finely decorated enamelware was handpainted before firing $300–$400. No. 3600 Series $200–$250 with white metal trimmings.

Page 133

No. 9200 "Stork in Bayou" pattern $250–$350. Mottled Grey $250–$350. No. 8000 Enameled Tea Pot $225–$275 with violets. No. 4000 series in Mottled Grey $275–$350.

Page 134

Water Coolers. Nos. 10, 20, and 804 all sell for about $65 to $125 each. The ornamental iron stands bring the same price. Nickel-plated Hotel or Restaurant Coffee Urns are still cheap for their size and form (and they make great fern stands). $25–$85.

Page 135

Stevens' Water Filters could process 15 to 30 gallons of muddy river water an hour into cool clear spring water (I think they were stuffed with layers of charcoal, sand, and gravel.) $40–$65 each. Centennial Cooler and Jewett Filters $30–$65 each. No. 73 Filter/Cooler with portrait panel $75–$150.

Page 136

Refrigerator/Sideboard $200–$500. Cedar Chests $200–$400. Nos. 2, 3, 4, & 5 Ice Boxes $200–$350. Nos. 0 & 1 Alaska Refrigerator $250–$500 each.

Page 137

Centennial Ice Chests $100–$200. Centennial Refrigerators $200–$450.

Page 138

Metal Ice Cream Moulds (not pictured, Animals $45–$200). No. 589 & 365 $30–$45. No. 310 Shell $30–$50. No. 325 Masonic $30–$50. No. 562 Auto $65–$85. No. 580 Grapes $30–$45. No. 37 Battleship without flags and figures $75–$125. The composition figures are scarce, especially the black people at $200 & up. Boat/Cupid $65–$95. Mother Goose $75–$100. Female Diver $65. Cupid w/bow $55.

Page 139

Ice Cream Freezers $35–$85 (Condition and label designs are important). Economy Ice Cutter $20–$45. French-style Ice Pick $10–$18. Four-prong chipper, 1873 patent $10–$20. No. 33 Cup-shape ice shaver $20–$35. Steak Pounder/Ice Shaver $25–$45. Crown Ice Chipper $20–$35. Gem Shaver $20–$35. Saw $15–$25.

Page 140

White Mountain Ice Cream Freezer $35–$55. Peerless Ice Cream Freezer $35–$55. Hotel Size Ice Cream Freezer $45–$85. Ice Cream Dishing Spoons $10–$25. Keywound Cornet Dishers $20–$45.

Page 141

American Machine Co. Freezers $35–$85. Triple Motion (with flywheel) $65–$110. Wonder 5-Minute Freezer $45–$85 (top price if wonderful label is intact).

Page 142

Champagne Tap $15–$30. No. 10 Nutcracker $60–$95. Nos. 1 and 11 Cork Pressers $65–$100. Nos. 2 and 12 Cork Pressers $85–$150.

Page 143

Cork Screws from an 1881 catalog. No. 25 Folding Pocket $35–$65. No. 45 Detachable Wire $10–$20. No. 42 French Level $55–$85. No. 30 Combined Ice Pick $15–$30. Clough's 1876 Patent $10–$20. No. 35 Spring Steel $10–$20. Champagne Cutter $5–$15.

Page 144

Cork Screws (top row) $4–$10 each. An advertising imprint on wood case adds value. Sperry's Patent $50–$100. Rodger's Champagne Nippers $45–$95.

Page 145

English Cork Screws from a circa 1900 catalog. No. 6950 $10–$20. No. 6951 $7–$15. No. 6952 $7–$15. No. 6953 $10–$20. No. 6954 $25–$60. No. 6956 with brush to clean off wine cellar dust (brushes are rarely found intact) $25–$65. No. 6957 $15–$25. No. 6960 $55–$100. No. 6962 $45–$100. No. 6963 $4–$10. No. 6966 $15–$25. No. 6969 $4–$10. No. 6971 $10–$15. No. 6972 $10–$15. No. 6972½ $10–$15. No. 6973 $35–$85. No. 6975 $15–$22. No. 6978 $25–$45.

Page 146

Cork Screws from an 1886 catalog. No. 45 $20–$30. Magic $20–$30. Globe $25–$50. Hercules $45–$75. Stag Handle $35–$45. Power $45–$65.

Page 147

No. 51 Cork Screw $7–$15. No. 52 $10–$20. No. 66 $10–$18. No. 123 Schinnocks' Patent $35–$50. Walker' Self-Puller $30–$45. No. 2 with brush, and a knife for cutting wires that held champagne cork in place $25–$45. No. 32½ Solid Metal Duplex $20–$35.

In the 1600's, corkscrews were called "steele worms." Samuel Henshall was the first inventor who bothered to patent one (in 1795). Prior to 1981, when the Watney and Baddige landmark book *Corkscrews for Collectors* was published, there were not enough collectors to fill up an average wine cellar. Since then, several museum exhibits and antique shows have featured this screwy collectible on a grand scale. Some high-priced corkscrews not illustrated here are: Folding ladies legs with stripes $300–$600. "Old Snifter" $55–$220. French bronze with cherubs $1,785. Two-pronged Mumford $300. W.W. Tucker patent 1878 fulcrum arm $715. Scrimshaw snake $275. 1905 Plant's patent $240.

Page 148

Table Manners from an 1887 publication.

Victorian Silverplate is emerging from a long-neglected period and prices have not yet stabilized. Collectors are looking for very unusual pieces with birds, bells, wheels, cherubs, and animals. (Art Deco hollow ware is rising rapidly, too, although not covered in this book.) Replating is an expensive process, so bright, mint pieces are going to command top dollar. Now is the time to buy up all those gargoyle and dragon mounted tea sets, cow creamers, and butter dishes, bulldog ash trays, and topless nymphs of all kinds. Victorian silverplate has a long way to go, and prices on the East Coast are still very low.

Page 149

Caster Sets of glass condiment bottles in silverplated frames were the standard centerpieces on Victorian dining tables. These 4, 5 or 6 bottle sets evolved from earlier groups of up to 10 containers on a flat tray. (Rancid meat needed a lot of seasoning?) The circa 1880 sets on this page sell for $150–$250 in bright, undented condition. Colored or fancy cut glass raises value.

Page 150

Caster Bottles are generally more valuable than their frames (which have survived in much greater numbers). These examples would fetch from $8 to $30 each. Most all have been reproduced, often in ruby red or cobalt blue. Reproduction bottles are thicker and clunkier looking than the fragile originals.

Page 151

Circa 1890 five-bottle Caster Sets $150–$250 in bright, undented condition. Call bells and cupid figures usually add value, as do famous maker names, or unusual glass.

Page 152

Circa 1892 ***Tilting Water Pitcher*** with many figural design elements, and two goblets. On the East Coast these can still be purchased for $200–$325 each. But a San Francisco silverplate specialist wouldn't take less than $1,000 for this Victorian delight.

Page 153

Electro Silverplated items circa 1880–1890. Lunch Caster Sets $15–$50, depending on design and glass type. Highest prices are on West Coast. ***Cake Baskets*** with medium to heavy decoration $125–175. Salt shakers $10–$18 each. Plain Knife Rests $2–$5.

Page 154

No. 821 Tilting Water Pitcher with cup $150–$350 for this common design with no figural elements. Nos. 88 and 99 are typical 1880's mass-production silverplate $125–$250 in good condition. (Much less if dull or damaged, and twice the price if resilvered.)

Page 155

Silverplated Flatware of the Victorian period came in hundreds of patterns, from plain to extremely ornate. With sterling silver pieces now selling in the $25–$250 range, collectors are turning to electroplated pieces, and prices are rising. Yes, the $1-an-item mixed assortments still abound in thrift shops and flea markets, but you can't buy fancy serving pieces from a "silver matching service" for less than the prices we list on the following pages. The low end of range reflects the East Coast market. Highest prices are found in California shops and shows. (Prices are for bright, unworn plate.)

Sugar Tongs $5–$15. Sugar shells $5–$15. Cream Ladles $5–$10. Table Spoons $5–$10. Dessert Spoons $4–$6. Tea Spoons $3–$6.

Page 156

Silverplated Butter Knives $5–$15. Medium Forks $4–$10. Salt Spoons $3–$8. Pie Knives $10–$25. Coffee Spoons $3–$10. Dessert Forks $5–$10. Mustard Spoons $6–$12.

Page 157

Silverplated Fish Knives $12–$25. Fish Forks $10–$20. Crumb Knives $15–$30. (Prices are for fancy patterns in bright, unworn condition.)

Page 158

Table Knives and Forks with inlaid cocobolo wood (obtained chiefly from the Isthmus of Panama) handles. $5–$12 each.

Page 159

Silverplated Table Knives can be inexpensive replacements for those missing from a sterling silver flatware set. Plain ones go for $2 to $5, and fancy patterns are from $10 to $20 each. One Arm Man Knives were developed following the Civil War $12–$18 each. Butter Knife $10–$25, depending on handle. Fruit Knife $4–$8.

Page 160

Midwestern and California shops have $15–$45 price tags on Bread Knives with B-R-E-A-D carved in their handles. (Reproductions and re-carved abound.) Plain-handled versions are $5–$15.

Page 161

Carving Sets are not yet highly collectible, but sportsmen and gun collectors have started to accumulate stag-handled pieces. Individual "steels" (the sharpener in the 3-pc. set) bring from $5–$20. Nicer Knives sell for up to $25 each. Forks seem to bring less, $5 to $8. The sets on this page would normally fall in the $15–$30 range.

Page 162

Butter Knife, bent $2–$5. Butter Knife, twist $2–$5. Saratoga Cake Knife $8–$20. Saratoga Pie Knife $8–$20. (A lot less money in a Pennsylvania flea market.)

Page 163

Dessert Fork, plain $1–$2. Dessert Fork, fancy $3–$8. Sugar Shell, fancy $4–$10. (Top prices are Southern California.)

Page 164

"Royal" Coffee Spoon $3–$8. "Mayflower" Coffee Spoon $3–$8. "Belmont" Coffee Spoon $3–$8. "Savarin" Coffee Spoon $3–$8. "Belmont" Pie Knife $8–$20. "Mayflower" Sugar Tongs $5–$10. "Mayflower" Fish Fork $6–12. Why didn't we buy a ton of these ten years ago, instead of IBM?

Page 165

Children's Knife, Fork, and Spoon Set $15–$30. (The original 1890's advertisement would fetch about the same price.)

Page 166

Silverplated Pie Knife $6–$15. Siren-pattern Cheese Scoop $10–$24. Medium Fork,"Assyrian"$3–$6. Table Spoon,"Newport"$3–$6. (Again, lowest prices are East Coast.)

Page 167

Salad Fork with Knife $10–$20. Sugar Shell $3–$8, with gilt bowl $8–$15. Gravy, Cream, or Oyster Ladle $4–$12. Magnolia Berry Spoon $8–$15.

Page 168

Souvenir Spoon $5–$12. Cream Ladle $4–$8. Bon Bon Scoop $3–$8. Sugar Shell $4–$8. Sugar Sifter $4–$8. Oyster Fork $2–$4. Pickle Spear $3–$5. (These are not great patterns.)

Page 169

Bar Spoons $1–$3. Nut Cracker $5–$8. Julep Strainer $5–$10. Nut Pick $2–$5. Sugar Tongs $5–$15.

Page 170

Silverplated Tea and Coffee Sets. The chance of finding an 1880's group in fine unrepaired condition on an original tray is remote. Replating costs about $30 a piece, per inch of height. A value range for this set, in acceptable condition, might be $250–$500. If the complete gold wash remains inside any of the following pieces, they would bring top price shown.

Page 171

Silverplated Syrup Pitchers. No. 0812 $15–$30. No. 39812½ $25–$60. No. 91912 $25–$60. Butter Dishes No. 16945 $30–$75. No. 62912 $25–$65. No. 64945 with cow finial on lid $60–$125. (These are all circa 1880.)

Page 172

Silverplated Butter Dishes. No. 2092 $45–$65. No. 87 $50–$95. No. 71 $40–$55. No. 201 $50–$95. No. 176 100–$150 (figural top). The following (15) pages are all from 1891-1893 catalogs.

Page 173

No. 1872 Combination Sugar Bowl and Spoon Holder $55–$75. No. 30 Spoon Rack with (12) Silverplated Spoons $50–$65. No. 75 Celery Stand with clear glass insert $50–$65. No. 66 Celery Stand $20–$30. No. 79 Celery Stand with rose-tinted Venetian glass insert $100–$175.

Page 174

Silverplated Spoon Holders. No. 20 (with rabbit handles) $75–$125. No. 3001 $20–$40. No. 28 Tray $15–$25. Straight-sided Spoon Holder $25–$40. No. 2093 $20–$50. No. 23 $45–$75. No. 9 $40–$55. (Top prices in each range are for gold-plated finish.)

Page 175

Quadruple Plate Silverware. No. 32 Bon Bon $30–$50. No. 31 Bon Bon $20–$35. No. 176 Two Salt Celler with Spoons $12–$30 set. No. 191 Table Salt Celler (Master Size) $30–$45. No. 2 Ink Eraser $5–$8. No. 100 Blotting Pad $10–$20. No. 27 Bon Bon Tray $30–$50. No. 2542 Ink Well $25–$35. No. 5 Trinket Tray $20–$35. No. 1783 Card Tray $15–$25.

Page 176

1890's Rustic Motif Silverplated Cups $15–$20 each. Gilt-lined $20–$30 each.

Page 177

No. 5 Water Lily Salt Celler $15–$25. No. 30 Pepper $5–$20. No. 28 Pepper (plain base) $5–$10. No. 8 Owl Pepper $65–$100. No. 31 Pepper $5–$15. No. 15 Pepper $75–$125. No. 14 Knife Rest $55–$85 each. No. 125 Napkin Ring $3–$8.

Page 178

Tea Set (6) pieces $150–$300 with no tray.

Page 179

No. 400 Square Waiter (Silverplated Tray) $65–$95 on East Coast, $100–$150 in California.

368

Page 180
Silverplated Crumb Sets $15–$25. Brandy Flask, with leaping trout that makes it worth $55–$75. No. 0252 Napkin Ring $4–$10. No. 78 Bird Napkin Ring $40–$60. No. 51 Match Holder $10–$20. No. 079 Toothpick Holder $20–$30. No. 0265 Napkin Ring (with space for monogram) $20–$30. No. 050 Stamp box $8–$22.

Page 181
Pickle Casters became popular table accessories just after the Civil War. The most valuable have colored, cut, or art glass inserts and figural finials. Those pictured are from Marshall Field's 1893 catalog, and would range from $75 to $200 each. The market price for cranberry, blue, etc., with enamel work, can quickly reach $400 or $500.

Page 182
Quadruple Silverplated Items. No. 4100 Mucilage Bottle $25–$35. No. 1994 "Dog" Inkstand $250–$350. No. 445 Baby Mug $8–$15 ($25–$35 if gold-lined). No. 0803 Cup $10–$15 ($25–$35). No. 0805 Cup $10–$15 ($25–$35). No. 0804 Cup $10–$15 ($25–$35). No. 01311 Syrup Pitcher $15–$20. No. 01309 Syrup Pitcher $15–$25. No. 1825 Gravy Boat $20–$30.

Page 183
Quadruple Plate Dessert Sets. Rows 1, 2 & 3 are $50–$80 per set. Solid Sterling Silver Tea and Coffee Sets vary widely in value. Weight, design, date, and maker's marks are all determining factors. This 1890's Gorham set could fetch from $3,000 to $5,000 without a tray.

Page 184
Napkin Rings were at their peak of popularity between 1870 and 1900. (Washing machines made them obsolete?) Each member of a family supposedly had their own individual ring, and most table napkins got more than one meal's usage. Reproductions of figural rings are available. They have an overall evenness to the finish, while antique examples have a strong contrast from light to dark patina. Prices for desirable old napkin rings range from $175 to $400 each. Cherubs, children and animals are the most popular subjects. Prices for those shown are No. 77 Bulldog $95–$175. No. 47 Squirrel $100–$200. Baseball Player $350–$450. Little Girl $125–$150. No. 335 Bird and Wishbone $50–$65. No. 68 Pair of Hares $100–$200. Moving parts, wheels, etc., could add another $50–$100 to value.

Page 185
Silverplated Cake Baskets from the Victorian period $55–$125 each.

Page 186
Quadruple Plate Berry Dishes, also called Fruit Baskets. (Colored glass inserts are the value-adders.) No. 311 $200–$400. No. 307 $200–$300. No. 142 $150–$200. No. 170 $275–$350.

Page 187
No. 147 Berry Dish (Compote-style) $40–$65. No. 312 Crystal or Canary Glass $125–$200. No. 300 Pattern Glass Insert $50–$75. No 2199 in Ivory with blue or yellow edge $125–$200.

Page 188
Japanned Tinware. No. 4 Perforated Fork & Spoon Box $50–$95. Combination Knife, Fork & Spoon $45–$85. Bill Head Box $45–$65. Post Office Box $25–$65. Spice Boxes: Ceylon $50–$75. Novelty $55–$85. Square $65–$125. Round $65–$125. Desk Top $75–$125.

Page 189
Japanned Tinware. Spoon and Fork Box $75–$95. Fancy Flour Box, Round $50–$75. Small Cash and Deed Boxes (locks in working order) $15–$30. Box Grater $45–$65. Tea Canister $45–$65. Green Garden Sprinkler $75–$135. French $95–$150. Square Cake Box $40–$60. Round Sugar $45–$65. Round Cake Box $50–$75.

Page 190
Spittoons & Cuspidors. Assorted colors $50–$65 each. Brass $55–$75 (widely reproduced, beware of affixed nameplates). Grey Graniteware $65–$100 each. Low, wide mouth $35–$50. Chocolate and white granite (not shown) $350. Spit Cups in graniteware are rare! A grey one with a hinged lid just sold for $225. A pewter-trimmed lady's cuspidor in gray granite (2 x 3 inches) set an auction record of $4,700 last year. The plain styles shown here in tin are probably worth at least $50–$75 each. (Partygoers with the chewing habit carried them from room to room.)

Page 191
Dust Pans, Japanned. Corrugated $35–$50. No. 10 $35–$55. Half covered $40–$55. Brush Compartment in Handle $75–$100. "Handy" $35–$50. Crumb Tray and Brush sets, colors $35–$50.

Page 192
Old Carpet Sweepers are not big sellers. These are worth $25–$55 each. Mr. Bissell filed his first patent in 1876. By 1880 these much-improved rug dusters had swept the country from coast to coast. Anna Bissell took over the company at her husband's death, and in 1893 brought out a toy carpet sweeper that saved the firm from a runious depression.

Page 193
Bullard's Carpet Stretcher $20–$30. Hammer-handle style $15–$25. Noyes foot-operated $25–$40. No. 1 socket-handled $15–$25. Once a year all carpets were unticked and taken outdoors for a good beating. Restretching and retacking was often a family affair.

Page 194
Bird Cage Trimmings from the 1880's are scarce. Opal fount $30–$45. Flint glass $25–$40. Wall Hooks $10–$25 (large cast iron or brass bring more). Swing $3–$5. Wire Nest $10–$20. Opal Baths $30–$50. Brass Springs $3–$10. Fancy Cage on Stand $175–$200.

Page 195
Revolving Perch Bird Cages. Nos. 602 & 603 $125–$200. Squirrel Cages Nos. 6 & 7 $150–$300.

Page 196
Japanned Tin Bird Cages. No. 46 $175–$250. No. 62 $100–$125. No. 15 $100–$125. No. 41 $125–$175. No. 83 $125–$175. No. 21 $150–$200. These are all circa 1880–1890. Plain white arch-top wire cages from 1900–1930 sell for $125 with floor stands.

Page 197
Brass Bird Cages on this page are valued at $85–$160 each.

Page 198
Mouse Traps are collectible! The "Delusion" self-setting $50–$55. Circular wood mouse traps with up to 6 holes appear in 1860's thru 1900's catalogs, and sell for $15–$20. Carriage jacks made of iron $25–$40. Wooden "Champion" $40–$75.

Page 199
Rat Traps, circa 1885. Lovet's cross-shaped choker $35–$55. Wire Rat Trap 6 x 11 inches $20–$30. 12 inch Wood Bottom Round Wire Rat $25–$35. 12 inch Globe, Straight Wire Side Bars $35–$55. Small, Round Wire Mouse $55–$65. Slayer Wooden Rat Trap $20–$25.

Page 200
Mouse Traps circa 1870–1880. 4 to 6 Hole, Choker $15–$25. Revolving Wire $50–$55. Perpetual (with lid) $45–$50. Catch-Em-Alive (tin) $60–$75. Marty (rat size) $25–$35.

Page 201
Fly Traps and Swatters. Tin Roach Trap $40–60. Harper's Column-shaped Wire Fly Trap $40–$50. Balloon Shape (with tin bottom present) $45–$60. Wind-Up Fly Fan (the whirling shadow scared flies silly) $400–$600. 1880's Fly Swatters $10–$25 each.

Our contributing expert for the Trap Section is Boyd Nedry of Comstock Park, Michigan.

Page 202
Foot Scrapers, Japanned (dark brown or black varnished) Iron. Nos. 12 & 20 $50–$100. No. 25 $65–$150. First sold circa 1860–1880, but widely reproduced over the years.

Page 203
Boot Jacks. Cast Iron, or Brass Nos. 3 & 13 $65–$90. American Bull Dog $95–$250. Nos. 2 & 12 $100–$175. No. 25 $95–$150. No. 15 $95–$135. Ripley's Automatic $50–$75. No. 4 Beetle has been made off and on since the 1860's. Repros sell for $30–$75. Antique examples $90–$135.

Page 204
Japanned Toilet Ware. Umbrella Stand $50–$100. Towel Stand $50–$100. Toilet Stand with Pitcher & Bowl (plain style) $125–$150. Toilet Stand with Decorated Pitcher, Bowl, Pan & Cup $200–$400. Soap Bracket $15–$35. Toothbrush Stand $10–$20. Sponge Basket $15–$30. Brush & Soap Holder $65–$100. Closet Paper Rack $15–$30.

Page 205
Japanned Toilet Ware. Originally sold in sets, but few have survived as such. Most often found are the decorated water carriers. Condition determines price. Nos. 321, 310, 316 $85–$100 each. No. 303 $75–$95 each. No. 103 $125–$20 each. No. 104 $85–$125 each. (Bottom Left) Bowl & Pitcher with banded colors $125–$175 per set. No. 145 Foot Tub $95–$135. No. 325 Pitcher $85–$135.

Page 206
Child's Bath Tub $85 in tin, up to $200 for a copper one.

Page 207
Bath Tubs from the 1880's are great decorator pieces. We've seen them at prices ranging from $75 for a small, plain, foot tub to $250 for a wood bottom plunge bath.

369

Page 208
Turnbull's Family Scales $200–$350. Turnbull's Market Scales $150–$195. The Market Scales were popular for 50 years.

Page 209
Spring Balance Scales. No. 70 $20–$35. German "Crab," Hide, or Ice Scales $35–$75. Nos. 50, 51, & 81 $15–$20. No. 84 $20–$35. No. 160 $35–$70. Letter Scales $45–$75. Package Scales $95–$150. Tobacco, with Brass Scoop $100–$150. (No Scoop $75–$95).

Page 210
Circular Face, Spring Balance Scales (large numerals for market use). Advertising imprints, e.g., cattle feed, bring top dollar. Price range for all on this 1884 catalog page: $75–$195 each.

Page 211
Family Scales (circa 1877–1881). No 170 $65–$125. No. 192 $75–$135. No. 198 $125–$200. "Windsor" $50–$75.

Page 212
Grocers' Scales. Even Balance with funnel scoop $100–$175. No. 661 Fairbanks Even Balance $100–$150 (ditto with side beam). No. 554 Counter Scales $100–$175. No. 547 Butter Scales $125–$175. Note: Top of price range includes some weights and a brass scoop.

Page 213
Grocers' Scales. Nos. 132 & 133 $125–$175. No. 161 Tea Scales $125–$175. No. 11 Counter Scales $150–$225. No. 44 Platform Scales $95–$175. (Surviving weights, scoops, and original paint are important price criteria.)

Page 214
Columbia Family, Spring Scales $30–$75. Universal Family Scales $30–$75. Both are circa 1900. Brass faces would add value. N. 580 Tea Scales $100–$150. No. 260 2 lb. Scales $100–$150. No. 32 Turnbull's Market Scales $1090–$195. No. 261 4 lb. Balance Scales $150–$200. (All of these are circa 1860–1890).

Page 215
Chatillion's Market Scales with Marble Slab $100–$225. Fairbanks Grocers' Platform Scales, Double Beam, Brass Scoop $100–$225.

Page 216
Steelyards $25–$60. Scale Beams $35–$70. Both styles were easily portable for field crops, hardware, feed, etc.

Page 217
Fairbanks Platform Scales, on wheels $125–$250. Used in hardware and feed stores, mills, lumber yards, farms, factories, freight depots, etc.

Page 218
Champion Beef Shaver $50–$65. Enterprise Cheese Knife $75–$100. No. 1 Cherry Stoner $30–$45. No. 750 Bone, Shell and Corn Mill $40–$75. (Farmers made their own chicken feed and eggshell hardner mixes.)

Page 219
Sausage Stuffers. No. 1 Perry's Patent $60–$90. Railroad, Tin Cylinder $90–$125. No. 2 Silver's Patent $90–$125. Nos. 3 & 4 Double Geared $90–$135. Anyone who has ever seen the actual ingredients used in these gut-stuffers will never eat another hot dog! Nothing was wasted, ears, eyes and bung holes all found their way into the mix.

Page 220
Sausage Stuffers. No. 1 Wagner $100–$150. Brecht 20 lb. $100–$150. Brecht 35 lb. $100–$150. Birkenwald Lard Press $75–$125.

Page 221
Sausage Stuffers. Nos. 22 & 24 $85–$135. Perry's 1874 pat. Lever Action $90–$150. No. 100 Woodruff's $85–$135. The Champion $85–$135. No. 32 & 33 Meat Cutters $100–$150.

Page 222
Meat Grinders (or Food Choppers) are very common on the West Coast. Gem-type grinders, circa 1900, run $5 to $15 each in every antique mall in California. Eastern and Midwestern dealers price them at $30–$45 each. The older, larger, Enterprise meat choppers sell nationwide for $35–$55.

Page 223
Starret's Meat and Vegetable Choppers are the ultimate Kitchen Collectibles (and cheap at $125–$250 each). These Rube Goldberg contraptions really worked well and made a fortune for their 29-year-old inventor, Leroy Starret. Mr. Starret was a dairy farmer by trade and a machinist at heart. He conceived the "Hasher" by watching a walking beam steam engine at work. Leroy whittled out a wooden model with a racheting mechanism that slowly turned the food-holding container while rapidly raising and lowering a guillotine blade. After receiving a patent in 1865, he began peddling the choppers all over New England. By 1868 he had formed his Athol Machine and Foundry, which went on to become the nation's largest manufacturer of machinists' hand tools. No one has ever found one of these gadgets with its complete original label intact, which gave operating instructions.

Page 224
The first "Lawn Mowers" were flocks of sheep that manicured the park-like estates of the landed gentry. In the 1870's, patented hand-pushed and horse-pulled machines began to appear on the domestic market. Horse-drawn $100–$300. Excelsior Hand Mower $65–$125. Earlier model with roller $85–$150.

Page 225
Pitcher Pumps were often used indoors, mounted on the kitchen drainboard with a slop bucket underneath. On the East Coast these can still be picked up for $10 to $25. In Southern California antique shops they range from $50 to $85. Cistern Pumps are also cheaper in farm country ($10–$30). They are $75 to $100 in the big city. Brass-Barreled Force Pump (no tub) $15–$30. Well Bucket $20–$40. Chain Pump Fixture $10–$25. Well Wheel $10–$25.

Page 226
Well pumps. The box-enclosed variety were called Curb Pumps, and drew water up by cups or chain valves. They do not appear often for sale, but have great decorative value (try for $150–$250 in the big city). Iron well pumps are dirt cheap in the East, and $100–up on the West Coast (it's a freight factor).

Page 227
People don't collect Post Hole Diggers (yet). They just use them, and are willing to pay $20–$40 for a sturdy looking tool.

Page 228
Mounted Grindstones run from $75 to $195 on nice oak frames. (The pedal-powered variety on angle iron bases with sheet metal sets are more modern.)

Page 229
Axes and Hatchets, circa 1860–1880. Unmarked hatchets average $8 to $15 each. Winchester, KeenKutter or Marbles trademarks can up the price three times as much. Carpenter adzes average $25–$35. Short handled, handmade, adzes with curved blades for bowl- or chair-making are very rare. Factory-made broad axes rarely sell for more than $45–$65 unless found with original handle and label. By 1875 over 25 different shapes of felling and hewing axe heads had evolved from those favored in Pennsylvania, Michigan, and California.

Page 230
Hay Knives were used to saw up bales. Heaths $25–$50. Spear Point $10–$25. Scimiter Blade $10–$25. Lightning $20–$35. Gem $20–$40.

Page 231
Hay Rakes, wooden $35–$100. Grain Cradles, with scythes $75–$200. Coal Forks $25–$50. Potato Scoop $35–$65.

Page 232
Wood cased Corn Shellers are very decorative in original pinstriped paint with charming advertising imprints. We have seen them at flea markets for $85, and in big city shops for $250 (weathered ones bring much less). Burrall's cast iron corn sheller, on stand $100–$200. Hand-held cast iron model $45–$75. Seed Stripper $65–$135. Poultry Mills for corn or shells $40–$75. Eagle Cider Mills $85–$250.

Page 233
Corn Huskers. Halls' Patent $5–$10. Brinkerhoff's Patent $5–$10. Husking Gloves $10–$25. Buckeye $5–$8. Corn Planters are $15–$25 in farm country and up to $40 on the West Coast.

Page 234
Ox Shoes $5–$10. Bull Snaps $10–$20. Bull Rings $3–$10. Cattle Leader $5–$10. Ox pins $3–$8. Ox Yokes $75–$150 (more if good handcarved ornamental details, or if sold in Texas).

Page 235
Hog Tongs $5–$15. Animal Catchers $10–$25

Page 236
Garden Wheelbarrow $200–$300. With bentwood legs $250–$400. Common Railroad Barrow $150–$200.

Page 237
Pitch Forks (Barley Forks), Steel $25–$35. Hawkeye $25–$50. Three-prong Wooden Field Fork $25–$50.

Antique Hardware is collected by antiquarians, builders, restorers, and art lovers. Victorian hardware combines design elements from Japan, Egypt, Greece, Rome, Persia, France, England, and Germany. It is eclectic to say the least, and appeals to a broad spectrum of the population. Prior to the Centennial Exposition in Philadelphia in 1876 and the introduction of William Eastlake's design influence, most hardware was either plain or lightly embellished with leaves or flowers, or perhaps a patriotic motif comprised of flags or eagles. But designers of the pre-1876 period had not felt compelled to cover every single inch of surface with the proliferation of birds, bees, vines, acorns, dolphins, gargoyles, lions, bears, butterflies, and geometric patterns that we now associate with the late Victorian period. Reproductions abound, but mostly they are practical "replacement hardware" for early 1900's pieces such as hinges, knobs, handles, hooks, brackets and escutcheons.

The three guest experts who priced the next 40 pages of antique builder's hardware are: **Don Stewart**, publisher of 32 books and pamphlets on locks and keys. For a complete list of related books and old catalog reprints, send a 50¢ stamp with your name and address to Key Collectors International, Post Office Box 9397, Phoenix AZ 85068.

Next is *Liz Gordon*, of *Liz's Antique Hardware*. Specializing in "Door, Furniture, Bath & Lighting Hardware." Ms. Gordon's large van full of inventory can be seen at the Long Beach Veteran's Stadium Swapmeet and at the Rose Bowl Flea Market. Both are Sunday events. Write to her at P.O. Box 16371, San Diego, CA 92116, or call (619) 284-1075 for an appointment.

Last but not least is *Muff's Restoration Hardware and Antiques*, 135 South Glassell Street, Orange, CA 92666. You will find a store full of old and new hardware that staggers the imagination. Send $5.00 for Muff's catalog of "Reproduction Hardware for Furniture, Hoosier Cupboards, Roll-Top Desks, Doors, Trunks, Locks and Keys." They are open Tuesday–Saturday. Or call (714) 997-0243 and ask for Gary.

Both of the above dealers are wholesale buyers of antique hardware, by the box or store-full. Other reproduction hardware dealers can be found in the pages of *The Old House Journal* and *Preservation News*, or in most collector publications listed in the bibliography.

Some of the prices that follow may astound those of you who live in areas where Victorian buildings are still being demolished, and old hardware can be bought for a song. Now is the time to pack it up in barrels and "Go West, young man."

Page 238
Shelf Brackets, circa 1870–1885. All are $15–$35 sellers (old or new). Reproduction iron is less at $8–$15 each. New cast brass brackets are $15–$35. Old brackets are worth more in sets or pairs. Individual pieces are hard to sell.

Page 239
Coat and Hat Hooks, circa 1880. All prices are per hook. Top end of range is for sets of 4 or more. Single items are slow movers. No. 640 $5–$10 each. No. 339 $12–$20. No. 350 $15–$25. Nos. 61 & 62 $8–$12.

Page 240
Coat and Hat Hooks. No. 57 $12–$18. No. 93 $6–$10. No. 7½ $6–$10. No. 3½ $8–$12. No. 490 $20–$40.

Page 241
Coat and Hat Hooks, circa 1875–1887. No. 345 $15–$30. No. 66 $15–$30. No. 600 $14–$25. No. 20 $5–$10. No. 500 $8–$15. No. 113 $6–$12.

Page 242
Furniture Feet. Nos. 050, 060 & 77 Ball & Claw. Old ones are $8–$12 each. Reproductions $5–$15. No. 4395 brass paw $10–$20, iron $5–$10. No. 6190 $8–$15. No. 4586 $10–$25. No. 34 $18–$35.

Page 243
Mixed Hardware from a 1915 catalog. Lion head with leaves $6–$12. Plain lion head escutcheon in cast brass $5–$9. All of the oval and round key-surrounds are $6–$8 each in old cast brass, and $2–$3 for lightweight stamped repros. French-curved double hooks are all $15–$30 each if old, and $6–$12 in similar reproductions. Gargoyle heads, single $12–$24, double $18–$35. These have also been reproduced. The old originals are cast iron with a thin copper plate finish.

Page 244
Sargent's Mortise Knob Lock. Full set with keys, knobs, and fancy plates $185–$245. The locks alone can be picked up at flea markets for $5–$8. Reconditioned, with new springs $25–$40. Sargent keys for above $1–$2. Special cut-to-fit costs extra. Fancy Victorian door key, 4 to 5 inch, brass $12–$16, iron $8–$10. This style made from 1850–1930. Common door keys $1–$4 in iron. Barn door keys $6–$8 each. Folding keys for pocket doors and store doors $8–$10 iron, $12–$18 brass.

Page 245
Plain Bow Cabinet Keys have little value ($1) to collectors, but when you need one, expect to pay $2–$5 each at an antique shop.

Page 246
Taper, Sprig and Double Bit Cabinet Keys run from $2 to $6. (I purchased a two-ton lot of these a couple of years ago and sold them all to dealers for 50¢ each. The original owner was a scrap metal yard in Great Britain. Those days are over.)

Page 247
Fancy Bow Cabinet Keys $2–$6 for average iron key. No. 1175 $8–$10. No. 5201 $8–$10 in gold plate. No. 1106 $6–$8. No 1010 Dolphin head is probably the most widely reproduced of all cabinet keys, $5 new. No 5084 is the rarest on this page at $14 in gilt.

Page 248
Mixed Escutcheons 1875–1900. (Row 1) No. 423 $4–$7. Oval, old $4–$6, repro $2–$4. No. 870 $4–$7. No. 890 $5–$8. Shield $8–$10. Horizontal pierced, old $5–$8, repro $3–$5. Filigree, old $6–$8, repro $2–$4. No. 131 $6–$10. Vertical pierced, old $5–$8, repro $2–$3. Double hole, engraved $7–$14. Double hole, plain $8–$12. No. 533 $3–$6. No. 506 $3–$6. No. 508 $5–$7. No. 515 $6–$10.

Page 249
Rose and Escutcheon Combined. (A rosette is the ornamental round plate surrounding the door knob shaft; a collar of sorts.) Top row are all $12 to $18 each in brass. Escutcheons on bottom row are $6–$9 each.

Page 250
Ornamental Flush Escutcheons were made for the sliding "pocket" doors used in Eastern United States. They would bring $20–$35 each on East Coast, and only $12–$15 in California. Sun Rim Latch Set could go anywhere from $50–$75 for 3 pieces. Unembellished, plain cast iron, rim locks are common at $5–$10 each.

Page 251
A 15-inch Bronze Store Door Lock and handle set would command $225–$350 from a hardware specialist. I've seen only one of these in 20 years of dealing.

Page 252
Bronze Bank Door Pulls $85–$150 each. These are from Sargent's 1884 catalog.

Page 253
Bronze Push Plates for swinging doors $35–$75. Butterfly and bird design should bring more.

Pages 254 and 255
Bronze and Plated Door Hinges (Butts) circa 1880. One dealer said she wouldn't take less than $100 a pair for this quality. Another said he did not have them often, but sells all for $15–$35 each. Reproductions of varying designs sell for $30–$50 a pair, and up.

Page 256
Mineral or "Bennington" reddish-brown door knobs sell in restortion shops for $18–$45 a pair with collars and shafts. Individual knobs can be found at flea markets for $5 each. White Porcelain and Black Porcelain sets sell for $18–$25, but are very common and usually cheaper in general line shops, often $10 for a knob and an old black rim latch. No. M1542A Bronze Knob and rosette $15 to $30. No. 1872P Light Bronze is rarer, but not as pretty, so $6–$12 with collar.

Page 257
Ornamental Door Knob. Nos. M1814A, K1434A and M1832A are White or Black Porcelain knobs with bronze rosettes. The combination is $9–$12 each, or $18–$24 a pair. (Rosettes alone are $5 each.) Nos. 1452, 1872, 1882, etc. are solid Bronze Knobs and Collars at $18–$35 each or $36–$65 a pair, including 2 knobs, 2 rosettes, and connecting shaft. Similar reproduction sets in brass are $12–$28.

Page 258
Rural Cupboard Catches are scarce, and worth $12–$15 per set of 2 pieces. No. 318 has been reproduced, $4–$8 new. Brass Engraved styles are also hard to find, and worth from $12 to $20 each. No. 1260 has been reproduced, $6–$10 new.

Page 259
Cupboard Turns from 1884 catalog would be rare finds easily worth $15–$25 each. No. 4240 has been copied, $9–$12 new.

Page 260
Ornamental Wall and Corner Brackets from an 1875 catalog. All would be rare finds and worth from $30 to $95 each. The double-decker, at 21 inches high, could bring $125.

Page 261
Flower Pot Brackets. Double arm $35–$50. Single arm $22–$35. Wire Floor Stands from the 1880's are worth $350–$800, depending on size and condition. Similar reproductions are currently being offered to the decorator trade for $850–$1,000 each.

Page 262
Victorian Door Bells came in chain-pull, lever-action, wind-up, and turn-button styles. Complete sets are worth $50–$95 or more. A plain non-embossed repro is available at $18–$28.

Page 263
Door Bell Lever Pulls. All are $15–$35 retail, and circa 1885.

Page 264-65
Letter Box Plates are rare survivors and worth $35–$75 each. The inside door plate is also valuable, $15–$30. A medium quality reproduction of No. 126 is available at $18–$28. Check with your postmaster before installing a vintage mail slot. Size requirements have changed since 1880.

Page 266
Star Padlocks from the 1870's are indeed rare. I have never seen No. 124, 3¾ inch size, but it is worth at least $160 or more. Little ones go for $25–$40. Miller's Patent Brass Padlocks are $40–$50.

Page 267
Scandinavian Padlocks were made by a dozen manufacturers in the U.S. between 1850 and 1950. They were used on jail cells, strong boxes, barrels, store doors, farm buildings, and chuck wagons $10–$50. Big brass ones bring top dollar. Key style 304 $6. Others $2. Horse Shoe Padlock $45–$60. Barnes is the common one. Brown's Patent Padlocks are so scarce God only knows what they are worth. Maybe $100 to $300 if circa 1871.

Page 268
Eagle Lock Company Padlocks. No. 4010 $25. No. 4002 $100. No. 4011 $20–$40. No. 4008 $10–$20. No. 4012 $20–$40. No. 4004 $20–$40. Some more, some less. Condition is everything.

Page 269
Hammacher Schlemmer's 1915 Padlock assortment. Mastodon 8-Lever $8–$15. Corbin 2-Lever $5–$10. Eton $4. H.S. & Co. 6-Lever $60–$80 for lock, $6–$10 for scarce key. Railroad Locks (lower right) require a price guide of their own. Some examples: Adlake steel $15. A & CAL $45. BH & ERR brass $55. B.R. & P. $65. CM & STPRR $60–$90. C & ORY $85. CI & LRR $90. CRI & PRR $125. D & RG $75–$125. D & HC Co. $175. DL & WRR $90. FFW & WRR $150. FRISCO $18. GM & NRR Co. $110. GPCO $90. HVRYCO $125. KCFS & MRR $125. LACKAWANNA $45. L&N $32. SO. PACIFIC $50–$90 and on, and on.

Page 270
Small Padlocks for dog collars, deed boxes, etc. are getting very scarce, Nos. 30 & 72 $8–$20 each. Wrought Iron Tumbler Padlocks are commonly sold for $5 or $10 in rural areas where they were used on every smokehouse and tool shed. But I have seen Nos. 201 and 404 in antique shops for $15–$20 each and separate keys at $2.

Page 271
Trumbler Trunk Locks from 1860's and 70's should bring $25–$35 if found intact. The example at lower right has been reproduced, without date, and retails at $22.

Page 272
Pendant Pattern Drawer Pulls. Engraved bronze, circa 1875 $8–$15 each.

Page 273
Bronze Plated Iron Drawer Pulls Nos. 52 & 61 $8–$12. Nos. 42 & 43 $10–$20 each.

Page 274
Drawer Pulls, Nos. 516, 1903R, 420½, 636, 501R, 455, 475R, 61, 150 and 71 are all quite scarce and worth from $8 to $12 each. Sets increase value over singles. Those Nos. followed by an "R" have been reproduced in brass and sell for $3–$6 new.

Page 275
Druggists' Drawer Pulls, old $8–$12. Similar repro $3–$5 new. No. 50 Household Drawer Pull $8–$12 old, similar repro $4–$6 new. No. 61 $8–$12. No. 96 $10–$15.

Page 276
Bronzed Iron Chest Handles and Fancy Animal Head Drawer Pulls from 1875 catalog have not yet been reproduced. Estimated value $18–$36 each.

Page 277
Furniture Pulls of the Late Victorian Period from a 1915 replacement hardware catalog. "Colonial" Lion Heads come in many sizes $8–$35 each, Repro $4–$6. No. 864 is typical of 1,000's of similar designs $6–$12 each, Repro $3–$6. Nos. 775, 1952 & 4417 $6–$12 old stamped, $8–$15 old cast brass. Repros. $5–$8. No. 5180 is a Sewing Machine Drawer Pull $8–$20 (old ones are hard to find). Similar repro $4–$8.

Page 278
Chandelier Hooks have not been reproduced because of limited demand. No. 37 $6–$10. No. 450 $8–$15. No. 240 $6–$10. No. 470 $10–$20. No. 460 $6–$10. Dolphin, 4 inch screw $15–$25.

Page 279
Thermometers, circa 1885. (Row 1) $5–$20 each. (Row 2) Cabinet $5–$10. Cottage $15–$30. Fancy Bronzed $30–$50. Distance $15–$30. Dairy $4–$8. Thermometers have recently become hot collectibles. Desktop models from the turn-of-the-century bring up to $200. Names to watch for are Pairpoint, Tiffany, Bearskin, Bradley and Hubbard.

Lamps & Lanterns are among the most popular Victorian factory-made collectibes. Candles, lard lamps, whale oil, and gas all preceded kerosene (coal oil) lighting devices, which came into popular use in the late 1850's. The last American oil lamp manufacturer, Aladdin Industries, began operation in 1908 as the Mantle Lamp Company, and continues in business today. We have not included their extensive line because two books have already covered the subject, and it is pretty late to be included in a Victorian Guide, even though many Aladdin lamps could pass for 1890's products.

The contributing expert for this section is *Ms. Bobbie Mongarro*, of *J. Camp Antiques*, 8395 La Mesa Blvd., La Mesa, CA 91941. Bobbie has been buying, selling, and restoring antique lamps for 30 years. She is frequently called upon by historical groups to recreate Victorian era lighting installations, and has a personal collection of 300 antique lamps and thousands of spare parts.

Page 280
Hand Lamps (for finding your way upstairs to the bedroom, or downstairs to the john) were produced by the millions between 1858 and 1940. They measure between 3½ and 7 inches tall, not including chimney. prices vary widely. Some with art glass fonts and shades run up to $1,000. So you must consult books on Victorian glass to appraise finer miniature oil lamps. Most, however, were of the dime store quality pictured here. Nos. C467, 468, 469, 448, 450, and 451 are from a 1905 catalog and range from $8 to $24 in value. The circa 1880 "Magic Hand Lamp" pictured with 4 inch dish is worth $25–$45. Other patented brass hand lamps in various shapes run about the same money. Neville Patent, Canadian lamps start at $75. Lanterns: Gypsy $30–$50. Great Western $30–$50. Square Lift $25–$45.

Page 281
Lanterns, circa 1880. No. 1 $25–$45 complete with font. No. 6 $35–$65. No. 4 $25–$45 (up to $95 with railroad logo). Nail City Senior $85–$125 complete with double glass globes. Gem $65–$85. No. 0 Tubular $24–$35. Buckeye Senior $85–$125 complete with font and double globes. Police Lantern $45–$65 with font inside.

Page 282
Dash Lanterns came in pairs with right and left-hand brackets. No.0 $45–$65 (pair $95–$150). No. 18 $65–$85 (pair w/brackets $125–$180. Searchlight $65–$95 (dated $75–$125). Star Kitchen lamp $65–$95 (dated $75–$125).

Page 283
Globe Street Lamp $85–$150 (dated $95–$165). No. 1 Ham's Cold Blast, Tubular $24–$38 (dated $35–$65). Mill lantern, 14 inch $65–$85 (dated $75–$95). Gem Cold Blast $24–$38 (dated $35–$65). No. 39 Railroad Lantern, for lard oil $45–$65 (with railroad logo and embossed logo on chimney glass $65–$120). Ham's No. 0 Clipper lift tin lantern $25–$45 (dated $35–$65). Embossed chimneys and colored glass puts any lantern nearer top dollar, as does a specific patent date or a railroad's name. Solid brass construction can triple value.

Page 284
Railroad Lanterns. No. 3 with both globes (double) intact $85–$125. No. 43 $45–$65 (with railroad logo on lamp and globe $65–$95). No. 43½ ditto. No. 13 Dash Board Tubular $65–$85 with embossed chimney. No. 1 Dash Board Queen complete with font and chimney $85–$125 (a very scarce lantern). No. 2 Police $45–$65 complete with font.

Page 285
Hand Lamps. No. 8440 $15–$25. No. C465 & 466 round wick $10–$15. Iron Lamp with double wick $25–$45. Lamp Filler, conical $18–$25. Nursery Lamp with Tea Kettle that fits opening when lid is removed $125–$175 set. Jacket Lamp with tubes and feeder $25–$45. Little Will 5 gal. oil can on wire stand $95–$150. Queen, with glass insert $30–$50. No. 8430 Side Lamp $35–$50.

Page 286
Oil Lamps from 1893 Marshall Field Catalog. No. 719 $85–$125 ($125–$165 with shade). No. 2 $65–$85 ($85–$145 with shade). No. 711 $145–$225 with bracket and shade. No. 2 embossed $125–$165 ($200 and up with silk shade). No. 32 student $225–$350 with shade & chimney (if dated $245–$450).

Page 287
Student Lamps, circa 1880–1900. The unique side-mounted oil reservoir eliminated shadows on reading material. Few have survived with original shades. Repro

shades fit only a few old frames, so measure before you buy. Nos. 1, 2, 3 & C1356 are $225-$350 (if dated $245-$450). No. 5 Double $450-$800.

Page 288

Parlor Table Lamps, circa 1895. Nos. 8340, 8342, 8346 & 8347 $225-$345. No. 8341 figural base with handpainted globe $265-$375. No. 8345 cherub $350-$550. No. 8350 $450-$650. No. 8351 $450-$800. No. 8353 $225-$400. No. 8355 $250-$500. No. 8356 $295-$600. No 8357 $295-$600. These are not exotic name brands such as Handell, Crown Milano, etc. Lamps that sell for thousands look like thousands. The glass will be art glass and the bases will be signed bronze.

Page 289

Parlor Lamps from a 1905 Butler Bros. catalog. Today they are called Gone-With-The-Wind Lamps, but Civil War oil lamps did not look like these. No. C1498 $400-$800 each. C1500 Indian Chief $800-$1,200. C1502 $550-$850 in red or satin glass. (Row 2) C1493-1, 2 & 3 $450-$600. Moose or Elk $800-$1,200. Nos. 5 & 6 Child at Sea $600-$900. (Row 3) C1504-1 $600-$900. No. 2 Smaller, Roses $450-$700. No. C1481 $250-$450 each. No. C1499 $600-$900. NOTE: All prices on these pages assume original shades or handpainted matching replacements. A damaged foot, missing handle, etc. reduces value greatly.

Page 290

Oil Lamps from Butler Bros. 1905 catalog. prices quoted here are for what you see on page. If no chimney or font, price reflects same. No. C460 $35-$60. Mammoth Stove Lamps "Royal" $200-$350 if all orig. "Banner" $300-$450. C1348 Dime Leaders $35-$85 each. C1326 Fancy Hand Lamps $45-$85 each. C1329 Crystal $80-$125 each. C1328 Defiance $45-$95 each. C1339 Tinted $85-$150 each. C1325 Pioneer $45-$85 each. Bracket Lamp $150-$250. C1347 Sewing $45-$125 each. C 1338 Big & Safe $85-$150 each. C330 Tall Pinwheel $65-$95. Short $65-$95. Short, Ribbed $45-$65. C1347 Sewing $45-$125 each. C1331 Table $65-$125. C1343 Decorated $125-$250. C1327 $75 each.

Page 291

Hanging Library Lamps were widely reproduced in the 1970's. Look for maker's name, date, and other signs of age. (Row 1) Round Shade is rarest on page $850-$1,200. (Upper right) $300-$600. (Row 2) Metal font $250-$500. (Center) Prism drops $550-$800. (right) $300-$600. (Bottom left) Animal Scene $800-$1,000. (Bottom right) $550-$800.

Page 292

Piano Lamps. No. 2598 Table style $350-$600 shade & chimney would add value). No. 2597 Brass Legs $300-$500.

Page 293

Pittsburgh lamps from 1893 Marshall Field Catalog. Nos. 3409, 3442 & 3467 $250-$400 each.

Page 294

Silk Lamp Shades from the 1890's are a one-in-a-million find. The fabric and frames just did not hold up. Reproduction shades have been made since the 1940's. They run from $100-$250 each. Ask a custom lamp shade shop for a quote from these illustrations. Specify a brass frame.

Page 295

Pittsburgh Lamps, tall metal bases in gilt or silver. $250-$450 each.

Antique Clocks are difficult to price. Traveling from state to state and shop to shop you will find identical old clocks priced 100% apart. Many dealers will not put a price tag on their rare finds, waiting instead for an offer they can't refuse. Just a few short years ago West Coast collectors could travel East and buy for half of California prices. Nowadays, Eastern dealers are coming out West to recapture their locally made clocks for resale back home.

The contributing expert for this section is *Jim Simpson*, owner of *Time and Treasures*, 8290 La Mesa Blvd., La Mesa, CA 91941. For the past 20 years, Mr. Simpson has bought, sold, appraised and repaired old clocks. Jim makes an annual scouting trip to Pennsylvania and his prices reflect this exposure..

Page 296

Fine Mantel Clocks. Cupid-adorned Ansonia, 8-day $650-$1,200. Antique oak with brass trim 12 inch by 23 inch $750-$850. Seated Knight, visible escapement $1,000 and up.

Page 297

Novelty Clocks with cheap 1-day movements. No. 2 Lawn Tennis $75-$100. Hand holding a 2-inch dia. clock $65-$95. Dog on platform $75-$100. No. 11 on brass chain $100 as shown, up to $400 for a 30-day calendar movement. No. 33 Parrot finial $125 and up. Soubrette with fancy dial $125. Fieldhand with wheelbarrow $100-$150.

Page 298

Don Caesar and Don Juan $900 and up. "Commerce" figure $125. Agriculture figure with clock $650-$800.

Page 299

Enameled iron clocks command a slight premium over wooden mantel clocks of the same design. The styles shown here are more valuable than the common straight-sided variety with marbleized pillars or columns, which average about $100 each in the Midwest. Open escapements are also value-adders. Pompeii $495. Neptuno $249. Otranto $295. Werra $249. Nubia $395. Louis $295.

Page 300

Waterbury Shelf Clocks, circa 1890. (Not shown are double-dial shelf clocks with simple calendar movements which sell in the $500-$650 range. Perpetual calendars are twice the price.) Chester $250. Sussex $250. Homer $250. Wayne $250. Lyons, simple one-dial calendar $300. Merwin $300.

Page 301

Seth Thomas Clocks. Concord $350. Atlas $550. Tampa $300. Walnut case $300. Octagon short-drop wall clock: 8-day spring $325. 8-day strike $345. 8-day simple calendar $375. 8-day calendar with strike $395. O.G. (ogee) 1-day weight-driven strike $200. 8-day version $300. Misc. Steeple Clocks (not shown) $100-$250.

Page 302

Novelty Alarm Clocks. These 1890's pin-ups are very scarce, but can still; be found for $75 and up. The plain roman numeral alarm clocks with visible balance wheels sell for $75. Later alarm clocks from 1904-1930 by Ansonia, Gilbert, Ingraham, National and Westclox start at $35 each. Cupid $75-$95. Lodge $100.

Page 303

Circa 1900 carved wood clocks, all $200-$250 (except upper right $300-$350).

Page 304

Cupids by New Haven and other 1890-1905 makers run about $175 if they have all their fingers, toes, and wing tips. The "Florist" clock is 8½ inches tall and worth $200 or more.

Page 305

Office Regulators, circa 1880-1905. No. 2643 $450. No. 2644 $1,250. No. 2645 $1,750. No. 2646 $400. No. 2647 $375 with calendar ($250-$325 with plain face).

Wicker Furniture was a Victorian favorite. Equally at home on the front porch and in the parlor, too, it blended beautifully with the lacy ferns and potted indoor plants of the period. *Linda Franklin* found these retail prices in her travels from Virginia to Pennsylvania. They pretty much match those here on the West Coast, and also parallel current price guide quotations in wicker books.

Page 306

No. 1200 Wicker Lounge Chair (circa 1900) $650-$950. Nos. 2123 & 2124 Fancy Reed Sideboards $450-$650. Natural shellac finish brings highest price. Heavy, gunky, paint layers lower value, but wicker can be carefully stripped and repainted. Reed wicker is more valuable than the twisted fibre variety. Original brass feet are a plus, and so are manufacturer's labels. Solid wooden frames are important and should be eyeballed for warpage.

Page 307

Wicker Chairs. Nos. 1169, 1174, 1178 and 1113 are worth from $350 to $600 each because of size or especially nice heart motifs. The other chairs on this page could bring anywhere from $175 to $450.

Page 308

Fancy Reed (Wicker) Parlor Stands. Nos. 2108, 1149 & 2106 $175-$250. (Row 2) Ladies & Gents Rockers $250-$450 (Heart shape is top dollar). (Row 3) Arm Chairs $200-$375. Rockers $250-$400. Settee $400-$595. All five pieces in a set $1,500 and up.

Page 309

Wicker Chairs (Row 1) $200-$450 each. Settee $500-$750. Three-piece set $1,200-$1,500. Five-piece set $1,750-$2,000. (Row 2) Chairs $175-$250 each. Three-piece set $900-$1,200. Five-piece set $1,200-$1,500. (Row 3) Chairs $175-$250 each. Three-piece set $800-$1,000. Five-piece set $900-$1,250.

Page 310

Fancy Wicker Work Stands. No. 353 $125-$250. No. 354 $150-$300. No. 355 $175-$350. Infant's or Work Stands No. 356 $125-$250. No. 357 $135-$275. No. 359 $150-$300. No. 358 $135-$275. Wicker Center Table (not shown) 28 inch diameter $450.

Page 311

No. 1187 Child's Deluxe Wicker High Chair $250-$350. No. 900 Budget High Chair

$175–$225. No. 910 Better Quality $195–$250. No. 1186 Very Best Child's High chair $295–$400. (Row 4) Fancy Willow Nursery Chairs. All are $175–$225. (Row 3) ditto, except Fancy Doll Hood Cradles $100, $150 & $200. Child's Crib (not shown) on stand $600–$800. Child's Wicker Rockers (not shown) $200–$275.

Golden Oak Furniture was a favorite in middle class American homes from 1890 to 1930. It was easily milled, pressed, formed and needed only a cheap gloss finish. In the 1940's, 50's and 60's, "modern" homemakers threw it out in favor of mahogany, maple, pine, and even blonde-stained ash. Imported teak wood also won a large market share. In the mid 1970's, newly married couples often started housekeeping with a few pieces of grandma's golden oak and its popularity staged a comeback that has persisted for two decades. Kitchen chairs and tables have been reproduced by the hundreds of thousands, and so has much old office furniture. The pieces pictured on these pages are from Simmons Hardware Catalog of 1905.

Our current national retail prices were provided by **Mrs. Sonia Nash**, co-owner of **The American Oak Co.**, 2946 Adams Ave., San Diego, CA 92116. Mr. & Mrs. Nash completely restore and refinish every piece of furniture they offer for sale.

Page 312

Children's High Chairs Golden Oak, circa 1900. Refinished and recaned $295–$595. Children's Pressed-Back Rockers $295–$595. Plainer styles (not shown) bring lower prices.

Page 313

Oak Rockers, Cane Backs and Seats $250–$350. Double-Panel Pressed Back with Turned Spindles $450–$650. Veneer Seat and Back, Side Chair $75–$85. Matching Rocker $250–$350. Matching End Chair $125–$185. (Row 3) Mahogany Veneer, Round-Bottom Parlor Chairs $95–$150.

Page 314

Ladies Fancy Rockers, circa 1900. (Top Row) $450–$550 refinished and reupholstered. (Bottom Row) Pressed Backs, Bent Oak Arms $350–$450 refinished.

Page 315

Morris Chairs were first made in England by William Morris, an artist and designer contemporary with Charles Eastlake, working circa 1860–1890. Americans fell in love with the new reclining easychair and it was mass-produced here well into the 1900's. The Sears and Roebuck quality shown here are worth from $275 to $550 each. Lion's heads up the value at least a hundred dollars. A Gustav Stickely label could mean $2,000–$7,000. (Bottom Row) Ladies Rockers $350–$550 each.

Page 316

Pressed Back Oak Dining Chairs, refinished and ready to sell. (Row 1) $95–$150 each in sets, less for singles. (Row 2) Veneer (plywood) Oak Side Chairs $75. Arm Chairs $150. (Row 3) Recaned and Refinished Oak in Sets. Side Chairs $95 each. Arm Chairs $175–$225 each.

Page 317

Fancy Rockers and Arm Chairs. (Row 1) $275–$350 each, all styles. (Row 2, left to right) $325, $250, $300, $200. (Row 3, left to right) $275, $250, $325, $275.

Page 318

Golden Oak Office Chairs. (Row 1) No. 567 $200–$300. No. 467 $350–$550. No. 465 $250–$450. (Row 2) No. 437 $350–$550. No. 435 $200–$400. No. 17 $350–$550. No. 15 $200–$400.

Page 319

Veneer-Back Oak Office Chairs (Row 1) No. 647 $200–$275. No. 1807 $250–$350. No. 645 $125–$175. (Row 2) No. 646 $150–$195. No. 636 $150–$195. No. 626 $200–$275. No. 86 $200–$275.

Page 320 and 321

Rare Sewing Machine prices appear directly on pages 320 and 321.

Page 322

Howe Mfg. Co. 1876 product catalog with history of sewing machine included. Centennial exhibit issue $45–$75.

Page 323

Howe Sewing Machines, from an 1876 catalog. Nos. 1, 2 & 3 are common models with many surviving examples. Value $125–$250 each. The "Cabinet Case" is very rare and a sewing machine collector might pay up to $750 for a nice one.

Page 324

Singer Sewing Machine, first model (no case), $1,000. With the original wooden packing case, which coverted into a treadle stand $1,500. Grover, Baker & Co.'s 1851–53 patent portable $1,000. Blees Noiseless Treadle $350. Wheeler & Wilson machine is worth less than $100 to a collector, but an antique shop or decorator would charge $200 to $300 for this one. I don't know what a Bolgiano Mfg. Co. 1893 patent water motor would fetch. Do you?

Page 325

Sewing Machines, circa 1880. Wheeler and Wilson, and The Wilson Sewing Machines are worth from $100 to $200 each to a collector. But antique shops price them, complete with treadle and case, at $200–$300.

Page 326

Antique Sewing Machines, 1870–1880. Both Beckwith portables are rare collector's items at $500 each, but they look like $100 retailers. The little Heberling Shirring and Gathering Machine also does not give an outward hint of its $350 collector value. New Homes' attractive floor model was made by the jillions, and is not really worth its $175–$275 antique shop price tag. The dainty-looking Davis Vertical-Feed floor model has a collector value of $200 and lots of decorator appeal at $250–$400. The 1872 Florence looks to be worth its average price tag of $350–$500 to collectors and consumers alike.

Page 327

Antique Portable Sewing Machines. Perfection Automatic Nos. 1 & 2 $75–$125 each. New Queen and similar heavier countertop models have no collector value, but decorative appeal at $50–$100. Cute little toy "Singer 20" of 1915 $75–$135. The 1880's pinking machine appears regularly at $35–$50. The circa 1915 Spenser Automatic sells for $75–$125. No. H125 Midget and No. H100 Gem are $75–$100 machines. (Lower right) Imported toy sewing machines, circa 1925, are not as old as they look $50–$100.

Pages 328 and 329

These mass-produced Treadle Sewing Machines, circa 1897–1915 have no collector value. However, they sew well and look even greater, at $100–$200 each. The ever-popular Singer with its premium oak and highly decorated head sells for $175–$350.

Page 330

Tailor's Shears, Heinisch's 1863 patent $30–$85. Rochus Heinisch received his first scissor patent in 1829, just nine years after his arrival here from Austria. Prior to his invention, the old fashioned shears were torture instruments to a tailor's hands. Belmont Lamp Trimmers $45–$65. Seymour's Shears $18–$50. No. 30 Lamp Wick Trimmers $20–$35.

Page 331

Sewing Kit. No. 9166 (8) pc. set with pearl handles, circa 1880 $45–$95 in velvet-lined case.

Page 332

Star Scissors $8–$18. Geneva No. 24 twisted-handles $5–$15. Most all Button Hole scissors sell for $5–$10 each. A Keen-Kutter, or Winchester, trademark can double or triple the price..

Page 333

Scissors from a 1915 catalog. (Col. 1) No. B1 $1–$3. No. B8 $1–$3. No. B9 $5–$10. No. 2095 $5–$10. No. 925 $5–$10. No. 900 $8–$20. (Col. 2) No. B2 $1–$3. No. 2095L $4–$8. No. 925L $3–$5. No. 920 $5–$10. No. 915L $1–$3. No. 900L $5–$10. (Col. 3) No. 300L $5–$10. Nos. 301 & 105 Stork-Handle have been made for over 100 years. Most examples are worth only $5 or $10, sterling silver handles $25–$40. No. 20 $1–$3. No. 607L $1–$3. No. 271 $1–$3. No. 60 $5–$10.

Page 334

Scissors, circa 1890. No. 5153 $1–$4. No. 5158 $1–$6. No. 7247 $5–$10. No. 4337 $3–$8. No. 2420 $8–$20. No. 9020 $3–$8. No. 5153 $1–$2. No. 7420 $5–$10. Folding Pocket Scissors with 1872 patent $15–$25.

Page 335

Scissors, circa 1890. No. Rx 1900 $1–$4. No. Rx 3442 $2–$6. No. Rx 2096 $3–$8. No. Rx 2603 $5–$10. No. Rx 717 $5–$10. No. Rx 1822 $5–$10. No. Rx 70 $2–$6. Button hole $5–$10. No. Rx 1122 $1–$4.

Page 336

Scissor Sets (3 pairs) with plain bows $15–$25 set. Fancy bows $25–$50 per set. Gold Thimbles sell in a very broad range, from $50 to $120 each. Sterling silver and coin silver thimbles are worth from $20 to $40 each. Silver with gold bands are $30–$50. The plain patterns, at the end of each row, are the cheapest. Cupids or cherubs appear on more modern thimbles than these, but are highly sought after at $90–$100 each. Thimble prices are highest in large cities.

Page 337

Silver & Gold Timbles circa 1890–1915. Condition is very important, wear reduces value. 10 karat gold is worth less than 14 karat. 24 karat is pure gold. "$^1/_{10}$ 18k gold filled" means that the rolled-on surface plating is less than 10% karat gold, and the underlaying base metal is brass or copper. Gold filled thimbles are not worth much more than silver ones. Heavy sterling or coin silver thimbles are more desirable than lightweight examples. Those with scenic bands engraved with buildings or landscapes are $45 in silver. World's Fair thimbles sell for $200 and up.

Inlaid stones add to value. Advertising thimbles have a price range completely unrelated to the base material. (Row 1) $25–$40 avg. Perhaps $50 for the gold-banded one. (Row 2) Silver $20–$40. Gold $50–$100. (Row 3) Gold $50–$100. Silver with band $30–$50. (Row 4) Coin Silver $20–$40. Plain patterns are cheapest. (Row 5) Sterling Silver $20–$40 each. Paneled and scrolled examples are the nicest ones shown. (Row 6) Sterling Silver $20–$40 each. A cute cherub would bring more. (Row 7) No. 8995 is the nicest silver one at $30–$40. The very plain 10 karat thimbles might fetch $40–$50.

Page 338

Sterling Silver Handle Scissors, circa 1895–1905. All on this page are $25–$40 sellers, if handles are marked to indicate sterling silver. Electroplated examples are worth one third as much. Highest prices are found at shops and shows where dealers specialize in antique sewing tools.

Page 339

Sterling Silver Sewing Novelties, circa 1915. (Row 1) No. 96234 $12–$18. No. 96038 $12–$20. No. 96230 $20–$30. No. 96040 $20–$35. No. 96035 $15–$20. (Row 2) No. 96233 $10–$15. No. 96039 $10–$15. Tapes, Nos. 96032, 33 & 34 $20–$35 each. Figural shaped tape measures (not shown) are twice as much. (Row 3) Set No. 96220 $22–$35. No. 96221 $42–$65. Shuttles, Nos. 96246, 47 & 48 $7–$12 each. (Row 4) Set No. 96223 $35–$55. No. 96219 $35–$55. No. 96213 $20–$40. No. 96216 $25–$45. Scissors, No. 96225 $20–$35. (Row 5) Set. No. 96222 $40–$85. No. 96201 $45–$95. No. 96031 $45–$95. (Row 6) No. 96203 $95–$135. No. 96205 $100–$165. No. 96206 $75–$120.

Washing Machines have been around since the early 1800's and no one really knows who invented the first one. We do know that by 1900 nearly 2,000 related patents had been filed, and over a hundred manufacturers were competing for a major market share.

Sears and Roebuck carried most of the designs shown on this page, and millions were sold. So don't expect more than the usual $75–$250 price for your old lint-catcher.

Most early (pre 1860) machines were more trouble to operate than the wooden wash boards of the period. Howard Snyder, a young mechanic with the firm of Maytag and Bergman, added an electric motor to their 1910 model, but basically it was still a peg-studded agitator contraption.

In 1922 Mr. Snyder developed a bottom-mounted agitator with fins which propelled water through the clothes, instead of vice-versa. He also changed Maytag's tub shape to round and made it out of aluminum instead of wooden staves. Maytag soon captured 60% of the market.

Page 341

prices: Wooden Wash Tubs $40–$100. Doty's Washer $125–$250. Universal Wringer $45–$70. Sargent's Wringer $150–$250.

Page 342

Universal Wringer $45–$70. Monitor Wringer $45–$70. Lamb's Washer/Wringer $50–$95. Ferguson Washer $150–$250. Clark's Washing Machine $150–$250.

Page 343

Wash Boards, circa 1878–1915. We've seen them for $15 to $150 in every antique mall in the land. Top dollar is a porcelain enamel scrubber, usually in some shade of blue. Even higher is the $350 asked for Bennington Pottery surfaced boards. Most common brass, glass, or galvanized boards are still under $35 each. A nice advertisement is a big selling point, otherwise you will be lucky to get more than $15 or $20 for your faded soap saver. The cone-shaped washing dolly with broomstick handle is common at $15–$20.

Page 344

Nos. 3 & 4 Folding Bench ringers $75–$150 (add $40 for each tub). Domestic Mangle and Wringer $75–$125 (widely used in hotels to wring sheets). Peerless Clothes Wringer, circa 1877 $45–$65. American Washing machine (wringer & tub) $50–$75. McDonald's Patent Washing Machine, circa 1880 $150–$250.

Page 345

Mr. Doty's Clothes Washer was a state-of-the-art machine in 1871 and perhaps the most popular brand in America for the next decade. 600 washers a day were shipped from his Middlefield factory. The main reasons for its success were a rustproof glavanized tub and the simple instructions printed in bold letters on the agitator handle: "Use Boiling Hot Suds."

Page 346

A circa 1900 Gravity Washer might clean your clothes in 6 minutes, but who is going to carry two 50-gallon loads of boiling hot wash and rinse water from the kitchen stove, to the washer, and out the back door? And who chops the wood to heat it? Woman's work is never done! Value $125–$195. The flat top makes a great geranium stand.

Page 347

Clothes Wringers circa 1900–1915 $35–$60 each, depending on the condition of rubber rollers and advertising message on the frame. (Notice the difference in design between these and those on page 342, which were produced in the 1870's).

Page 348

Copper Measures $30–$80 each. $200–$275 for 5-piece set. Wash Boilers, tin $60–$80. Copper $125–$175. (Wooden handles, set out from the ends, were an early 1900's improvement which saved a lot of burned knuckles.)

Page 349

Washing Machines circa 1905–1915. No. 5 American $100–$195. No. 22 Miracle $150–$295 (restored). No. 6 Advance $100–$195. Rapid Suction Washer, tin $10–$20. Copper $30–$60. Ironing Tables $18–$30 (see page 95 for earlier ironing boards, worth more).

Page 350

Mrs. Pott's Smoothing Irons are the most common detachable-handle style found in shops today. Individual bottoms sell for $5 to $8 each. You can fill them with sand and plant a cactus on top, or use 'em for door stops. Add an original bentwood handle and price jumps to $25 or $35. A set of three, with handle and trivet, ought to bring $60–$75.

Page 351

Most Patent Smoothing and Polishing Irons on this page can be found for $15–$35 each.

The Star and Geneva styles are the most common. The Keystone, Toilet Iron, and Troy Polishing Irons are harder to find.

Page 352

(Row 1) Common Sad Irons $8–$20. (Row 2) Troy Polishing Irons $15–$30. (Row 3) Very Plain-style Charcoal Irons $25–$55. Fancy embossing and figural top-catches would raise value considerably, but most of these would be European. (Reproduction alert: Rusty charcoal irons with dove or rooster finials, and saw-tooth lid undersides, are being made in Mexico today. I used to sell them wholesale for $20 each.)

Page 353

Flat Iron heaters (irons not included). No. 15 Stove $150–$200. Sun, Globe, and Improved $45–$65 each. Gas wall fixture $15–$30. Trivets $20–$40. Coffee Pot Stands $10–$20.

Page 354

Sad Iron Stands, circa 1865–1885 $25–$40. Coffee Pot Stands $15–$20.

Page 355

Toy Sad Irons, circa 1885. Perforated or Wood Handle, double-end, with Trivets $95–$150 (Repros are usually cadmium-plated). No. 75 Toy Sad Iron and Stand $75–$95. Toy Duck Sad Iron and Stand $125–$150. Common 2½ lb. Sad Irons about 4 inches long $8–$20.

Page 356

Sad Irons on this page are circa 1872–1887, and sell for $25–$55 each.

Page 357

Sensible Sad Iron Set of (3) irons, 1) stand, and (1) handle $65–$125. Single iron w/handle $40. No. 4 Gem Polishing Iron $25–$40. Polishing collars, cuffs and shirt-fronts went out the window when cheap, disposable, celluloid replacements were invented, but they kept on making polishing irons in England until World War II.

Page 358

Charcoal Irons, circa 1860–1890. Nos. 1, 2, 3 & 4 are rather plain, and worth $25–$55 each. Double Flue type are embossed and of a more intricate design $45–$65.

Page 359

Hair Curlers and Frizzers $8–$20. Nos. 4 & 6 Langtry Frizzing Irons $5–$15. No. 10 Pinching Irons (for spit curls, used with paper to prevent scorching) $8–$15. Five-prong Fluting Scissors $8–$15. Three-prong $5–$10.

The ancient Egyptians were the first to invent curling irons – to jazz up their straight hair. Wigs have been curled for centuries, and men have been using these tools since Assyrian warriors learned to curl their beards. The permanent wave was first wound on chicken bones (early 1800's), and heated until a little fat started to fry. In 1872 Marcel Crateau invented the modern curling iron and retired quite wealthy at the age of 45.

Page 360

Fluting Machines $125–$150. These evolved from hotdog-shaped goffering irons that made the ruffles and tucks in 16th century costumes. They crank out multiple pleats in up to 15 different sizes.

Page 361

Hand Fluters. Shepard's No. 85 $75–$100. Geneva $80 with iron base, up to $100 with brass base. Young's, or Universal Plaiters (pleating iron sets) $75–$100.

SEARS MOTOR BUGGY $395.00
12-HORSE POWER, 2-CYLINDER ENGINE

THE ONLY PRACTICAL LOW PRICED MOTOR BUGGY

ALL COMPLETE AS ILLUSTRATED, WITH SOLID RUBBER TIRES, TOP, LAMPS AND FENDERS. TESTED AND READY TO RUN. NO EXTRAS.

NOTHING FOR YOU TO BUY BUT GASOLINE.